Feature Papers for the Inaugural Issue of Applied Biosciences

Feature Papers for the Inaugural Issue of Applied Biosciences

Editor

Robert Henry

MDPI

Basel • Beijing • Wuhan • Barcelona • Belgrade • Novi Sad • Cluj • Manchester

Editor
Robert Henry
Queensland Alliance for
Agriculture and
Food Innovation
University of Queensland
Brisbane
Australia

Editorial Office
MDPI AG
Grosspeteranlage 5
4052 Basel, Switzerland

This is a reprint of articles from the Special Issue published online in the open access journal *Applied Biosciences* (ISSN 2813-0464) (available at: https://www.mdpi.com/journal/applbiosci/special_issues/Inaugural_Issue_Applbiosci).

For citation purposes, cite each article independently as indicated on the article page online and as indicated below:

Lastname, A.A.; Lastname, B.B. Article Title. *Journal Name* **Year**, *Volume Number*, Page Range.

ISBN 978-3-7258-1709-2 (Hbk)
ISBN 978-3-7258-1710-8 (PDF)
doi.org/10.3390/books978-3-7258-1710-8

Contents

applied biosciences

Review

RNAi-Based Biocontrol of Pests to Improve the Productivity and Welfare of Livestock Production

Pia S. Menezes [1,†], Yakun Yan [2,†], Yunjia Yang [2,†], Neena Mitter [2], Timothy J. Mahony [2] and Karishma T. Mody [2,*]

1 School of Chemistry and Molecular Biosciences (SCMB), The University of Queensland, Brisbane, QLD 4072, Australia
2 Queensland Alliance for Agriculture and Food Innovation (QAAFI), The University of Queensland, Brisbane, QLD 4072, Australia
* Correspondence: k.mody@uq.edu.au; Tel.: +61-7-33462318
† These authors contributed equally to this work.

Abstract: Insects and ectoparasites are causes for major concern throughout the world due to their economic and welfare impacts on livestock agriculture. Current control measures involve chemicals such as acaricides which pose challenges like chemical resistance and longer withholding periods. To enable more sustainable agriculture practices, it is important to develop technologies that combine targeted effectiveness with minimal environmental footprint. RNA interference (RNAi) is a eukaryotic process in which transcript expression is reduced in a sequence-specific manner. This makes it a perfect tool for developing efficient and effective biological control against pests and pathogens. Double-stranded RNA (dsRNA) is the key trigger molecule for inducing RNAi; this concept is widely studied for development of RNA-based biopesticides as an alternative to chemical controls in crop protection for targeting pests and pathogens with accuracy and specificity. In this review, we discuss key advances made using RNAi technology and how they can be applied to improve health in livestock industries. This includes research focused on different delivery mechanisms of dsRNA, important developments in regulatory frameworks, and risk identification, that will enable the future adoption of RNAi technologies to improve animal health.

Keywords: RNA interference; dsRNA delivery; biological control; animal health

Citation: Menezes, P.S.; Yan, Y.; Yang, Y.; Mitter, N.; Mahony, T.J.; Mody, K.T. RNAi-Based Biocontrol of Pests to Improve the Productivity and Welfare of Livestock Production. *Appl. Biosci.* **2022**, *1*, 229–243. https://doi.org/10.3390/applbiosci1030015

Academic Editor: Demetrios A. Arvanitis

Received: 25 August 2022
Accepted: 12 October 2022
Published: 14 October 2022

1. Introduction

Pests and pathogens have threatened primary production since farming began. According to the Food and Agriculture Organization, parasites affect 80% of the world's livestock population [1–3]. In livestock production systems, chemical pesticides such as insecticides are commonly utilised to control and combat pests like insects, because of their low prices, high availability, fast mode of action, and perceived reliability. However, the extreme dependence on the use of these compounds has resulted in ubiquitous, low-level exposure, which is potentially harmful to both human health and ecosystems [4]. Besides risks to human health, widespread and repeated insecticide application can also result in environmental concentration build-up in soils [5]. Furthermore, direct application of these chemical pesticides have other important limitations like, poor specificity, short duration of protection, poor water-solubility, difficulties with site-specific uptake by the targeted pest, and inducing chemical-resistant pathogen strains [4,6]. Some novel applications have been investigated to increase the efficiency of insecticides, such as site-specific spraying and human–robot collaborative sprayer. However, these methods require significant investment and expensive maintenance cost, making them less practical in many agricultural applications [7,8]. Thereby developing safe, environmentally friendly, effective, and sustainable pest management strategies has become a much-needed requirement to improve the productivity and sustainability of global agriculture.

RNA interference (RNAi) has emerged as a promising candidate for the development of biological based control strategies, offering excellent target species specificity and sustainability management of pests and pathogens affecting livestock agriculture. Unlike conventional pesticides used to control pests, RNAi based control uses the target pests' own molecular mechanisms to initiate silencing of essential genes. Ideally, the RNAi approach drives mRNA degradation to block targeted protein production and inhibits pest growth, depending on the developmental stage of the pest and gene of interest targeted. The RNAi mechanism is a conserved, nucleic acid metabolism, which can be initiated by exogenously applied or endogenously expressed double-stranded RNAs (dsRNA) in many species [9,10].

RNAi technology has shown great potential in crop protection for controlling hemipteran pests, including aphids, whiteflies, planthoppers and psyllids [10]. The first commercial application of RNAi was done using SmartStax-PRO, developed by Monsanto and Dow AgroSciences to protect maize against corn rootworms. Transgenic maize MON874411 expressed dsRNA targeting DvSnf7 in Western corn rootworm (WCR), which was used on-field for controlling WCR, and other corn rootworms, lepidopteran pests and weeds [11,12]. In 2011, the Monsanto company published a patent for using polynucleotide molecules for gene regulation in plants [13]. However, fewer studies have focused on using this innovative technology for controlling pests affecting animals in agriculture. Recently, there have been an increasing number of reports on the application of RNAi to improve the control of livestock pest, that have proven to be recalcitrant to conventional control strategies. For example, limited research on nematode parasites in pigs, such as *Trichinella spiralis* and *Ascaris suum*, and ticks in cattle have been reported to have potential to be controlled efficiently with RNAi technology [14–17].

In this review, the potential advantages that RNAi technology offers are detailed, along with the scope and significance of using RNAi to control pests impacting the health of livestock. Additionally, key attributes of RNAi target identification, effector molecule design and different delivery methods used for the delivery of dsRNA are reviewed. In addition, the challenges associated with the use of this technology in field are also discussed.

2. RNA-Interference (RNAi)

RNAi is an endogenous, post-transcriptional gene regulation mechanism which has been identified in almost all eukaryotes; from plants, fungi, algae, protozoans, invertebrates, to vertebrates [18,19]. It was first reported by Napoli and Jorgensen in 1990 when they observed endogenous gene co-suppression while studying pigmented petunias [20,21]. In animals, RNAi was first documented in the nematode *Caenorhabditis elegans* [22,23]. While the intracellular components of RNAi are similar across species, it was initially known as quelling in fungi, and post transcriptional gene silencing in plants [24]. It is a highly conserved mechanism, which is highly sequence specific and selective in its activity [25].

2.1. RNAi Mechanism

While a comprehensive review on RNAi mechanism discussing all the different facets of this technology is outside the scope of this review, the basic mechanism of RNAi when initiated via delivered exogenously applied dsRNA are well described and are illustrated in Figure 1.

RNAi-mediated post-transcriptional gene silencing is triggered by the processing of a dsRNA precursor into short single stranded RNA effector molecules [18]. There are three RNAi pathways, depending on the RNA class of effector involved, small-interfering RNA (siRNA), piwi-interacting RNA (piRNA), and microRNA (miRNA) [18,25–27]. Of these, siRNA is considered to be the 'classical' pathway, with dsRNA being the trigger molecule for gene silencing in insects and other species [27]. In the siRNA pathway, once taken up by cells dsRNA is cleaved by an endonuclease, Dicer, into pre-siRNA duplexes [24]. The pre-siRNA is a 21–23 nucleotide (nt) long duplex with 2 nt overhangs at each of the 3′ termini. The pre-siRNA duplex is bound by the RNA-induced silencing complex

(RISC), with one strand, the guide strand, being retained within the complex, while the complementary or passenger strand of the pre-siRNA duplex is degraded [4,28]. The guide siRNA strand with the Argonaut proteins (Ago) within the activated RISC, identify matching mRNA in a sequence dependent manner, resulting in suppression of translation or mRNA degradation [29]. This results in loss of protein function which may lead to lethality or stunted growth of the target organism [4,27].

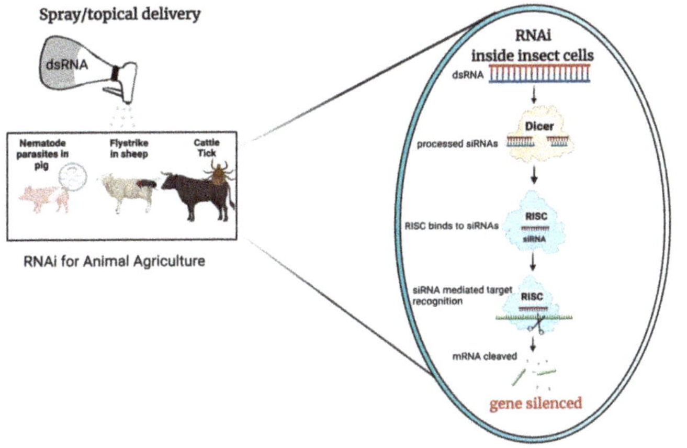

Figure 1. Schematic example of RNA interference (RNAi) delivery via spray/topical application for animal agriculture; this image was made using BioRender. After application over the animal fur/surface, the dsRNA would enter the cells of the insect pest or pathogen. dsRNA is cleaved by dicer-2 into pre-siRNA duplexes, that are further processed into small-interfering RNA (siRNAs, 21–24 nt) effector molecules. The RNA-induced silencing complex (RISC) binds to the siRNAs and guides sequence-dependant degradation or translational inhibition of homologous mRNAs, which results in RNAi-mediated gene silencing.

2.2. RNAi Targets

Target gene selection depends on the class of RNA effector molecule to be used. While both siRNA and piRNA are generated from long, complementary dsRNAs, miRNA is endogenous in nature, usually processed from stem-loops, and siRNA are exogenous and are directly generated from cleaved dsRNAs [30]. The first step in developing an RNAi product is to identify potential target genes and design dsRNA targeting them. In 2010, a research study suggested that there are five factors which play a key role in influencing the efficiency of RNAi as biocontrol, namely, the concentration, nucleotide sequence, length of the dsRNA, effective period of the dsRNA induced silencing, and life stage of the target [31].

The perfect RNAi target gene should be transcribed into an mRNA with a high turn overrate, that is translated into a protein with a short half-life, and is transcribed at all stages of life cycle, with the loss of function leading to mortality or severe impairment in the pest or pathogen of interest. In addition, the target transcript should be poorly conserved across species to maintain maximum specificity of mRNA for minimal environmental effects on non-target species [27]. This is a challenging criterion to meet, as those genes that are essential to cell viability tend to be more conserved between species.

It has reported that the efficiency of dsRNA uptake is length dependent, requiring an optimal length and dosage to induce RNAi successfully [32]. Studies have shown that dsRNA as short as 30 and 60-bp can induce 30% and 70% gene knockdown in the red flour beetle (*Tribolium castaneum*), and the optimum length 200–520 bps for most insects [31,33]. This ideal dosage changes according to the organism's susceptibility to RNAi and their potential for systemic RNAi, formulation for delivery, gene expression abundance, life and development stage for gene expression [27]. Unfortunately, genes that often would lead to mortality if targeted by RNAi are highly conserved across nature, due to their evident importance for homeostasis [27].

Various selection methods can be used to develop efficient RNAi targets to minimize side effects and enhance expected silencing. Since RNAi relies on high gene specificity to the selected target species based on sequence divergence, dsRNA needs to be screened and designed to be specific to a target gene or to target genes on a broad spectrum, closer to related species [34]. To increase knockdown or have higher than the expected RNAi result, multiple targets can be selected for a single transcript to ensure variation within the target species is accounted for. A common application that can be used to evaluate off-target genes is the Basic Local Alignment Search Tool (BLAST), and it has been used to find contiguous matching sections of 17 nt or more in the genomes of interesting organisms [34]. BLAST is a common approach for finding regions of identity, and for identifying the functionality of siRNA. Thereafter, a wide range of specialised programs can be used to design the sequences. These include siRNA-Finder (si-Fi), siDirect, dsCheck, and RNAi Designer by ThermoFisher [35].

Amongst insect researchers it is widely accepted that several dsRNA should be screened as a combination as each gene is highly unique to the target gene and the insect species [33]. Whyard et al. demonstrated species specific insecticides could be achieved by targeting the variable 3′-UTR regions of the *y*-Tubulin transcript, allowing for a specific species knockdown between four closely related *Drosophila* species [36]. The results of this study clearly demonstrate that with sufficient sequence information and thorough interrogation of this available data, the development of highly specific RNAi based insecticides is highly feasible.

While the application of RNAi animal agriculture is still in its infancy, there are more dosage trials in plant agriculture, such as efficiency of dosage needed per acre regarding insect pests. Despite the fact that sufficient field experimentation is lacking, it is estimated that near 2–10 g of dsRNA is needed per acre of plants [30]. As this depends on the organisms response efficiency and target gene, this number is open to high levels of variability [30].

2.3. dsRNA Uptake Mechanism

The use of RNAi to protect plants and animals by suppressing essential gene function in pest species has been well documented in model invertebrates such as *Caenorhabditis elegans*, and *Drosophila melanogaster* [37]. While most of the reported experiments involved injecting dsRNA into the target organism, this is not a viable delivery approach for in-field applications in animal production, thus creating the need for autonomous dsRNA uptake [31,37]. Studies have reported two dsRNA uptake mechanisms, the transmembrane channel-mediated uptake and an alternate endocytosis-mediated uptake [37].

Transmembrane channel-mediated uptake mechanism has been best studied in *C. elegans*, more specifically with worms carrying mutant *sid* genes that lack systemic RNAi capacity [31]. This pathway involves two proteins, SID-1 and SID-2, with changes being made to the protein molecule affecting other cells and their activity. SID-1 is a hypothesized transmembrane protein which passively transports dsRNA into the cells [31,38]. The role played by SID-2 has been hypothesized to help facilitate RNAi by coordinating with SID-1 either by modifying the molecule to activate it or binding to dsRNA for delivery or by inducing endocytosis [31]. It has been noted that SID-1 is considered to be a 'cell surface conduit' for dsRNA uptake [32]. Meanwhile, endocytosis-mediated mechanism was first studied in *D. melanogaster*, since it had no *sid* gene orthologues and the mechanism was later confirmed in *C. elegans* [31]. This is a cell-autonomous pathway wherein dsRNA is transported intracellularly via vesicles [31,38]. Drosophila S2 cells were able to take up dsRNA and initiate an RNAi response by receptor-mediated endocytosis uptake which is an active process involving vacuolar H^+ (V-H) ATPase [31,32]. This pathway is active, specific and is conserved in other animal cells [32].

Some organisms can take up exogenous dsRNA and trigger an RNAi response in the entire animal systemically, like *C. elegans*, while others cannot transmit this RNAi response in another cell [32]. In *Drosophila*, it has been confirmed that transposon inversion can activate RNAi, and the homologous gene silencing caused by it is similar to co-inhibition in plants [39]. In a *C. elegans* assay, it was found that the quelling defective protein, QDE-1, a necessary protein for RNAi process, is homologous to RNA-dependent RNA polymerase (RDRP), supporting RNAi process potential involved in RNA replication and regulation [40]. Similarly, dsRNA injection into phloem of the plant can spread throughout the whole plant to initiate systemic RNAi [41]. Interestingly, RNAi can also be induced by immersing nematodes in dsRNA-containing liquids or by feeding nematodes with *E. coli* bacteria engineered to express dsRNA [42]. The pathway for entry of dsRNA into the target cell determines RNAi efficiency and provides a starting point for the development of dsRNA delivery systems [32].

2.4. dsRNA Delivery

RNAi efficacy after cellular uptake had been a concern since the immune system and gut enzymes, mainly intestinal nucleases, digest dsRNA. Thereby affecting dsRNA activity as a pest control [43,44]. This identified the need for a carrier system to protect the dsRNA from degradation and thus, the delivery system for dsRNA can play an important role in the success of RNAi activity. There are several delivery systems, such as microinjection, transgenic crops expressing dsRNA, and oral feeding [45].

2.4.1. Injection

The first microinjection experiment was reported in 1998 by Kennerdell and Carthew in *Drosophila melanogaster* where they showed successful downregulating of frizzled (fz) and frizzled 2 (Dfz2) genes [46,47]. dsRNA corresponding to Wnt ortholog wingless (wg) gene was injected into drosophila embryos, resulting in wg-like phenotype at the site of injection with the RNAi effect being localized. Apart from *D. melanogaster*, dsRNA was injected in *T. castaneum*, which showed RNAi response in larval, and pupal tissues, and phenotypical defects in adults [48]. This was considered to be a robust system since the previous experiments only reported changes seen in embryos and embryonic RNAi does not persist in the later stages of the insect life cycle [47,48]. Overall, there appears to be a wide range of experimental success rates for different species, with larvae, having injections performed on or between segments, while it was convenient to inject adults under the wings [49]. Although there are many successful cases of injection dsRNA delivery triggering RNAi, it cannot be denied that different microinjection protocols can significantly affect the experimental success rate. There is also the potential to injure the cuticle at the injection site which has necessitated the shift to oral administration of dsRNA [50]. It has been speculated to be a combination of several elements, including the insect's life stage, injected concentration, and injected dsRNA quantity [49].

2.4.2. Oral Feeding

Ingestion in pests include oral feeding and transgenic plant intake [47]. In 2009, Whyard et al., successfully utilized a liposome carrier for dsRNA and triggered RNAi in four insect pest species; *D. melanogaster*, *T. castaneum*, *A. pisum*, and *M. sexta* [36]. The first demonstration of dsRNA ingestion in rootworms was been studied and documented, with *Dvsnf7* dsRNA being fed to western corn rootworm (WCR—*Diabrotica virgifera virgifera*). At 24 h post feeding of larvae resulting in them showing high absorbance to 60 bp dsRNA and the rootworm began to die after showing developmental retardation [49]. Similar results had been found in Sri Lanka weevil, a highly polyphagous pest [51]. When *Pros2* and *Snf7* dsRNAs were orally administered to the target, large decline in gene transcript levels were seen along with significant mortality rates (78.6 and 92.7% respectively) [51]. Collectively, these studies indicate the potential of oral delivery as a more practical approach.

2.4.3. Spraying

Considering that among these delivery systems, microinjection has proven to be not feasible for in-field application as it can develop ecological risks, and feeding on plants tends to show low efficacy due to degradation of dsRNA by enzymatic activity within the digestive system of the target pest, alternate delivery systems are required [52]. One such method is the spray-induced gene silencing (SIGS) delivery system which involves spraying of dsRNA that has been loaded onto plant or animal surfaces [52,53]. Dalakouras et al., reported that applying high-pressure spraying of GFP gene to plant can result in locally silencing of GFP gene [54]. In green fluorescent protein (GFP)-expressing transgenic *Nicotiana benthamiana* plants, merely spraying and injecting siRNA failed to induce silencing, while high-pressure spraying of multiple sizes of siRNAs (21, 22, and 24 nt) induced localised GFP silencing [54]. These studies supported the idea that a specifically designed application system to different host or targeted gene could be required in some situations.

2.4.4. Nanoparticle Delivery System

The first evidence for using nanoparticle carriers for dsRNA was published [55]. In their study, naked dsRNAs were shown to be ineffective with direct feeding and showed more effective results (95.4% RNAi efficiency and 80.5% reduction in target population) with a nanocarrier-based delivery method [55,56]. The nanoparticle carrier used in the study showed negligible toxicity to non-target cells and aphids [56]. Additionally, nanoparticles have enabled penetration and uptake of dsRNA into the cells using transdermal dsRNA fluorescent nanocarrier delivery system [30,57]. A scientific publication reported significant knockdown (from 86.86 to 58.87%) using this delivery method for *Aphis glycines*, a soybean pest [57]. The use of nanocarriers is being researched for overcoming such limitation across different delivery systems, especially since it reduced dependence on viral-vectors and is safer and more convenient alternative for mass production [44,57].

One of the most commonly used nanomaterials is chitosan, which is a biodegradable polymer made by deacetylation of chitin [24]. Chitosan was used as a nanocarrier to deliver dsRNA in *Anopheles gambiae* larvae and was found to protect dsRNA and help stabilize it. This was fed to the mosquito larvae via agarose gel coated food which triggered RNAi and led to increased mortality [58]. Recent studies on nanocarriers which have shown promising results are layered double hydroxide (LDH) clay nanosheets. It was demonstrated by Mitter et al., in 2017, where non-toxic LDH were adsorbed with dsRNA (dsRNA-LDH complexes) to form BioClay™ which can be sprayed onto plant/leaf surfaces and induce an RNAi reaction in the target pest [59]. The dsRNA was protected from by the LDH nanosheets effectively against nucleases, and showed strong adhesion even after rinsing [59]. The LDH nanosheets enhanced the stability of dsRNA under environmental conditions for a longer period, making BioClay™ suitable for spraying on-field [59,60]. Apart from nanocarriers, cationic lipids are another medium for transferring dsRNA into target cells [61]. Lipid-based transfection reagents can also increase RNAi efficiency, these vesicles are referred to as liposomes. Electrostatic interactions between positively charged cationic lipids and negatively charged dsRNA can produce liposomes [62].

More research was done using similar technology, but with a controlled dsRNA release. One such approach is by making dsRNA release pH dependent, with the clay nanocarrier releasing dsRNA either under acidic or basic conditions, based on the target site pH level and the environment conditions to maintain stability [63]. This was investigated in our lab, where the stability of dsRNA was tested after loading on two types of biodegradable clay particles: Clay 1 (releases dsRNA under acidic conditions) and Clay 2 (releases dsRNA under alkaline conditions) on cattle hide. Cattle skin was treated with Cy3 labelled dsRNA alone and Cy3 labelled dsRNA loaded on Clay1 or Clay2. The treated skin samples were imaged using confocal microscopy for the presence of fluorescence signal. We found that the dsRNA loaded on clay particles was stable unlike naked Cy3-dsRNA which degraded and was not visible after washing. This study showed that the clay particles increased inherent stability of the dsRNA molecules, offering promise to provide a sustainable solution for animal health [63]. A study in 2020 compared various inorganic nanomaterials as potential carriers for dsRNA in *C. elegans*, of which only dsRNA adsorbed to a synthetic cationic nanocarrier, Mg-Al LDH showed effective knockdown with gene expression at the lowest concentration levels being brought down to 66.8% when compared to the control [64].

3. Application of RNAi for Animal Health

Over the past ten years, there has been a significant increase in published papers regarding application of RNAi to control in parasites affecting livestock, the data of which has been figuratively described in Figure 2A. This section highlights those RNAi studies which have been focused on the control of pests that affect pigs, sheep, and cattle pest.

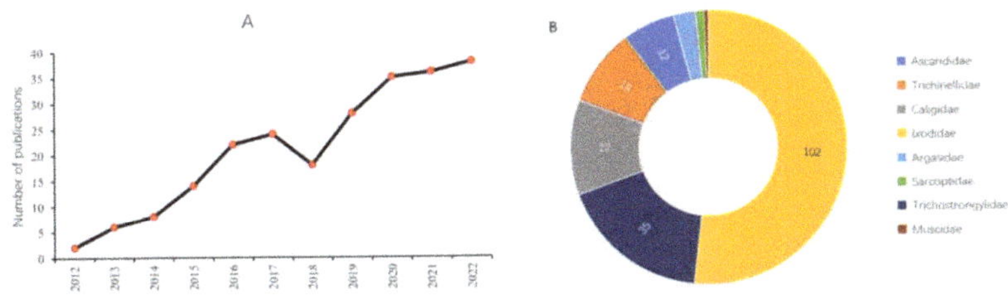

Figure 2. Recent publications of RNAi based pest management associated with livestock. (**A**) Number of publications in the past 10 years based on a query-based PubMed search. (**B**) Recent quantitative distribution of RNAi studies in major livestock pest families. Based on PubMed search: More than 203 publications have been identified with queries "RNAi, gene silencing, specific species name, pest management, and specific livestock animals" by PubMed search.

RNAi technology has been described as a novel method for determining gene function and has provided an appealing approach to pest and disease management [65]. There has been abundant transcriptomic/genomic resources available for *Ascaris suum*, making it an experimental model among nematode parasites [66]. Apart from studies focused on nematode parasite management in pigs, there were studies done with regard to tick control in cattle. Ticks are ectoparasites of both animals and humans and are a significant disease vector affecting cattle husbandry globally [67,68].

With regard to human health, large-scale production of mRNA has been achieved for commercial production with the development of mRNA vaccines for COVID in the recent years [69]. Lipid nanoparticles (LNPs) which are composed of ionisable lipids, helper phospholipids, cholesterol, and PEGylated lipids are used to deliver mRNA vaccines developed by biopharmaceutical companies such as Moderna against COVID-19 and Zika viruses and BioNTech against COVID-19 spike proteins [69]. Over the past few years, research in pest management has further extended towards other livestock pest species. Table 1 enumerates some of the major experiments conducted in the past decade on the application of RNAi based pest control in the agricultural sector.

Table 1. A summary of the RNAi studies focused on the management of pest associated with livestock production.

RNAi Target	Parasites	Delivery	Livestock	Efficiency	Ref
As-eft-1, As-eft-2, As-gmpr, As-tnc-1, As-rab-3, As-hb-1, As-unc-29, As-unc-38	*Ascaris suum*	Soaking	Pig	Different targets in adult *A. suum* were tested for gene knockdown. All targets were significantly silenced in different region of adult *A. suum*. Furthermore, all genes indicate they are susceptible to RNAi in adults *A. suum*, especially neuronal genes.	[15]
As-enol-1	*Ascaris suum*	Soaking	Pig	Enolase gene expression was significantly silenced, and the mortality rate increased by 20.11% ($p < 0.01$) after soaking expressed sequence tag (EST) representing *As-enol-1* dsRNA derived from *A. suum* for 72 h.	[65]
EST-06G09	*Ascaris suum*	Soaking	Pig	The gene of EST 06G09 was silenced after 72 h soaking. The survival rate achieved highest peak at 48 h (20.37%, $p < 0.01$). The gene 06G09 has function in larval development.	[70]
TsSPIs	*Trichinella spiralis*	Soaking	Pig	TsSPI mRNA and protein expression levels in larvae decreased by 68.7% and 68.4%, respectively ($p < 0.05$). RNAi-mediated suppression of TsSPI expression in *T. spiralis* drastically decreased larval infectivity and survival inside the host.	[71]
LsalMS	*Lepeophtheirus salmonis*	Soaking	Salmon Lice	Myosuppressin (LsalMS) knockdown resulted in reduction in the amount of muscle, with skeletal and visceral muscles both showing anomalies in the lice. Additionally, LsalMS down-regulation also had an impact on feeding, spermatophore deposition, and moulting.	[72]
Ubiquitin-63E	*Rhipicephalus microplus;*	Injection	Cattle	All ticks treated with dsRNA did not have offspring. Ubiquitin-63E knockdown was confirmed by microarray and qRT-PCR.	[73]
HBP	*Amblyomma mericanum*	Injection	Cattle	dsRNA specific for histamine binding protein (HBP) transcript was injected into the haemolymph. Less expression of HBP mRNA was shown by molecular evidence, resulting in lower histamine binding ability.	[74]
Subolesin	*Ornithodoros spp.*	Injection	Cattle	By injecting dsRNA to silence the subolesin gene (Silencing 70–78%), oviposition was prevented (88.6% and 95.4%).	[75]
Fhteg1, Fhteg5, Fhteg8,	*Fasciola hepatica*	Soaking	Cattle	92.9% transcript reduction of Fhteg1, 74.1% transcript reduction of Fhteg5. But no observable effect on phenotype.	[76]
SsGST-mu1	*Sarcoptes scabiei*	Soaking	Sheep	Compared to controls, knockdown caused a 35% decrease in transcription of the target gene.	[77]
Hyaluronidase	*Haemonchus contortus*	Oral feeding	Sheep	A significant drop in worm burden and significant decrease in hyaluronidase activity in siRNA treated group.	[78]

The data from Table 1 demonstrates that RNAi has the potential to be successfully applied for parasite management and improve animal husbandry practices. RNAi technology has been described as a novel method for determining gene function and has provided an appealing disease management approach in parasitic nematodes. Out of the examples illustrated, the most promising results were obtained for nematode management, highlighting its potential for improving pig health. More than half of the studies focusing on tick control were on *Ixodidae* and *Argasidae* families, covering the following species, *Amblyomma mericanum, Ixodes scapularis, Haemaphysalis longicornis, Rhipicephalus sanguineus, Dermacentor marginatus, and Boophilus microlpolu.*

The field application of RNAi product for pest control In animal is still in an early stage. One specific research example is the application of RNAi against *Haemonchus contortus* commonly known as barber's pole worm in sheep. Researchers successfully used siRNA treatment for 8000 barber's pole worms [78]. They used a dosage of ~1 mg/10,000 larval to observe a RNAi effect. For this study, the quantity of siRNA needed for RNAi effect was ~1 mg per animal [78]. In animal agriculture, each animal needs to be treated individually against pests and the exact quantity of RNA precursors required RNAi silencing can be very subjective depending on various factors like: (1) Size, animal types, animal density and living conditions of the farm. (2) Delivery system (nanoparticle/directly) and methods (injection/feeding). (3) Knockdown of genes with different function. (4) Species, stages of the pest. (5) Infection abilities associated with different pest and animal also need to be considered.

DsRNA has been delivered to the various tick species using different techniques, including microinjection, soaking, and feeding. As a major parasite of cattle, research studies focusing on RNAi strategy for tick species have achieved considerable gene knockdown or mortality increase [67,68]. Many parasites that affect livestock remain significant research subjects for flystrike in sheep, anaplasmosis in cattle, nematode infection in pigs and ectoparasites of salmon. While large-scale production of dsRNA has been expensive in the past, there has been a significant drop in production costs. This is due to the development of mass production systems, such as bacterial expression vectors for in vivo synthesis [79]. Thus, the use of RNAi as biopesticide has an excellent potential as control strategy for these pests affecting livestock industries.

4. Regulatory Framework

The excitement surrounding the development of RNAi for different agricultural applications has been building for years, with the full breadth of these potential benefits across many areas being thrown into sharp focus. Specifically, RNAi has the potential to replace conventional chemical pesticides with the ability to target genes selectively, and therefore only induce mortality in the target organism. As an emerging agricultural technology, RNAi has already shown its appealing capabilities in multiple areas. RNAi for developing sustainable agriculture has growing number of publications and patents (according to a search in the World Intellectual Property Organization database using the terms 'RNAi' and 'agriculture', ~1100 patent applications have been lodged), yet very few agricultural applications have made it to market. Despite the potential of RNAi there are still many challenges to be addressed to enable its application in the field.

One important limitation in the development of regulatory framework for RNAi include difference in the efficacy of each dsRNA under environmental conditions. RNAi-based products can be applied via some of the delivery methods described above. The exogenous spray application of RNAi as SIGS to improve plant health has the potential to become a fast-growing market. Despite existing framework available for regulatory assessment and approval of genetically modified plants, appropriate safety evaluations, and authorization procedures for SIGS-based products are less clear [80]. The degradation profile of each target dsRNA entering the environment needs to be extensively studied to assess the risk of unanticipated persistence [81]. The three environmental and highly researched microcosms of interest in this context are soil, surface water and sediment. Within soil foliarly applied dsRNA degradation rate for DT_{50} (Time to 50% degradation) and DT_{90} (Time to 90% degradation) was <30 h and <35 h, respectively, [81]. Interesting enough, insect bioassays demonstrate that the loss of functional capacity of dsRNA is unlikely to be caused by soil texture, clay content, pH, or other differences. Instead, it is commonly believed that microbial degradation is the primary driver of dsRNA degradation [81].Fischer et al., demonstrated that dsRNA remained stable in sterile water for three days. While this information lays a good foundation of dsRNA degradation in the ecosystem suggesting low risk of persistence [82]. As RNAi is a rapidly evolving field with most research focused on model organisms it is still not fully understood, making the design and implementation of regulations to be in constant flux.

Safe consumption of dsRNAs by humans and other vertebrates holds a long history; as dsRNA and its processed products are natural components of food and feed [83,84]. Oral uptake of dsRNA or siRNA by humans or farm animals has been reported to have negligible impact due to degradation and multiple barriers in the gastrointestinal tract of mammals [85,86]. The regulatory status of RNAi technology is different for each country. Topical dsRNA will be the likely treatment for animal agriculture, hence topical dsRNA regulatory approval is discussed below for selected jurisdictions. European Union (EU) regulations have been initially set on chemicals as active substances, no specific guidance documents defining the data requirements for the authorization of dsRNA-based application for plant protection are outlined. However, "the Commission may ... adopt or amend technical and other guidance documents such as explanatory notes, guidance documents on the content of the application concerning microorganisms, pheromones, and biological products, for the implementation of this regulation. The Commission may ask the Authority to prepare or to contribute to such guidance documents" [87].

The Environmental Protection Authority (EPA) in New Zealand regulates RNAi under the Hazardous Substances and New Organisms (HSNO) Act. The NZ EPA and Decision-Making Committee decided that due to topical dsRNA lacking the ability to alter the genetic code, it is out of the scope of HSNO, and instead may be classed as a hazardous substance. However, EPA databases of hazardous chemical databases does not list dsRNA, making the regulations on dsRNA almost non-existent [88]. In the United States, pesticide goods must be registered with the US Environmental Protection Agency (EPA) before they can be manufactured, transported, or sold [89]. In 2017, the EPA approved the first RNAi pesticide [90]. SmartStax Pro is a GM maize seed that will use both transgenic insecticidal proteins and RNAi to combat western and northern corn rootworm [11,90]. The EPA's standards in Title 40 of the Code of Federal Regulations (CFR) Part 158 lay out the data requirements for assessing pesticide products [91]. These data criteria aid in the evaluation of pesticide active components as well as pesticide products [91]. In addition to traditional chemical pesticide data requirements, 40 CFR Part 158 defines data requirements for biochemical and microbiological insecticides, which are often less toxic intrinsically than conventional chemical pesticides [91].

In Australia, as of the 8 October 2019, the Gene Technology Regulations 2001 legislation was amended, to include a provision that explained that gene technology does not include approaches involving the introduction of RNA into an organism, if the RNA cannot be translated into a polypeptide, an infectious agent cannot be produced, or the organisms' genomic sequence is not altered as a result. If these conditions are met, the organism treated with dsRNA is not classified as a GMO under the Gene Technology Act of 2000 and is thus not regulated by the OTGR [34]. However, topically applied dsRNA is still subject to regulatory oversight by the Australian Pesticides and Veterinary Medicines Authority (APVMA) under the category of agricultural chemical products. To be approved by the APVMA, appropriate evidence must be presented for human health and safety, worker health and safety, environmental fate and toxicity, target efficacy and crop safety, foreign trade implications, and chemistry and manufacturing processes [34].

RNAi for controlling pests (O: *Diptera, Hemiptera, Lepidoptera, Coleoptera*), which affect plants/crops are proving to be efficient, hence promoting the transition of the novel and innovative RNAi to the market [92,93]. This existing data on RNAi research available to regulators to identify, assess, manage, and mitigate risks can help expand on traditional laws, biosafety regulations, market-based or economic regulatory schemes to create a specific niche assessment for non-GMO RNAi-products globally. The legal regulatory framework surrounding RNAi is looking more promising than ever due to the completed assessment to date and the regulatory framework already in places such as Australia regarding SIGS-products. While these outcomes have been focused on the plant/crop sector, they will have a profound impact on our knowledge of gene silencing and inform the way we apply this technology to improve the productivity of livestock.

5. Conclusions

RNAi-based biological controls can offer high target specificity, non-GM, environmentally friendly strategy against pests and pathogens. As described above, the RNA-mediated management for animal agriculture is still in its infancy and substantial research focus will be required for it to reach its full potential. Additionally, RNAi applications also need to be explored and adopted from the regulatory and community acceptance aspects of this technology. The breadth and success reported for RNAi studies conducted to control pests affecting plants, suggest dsRNA has a promising future in addressing current deficiencies in the control of livestock pests. If development is conducted diligently and thoroughly, encompassing bioinformatic identification, in silico best-design of RNAi effectors, and laboratory- and field-based RNAi toxicity studies, RNAi-based biopesticides have the potential to revolutionize pest management in a safe, specific, and effective manner. For successful development of RNAi strategies to improve animal health this issue must be addressed by understanding the movement of the dsRNA within the pest of interest to improve the penetration and persistence in the environment.

Author Contributions: Conceptualization, K.T.M.; methodology, writing, review, and editing—Y.Y. (Yakun Yan), T.J.M., P.S.M., Y.Y. (Yunjia Yang), N.M. and K.T.M. All authors have read and agreed to the published version of the manuscript.

Funding: This research was funded by Advance Queensland Industry Research Fellowship with grant number CTS 25856/19.

Institutional Review Board Statement: Not applicable.

Informed Consent Statement: Not applicable.

Acknowledgments: Y.Y. (Yunjia Yang), T.M., P.M., Y.Y. (Yakun Yan), N.M. and K.M. would like to thank the support of The University of Queensland and Queensland Alliance of Agriculture and Food Innovation (QAAFI) and School of Chemistry and Biology (SCMB). This project is supported by K.M.'s Advance Queensland Industry Research Fellowship by Queensland Government, Queensland, Australia and YY's (Yunjia Yang and Yakun Yan) Ph.D. scholarship from The University of Queensland, Brisbane, Queensland, Australia.

Conflicts of Interest: The authors declare no conflict of interest.

References

1. FAO. *The State of Food and Agriculture 2001*; Food & Agriculture Organization: Rome, Italy, 2001.
2. Savary, S.; Willocquet, L.; Pethybridge, S.J.; Esker, P.D.; McRoberts, N.; Nelson, A.D. The global burden of pathogens and pests on major food crops. *Nat. Ecol. Evol.* **2019**, *3*, 430–439. [CrossRef]
3. Narladkar, B. Projected economic losses due to vector and vector-borne parasitic diseases in livestock of India and its significance in implementing the concept of integrated practices for vector management. *Vet. World* **2018**, *11*, 151. [CrossRef]
4. Worrall, E.; Hamid, A.; Mody, K.; Mitter, N.; Pappu, H. Nanotechnology for Plant Disease Management. *Agronomy* **2018**, *8*, 285. [CrossRef]
5. Chagnon, M.; Kreutzweiser, D.; Mitchell, E.A.D.; Morrissey, C.A.; Noome, D.A.; Van der Sluijs, J.P. Risks of large-scale use of systemic insecticides to ecosystem functioning and services. *Environ. Sci. Pollut. Res. Int.* **2014**, *22*, 119–134. [CrossRef]
6. Bai, G.; Shaner, G. Management and resistance in wheat and barley to Fusarium head blight. *Annu. Rev. Phytopathol.* **2004**, *42*, 135–161. [CrossRef] [PubMed]
7. Sala, S.; Vighi, M. GIS-based procedure for site-specific risk assessment of pesticides for aquatic ecosystems. *Ecotoxicol. Environ. Saf.* **2008**, *69*, 1–12. [CrossRef]
8. Karimzadeh, R.; Hejazi, M.J.; Helali, H.; Iranipour, S.; Mohammadi, S.A. Assessing the impact of site-specific spraying on control of Eurygaster integriceps (Hemiptera: Scutelleridae) damage and natural enemies. *Precis. Agric.* **2011**, *12*, 576–593. [CrossRef]

9. Kim, Y.H.; Issa, M.S.; Cooper, A.M.; Zhu, K.Y. RNA interference: Applications and advances in insect toxicology and insect pest management. *Pestic. Biochem. Physiol.* **2015**, *120*, 109–117. [CrossRef]
10. Jain, R.G.; Robinson, K.E.; Fletcher, S.J.; Mitter, N. RNAi-based functional genomics in Hemiptera. *Insects* **2020**, *11*, 557. [CrossRef]
11. Head, G.P.; Carroll, M.W.; Evans, S.P.; Rule, D.M.; Willse, A.R.; Clark, T.L.; Storer, N.P.; Flannagan, R.D.; Samuel, L.W.; Meinke, L.J. Evaluation of SmartStax and SmartStax PRO maize against western corn rootworm and northern corn rootworm: Efficacy and resistance management. *Pest Manag. Sci.* **2017**, *73*, 1883–1899. [CrossRef]
12. Rodrigues, T.; Sridharan, K.; Manley, B.; Cunningham, D.; Narva, K. Development of dsRNA as a sustainable bioinsecticide: From laboratory to field. In *Crop Protection Products for Sustainable Agriculture*; ACS Publications: Washington, DC, USA, 2021; pp. 65–82.
13. Sammons, R.D.; Ivashuta, S.I.; Liu, H.; Wang, D.; Feng, P.C.; Kouranov, A.Y.; Andersen, S.E. Polynucleotide Molecules for Gene Regulation in Plants. US Patent 2011/0296556 A1, 1 December 2011.
14. Maule, A.G.; McVeigh, P.; Dalzell, J.J.; Atkinson, L.; Mousley, A.; Marks, N.J. An eye on RNAi in nematode parasites. *Trends Parasitol.* **2011**, *27*, 505–513. [CrossRef]
15. McCoy, C.J.; Warnock, N.D.; Atkinson, L.E.; Atcheson, E.; Martin, R.J.; Robertson, A.P.; Maule, A.G.; Marks, N.J.; Mousley, A. RNA interference in adult Ascaris suum–an opportunity for the development of a functional genomics platform that supports organism-, tissue-and cell-based biology in a nematode parasite. *Int. J. Parasitol.* **2015**, *45*, 673–678. [CrossRef]
16. Yang, D.; Zeng, Y.; Sun, X.; Yue, X.; Hu, C.; Jiang, P.; Liu, R.; Ciu, J.; Wang, Z. Trichinella spiralis: RNAi-mediated silencing of serine protease results in reduction of intrusion, development and fecundity. *Trop. Biomed.* **2020**, *37*, 932–946. [PubMed]
17. Almazán, C.; Lagunes, R.; Villar, M.; Canales, M.; Rosario-Cruz, R.; Jongejan, F.; de la Fuente, J. Identification and characterization of Rhipicephalus (Boophilus) microplus candidate protective antigens for the control of cattle tick infestations. *Parasitol. Res.* **2010**, *106*, 471–479. [CrossRef] [PubMed]
18. Schuster, S.; Miesen, P.; van Rij, R.P. Antiviral RNAi in insects and mammals: Parallels and differences. *Viruses* **2019**, *11*, 448. [CrossRef]
19. Chen, J.; Peng, Y.; Zhang, H.; Wang, K.; Zhao, C.; Zhu, G.; Reddy Palli, S.; Han, Z. Off-target effects of RNAi correlate with the mismatch rate between dsRNA and non-target mRNA. *RNA Biol.* **2021**, *18*, 1747–1759. [CrossRef] [PubMed]
20. Napoli, C.; Lemieux, C.; Jorgensen, R. Introduction of a chimeric chalcone synthase gene into petunia results in reversible co-suppression of homologous genes in trans. *Plant Cell* **1990**, *2*, 279–289. [CrossRef]
21. Sen, G.L.; Blau, H.M. A brief history of RNAi: The silence of the genes. *FASEB J.* **2006**, *20*, 1293–1299. [CrossRef]
22. Guo, S.; Kemphues, K.J. par-1, a gene required for establishing polarity in *C. elegans* embryos, encodes a putative Ser/Thr kinase that is asymmetrically distributed. *Cell* **1995**, *81*, 611–620. [CrossRef]
23. Fire, A.; Xu, S.; Montgomery, M.K.; Kostas, S.A.; Driver, S.E.; Mello, C.C. Potent and specific genetic interference by double-stranded RNA in Caenorhabditis elegans. *Nature* **1998**, *391*, 806–811. [CrossRef]
24. Wytinck, N.; Manchur, C.L.; Li, V.H.; Whyard, S.; Belmonte, M.F. dsRNA uptake in plant pests and pathogens: Insights into RNAi-based insect and fungal control technology. *Plants* **2020**, *9*, 1780. [CrossRef]
25. Bramlett, M.; Plaetinck, G.; Maienfisch, P. RNA-based biocontrols—A new paradigm in crop protection. *Engineering* **2020**, *6*, 522–527. [CrossRef]
26. Darlington, M.; Reinders, J.D.; Sethi, A.; Lu, A.L.; Ramaseshadri, P.; Fischer, J.R.; Boeckman, C.J.; Petrick, J.S.; Roper, J.M.; Narva, K.E. RNAi for Western Corn Rootworm Management: Lessons Learned, Challenges, and Future Directions. *Insects* **2022**, *13*, 57. [CrossRef]
27. Vogel, E.; Santos, D.; Mingels, L.; Verdonckt, T.-W.; Broeck, J.V. RNA interference in insects: Protecting beneficials and controlling pests. *Front. Physiol.* **2019**, *9*, 1912. [CrossRef]
28. Nunes, C.C.; Dean, R.A. Host-induced gene silencing: A tool for understanding fungal host interaction and for developing novel disease control strategies. *Mol. Plant Pathol.* **2012**, *13*, 519–529. [CrossRef]
29. You, L.; Zhang, F.; Huang, S.; Merchant, A.; Zhou, X.; Li, Z. Over-expression of rna interference (rnai) core machinery improves susceptibility to rnai in silkworm larvae. *Insect Mol. Biol.* **2020**, *29*, 353–362. [CrossRef]
30. Zotti, M.; Dos Santos, E.A.; Cagliari, D.; Christiaens, O.; Taning, C.N.T.; Smagghe, G. RNA interference technology in crop protection against arthropod pests, pathogens and nematodes. *Pest Manag. Sci.* **2018**, *74*, 1239–1250. [CrossRef]
31. Huvenne, H.; Smagghe, G. Mechanisms of dsRNA uptake in insects and potential of RNAi for pest control: A review. *J. Insect Physiol.* **2010**, *56*, 227–235. [CrossRef]
32. Saleh, M.-C.; van Rij, R.P.; Hekele, A.; Gillis, A.; Foley, E.; O'Farrell, P.H.; Andino, R. The endocytic pathway mediates cell entry of dsRNA to induce RNAi silencing. *Nat. Cell Biol.* **2006**, *8*, 793–802. [CrossRef]
33. Joga, M.R.; Zotti, M.J.; Smagghe, G.; Christiaens, O. RNAi Efficiency, Systemic Properties, and Novel Delivery Methods for Pest Insect Control: What We Know So Far. *Front. Physiol.* **2016**, *7*, 553. [CrossRef]
34. Fletcher, S.J.; Reeves, P.T.; Hoang, B.T.; Mitter, N. A Perspective on RNAi-Based Biopesticides. *Front. Plant Sci.* **2020**, *11*, 51. [CrossRef] [PubMed]
35. Luck, S.; Kreszies, T.; Strickert, M.; Schweizer, P.; Kuhlmann, M.; Douchkov, D. siRNA-Finder (si-Fi) Software for RNAi-Target Design and Off-Target Prediction. *Front. Plant Sci.* **2019**, *10*, 1023. [CrossRef] [PubMed]
36. Whyard, S.; Singh, A.D.; Wong, S. Ingested double-stranded RNAs can act as species-specific insecticides. *Insect Biochem. Mol. Biol.* **2009**, *39*, 824–832. [CrossRef] [PubMed]

37. Katoch, R.; Sethi, A.; Thakur, N.; Murdock, L.L. RNAi for insect control: Current perspective and future challenges. *Appl. Biochem. Biotechnol.* **2013**, *171*, 847–873. [CrossRef] [PubMed]
38. Feinberg, E.H.; Hunter, C.P. Transport of dsRNA into cells by the transmembrane protein SID-1. *Science* **2003**, *301*, 1545–1547. [CrossRef]
39. Okamura, K.; Chung, W.-J.; Ruby, J.G.; Guo, H.; Bartel, D.P.; Lai, E.C. The Drosophila hairpin RNA pathway generates endogenous short interfering RNAs. *Nature* **2008**, *453*, 803–806. [CrossRef]
40. Mourrain, P.; Béclin, C.; Elmayan, T.; Feuerbach, F.; Godon, C.; Morel, J.-B.; Jouette, D.; Lacombe, A.-M.; Nikic, S.; Picault, N. Arabidopsis SGS2 and SGS3 genes are required for posttranscriptional gene silencing and natural virus resistance. *Cell* **2000**, *101*, 533–542. [CrossRef]
41. Price, D.R.; Gatehouse, J.A. RNAi-mediated crop protection against insects. *Trends Biotechnol.* **2008**, *26*, 393–400. [CrossRef]
42. Alper, S.; McBride, S.J.; Lackford, B.; Freedman, J.H.; Schwartz, D.A. Specificity and complexity of the Caenorhabditis elegans innate immune response. *Mol. Cell. Biol.* **2007**, *27*, 5544–5553.
43. Zhu, K.Y.; Palli, S.R. Mechanisms, applications, and challenges of insect RNA interference. *Annu. Rev. Entomol.* **2020**, *65*, 293–311. [CrossRef]
44. Wei, H.; Tan, S.; Yan, S.; Li, Z.; Shen, J.; Liu, X. Nanocarrier-mediated transdermal dsRNA-NPF1 delivery system contributes to pest control via inhibiting feeding behavior in Grapholita molesta. *J. Pest Sci.* **2022**, *95*, 983–995. [CrossRef]
45. Yan, S.; Ren, B.Y.; Shen, J. Nanoparticle-mediated double-stranded RNA delivery system: A promising approach for sustainable pest management. *Insect Sci.* **2021**, *28*, 21–34. [CrossRef] [PubMed]
46. Kennerdell, J.R.; Carthew, R.W. Use of dsRNA-mediated genetic interference to demonstrate that frizzled and frizzled 2 act in the wingless pathway. *Cell* **1998**, *95*, 1017–1026. [CrossRef]
47. Yu, N.; Christiaens, O.; Liu, J.; Niu, J.; Cappelle, K.; Caccia, S.; Huvenne, H.; Smagghe, G. Delivery of dsRNA for RNAi in insects: An overview and future directions. *Insect Sci.* **2013**, *20*, 4–14. [CrossRef]
48. Tomoyasu, Y.; Denell, R.E. Larval RNAi in Tribolium (Coleoptera) for analyzing adult development. *Dev. Genes Evol.* **2004**, *214*, 575–578. [CrossRef]
49. Bolognesi, R.; Ramaseshadri, P.; Anderson, J.; Bachman, P.; Clinton, W.; Flannagan, R.; Ilagan, O.; Lawrence, C.; Levine, S.; Moar, W. Characterizing the mechanism of action of double-stranded RNA activity against western corn rootworm (Diabrotica virgifera virgifera LeConte). *PLoS ONE* **2012**, *7*, e47534. [CrossRef]
50. Narva, K.E.; Siegfried, B.D.; Storer, N.P. Transgenic approaches to western corn rootworm control. In *Yellow Biotechnology II*; Springer: Heidelberg, Germany, 2013; Volume 163, pp. 135–162.
51. Pinheiro, D.H.; Taylor, C.E.; Wu, K.; Siegfried, B.D. Delivery of gene-specific dsRNA by microinjection and feeding induces RNAi response in Sri Lanka weevil, Myllocerus undecimpustulatus undatus Marshall. *Pest Manag. Sci.* **2020**, *76*, 936–943. [CrossRef] [PubMed]
52. Zhang, Y.H.; Ma, Z.Z.; Zhou, H.; Chao, Z.J.; Yan, S.; Shen, J. Nanocarrier-delivered dsRNA suppresses wing development of green peach aphids. *Insect Sci.* **2022**, *29*, 669–682. [CrossRef] [PubMed]
53. Wang, M.; Jin, H. Spray-Induced Gene Silencing: A Powerful Innovative Strategy for Crop Protection. *Trends Microbiol.* **2017**, *25*, 4–6. [CrossRef]
54. Dalakouras, A.; Wassenegger, M.; McMillan, J.N.; Cardoza, V.; Maegele, I.; Dadami, E.; Runne, M.; Krczal, G.; Wassenegger, M. Induction of silencing in plants by high-pressure spraying of in vitro-synthesized small RNAs. *Front. Plant Sci.* **2016**, *7*, 1327. [CrossRef]
55. Zhang, Y.; Li, C.; Wang, Y.; Hu, Y.; Christie, P.; Zhang, J.; Li, X. Maize yield and soil fertility with combined use of compost and inorganic fertilizers on a calcareous soil on the North China Plain. *Soil Tillage Res.* **2016**, *155*, 85–94. [CrossRef]
56. Zheng, Y.; Hu, Y.; Yan, S.; Zhou, H.; Song, D.; Yin, M.; Shen, J. A polymer/detergent formulation improves dsRNA penetration through the body wall and RNAi-induced mortality in the soybean aphid Aphis glycines. *Pest Manag. Sci.* **2019**, *75*, 1993–1999. [CrossRef]
57. Yan, S.; Qian, J.; Cai, C.; Ma, Z.; Li, J.; Yin, M.; Ren, B.; Shen, J. Spray method application of transdermal dsRNA delivery system for efficient gene silencing and pest control on soybean aphid Aphis glycines. *J. Pest Sci.* **2020**, *93*, 449–459. [CrossRef]
58. Zhang, X.; Zhang, J.; Zhu, K. Chitosan/double-stranded RNA nanoparticle-mediated RNA interference to silence chitin synthase genes through larval feeding in the African malaria mosquito (*Anopheles gambiae*). *Insect Mol. Biol.* **2010**, *19*, 683–693. [CrossRef]
59. Mitter, N.; Worrall, E.A.; Robinson, K.E.; Li, P.; Jain, R.G.; Taochy, C.; Fletcher, S.J.; Carroll, B.J.; Lu, G.; Xu, Z.P. Clay nanosheets for topical delivery of RNAi for sustained protection against plant viruses. *Nat. Plants* **2017**, *3*, 1–10. [CrossRef]
60. Worrall, E.A.; Bravo-Cazar, A.; Nilon, A.T.; Fletcher, S.J.; Robinson, K.E.; Carr, J.P.; Mitter, N. Exogenous application of RNAi-inducing double-stranded RNA inhibits aphid-mediated transmission of a plant virus. *Front. Plant Sci.* **2019**, *10*, 265. [CrossRef] [PubMed]
61. Lechanteur, A.; Sanna, V.; Duchemin, A.; Evrard, B.; Mottet, D.; Piel, G. Cationic liposomes carrying siRNA: Impact of lipid composition on physicochemical properties, cytotoxicity and endosomal escape. *Nanomaterials* **2018**, *8*, 270. [CrossRef] [PubMed]
62. Zuhorn, I.S.; Engberts, J.B.; Hoekstra, D. Gene delivery by cationic lipid vectors: Overcoming cellular barriers. *Eur. Biophys. J.* **2007**, *36*, 349–362. [CrossRef] [PubMed]
63. Mody, K.T.; Zhang, B.; Li, X.; Jain, R.; Li, P.; James, P.; Mahony, T.J.; Xu, Z.; Mitter, N. Topical RNAi for Sustainable Animal Health. *Multidiscip. Digit. Publ. Inst. Proc.* **2020**, *36*, 170.

64. Lichtenberg, S.S.; Laisney, J.; Elhaj Baddar, Z.; Tsyusko, O.V.; Palli, S.R.; Levard, C.; Masion, A.; Unrine, J.M. Comparison of nanomaterials for delivery of double-stranded RNA in Caenorhabditis elegans. *J. Agric. Food Chem.* **2020**, *68*, 7926–7934. [CrossRef]
65. Chen, N.; Xu, M.-J.; Nisbet, A.J.; Huang, C.-Q.; Lin, R.-Q.; Yuan, Z.-G.; Song, H.-Q.; Zhu, X.-Q. Ascaris suum: RNAi mediated silencing of enolase gene expression in infective larvae. *Exp. Parasitol.* **2011**, *127*, 142–146. [CrossRef]
66. Jex, A.R.; Liu, S.; Li, B.; Young, N.D.; Hall, R.S.; Li, Y.; Yang, L.; Zeng, N.; Xu, X.; Xiong, Z. Ascaris suum draft genome. *Nature* **2011**, *479*, 529–533. [CrossRef]
67. Peter, R.; Van den Bossche, P.; Penzhorn, B.L.; Sharp, B. Tick, fly, and mosquito control—Lessons from the past, solutions for the future. *Vet. Parasitol.* **2005**, *132*, 205–215. [CrossRef] [PubMed]
68. de la Fuente, J.; Estrada-Pena, A.; Venzal, J.M.; Kocan, K.M.; Sonenshine, D.E. Overview: Ticks as vectors of pathogens that cause disease in humans and animals. *Front. Biosci.* **2008**, *13*, 6938–6946. [CrossRef]
69. Fang, E.; Liu, X.; Li, M.; Zhang, Z.; Song, L.; Zhu, B.; Wu, X.; Liu, J.; Zhao, D.; Li, Y. Advances in COVID-19 mRNA vaccine development. *Signal Transduct. Target. Ther.* **2022**, *7*, 94. [CrossRef]
70. Xu, M.; Chen, N.; Song, H.; Lin, R.; Huang, C.; Yuan, Z.; Zhu, X. RNAi-mediated silencing of a novel Ascaris suum gene expression in infective larvae. *Parasitol. Res.* **2010**, *107*, 1499–1503. [CrossRef]
71. Yi, N.; Yu, P.; Wu, L.; Liu, Z.; Guan, J.; Liu, C.; Liu, M.; Lu, Y. RNAi-mediated silencing of Trichinella spiralis serpin-type serine protease inhibitors results in a reduction in larval infectivity. *Vet. Res.* **2020**, *51*, 139. [CrossRef]
72. Komisarczuk, A.Z.; Kongshaug, H.; Li, M.; Nilsen, F. RNAi mediated myosuppressin deficiency affects muscle development and survival in the salmon louse (*Lepeophtheirus salmonis*). *Sci. Rep.* **2019**, *9*, 6944. [CrossRef]
73. Lew-Tabor, A.; Kurscheid, S.; Barrero, R.; Gondro, C.; Moolhuijzen, P.; Valle, M.R.; Morgan, J.; Covacin, C.; Bellgard, M. Gene expression evidence for off-target effects caused by RNA interference-mediated gene silencing of Ubiquitin-63E in the cattle tick Rhipicephalus microplus. *Int. J. Parasitol.* **2011**, *41*, 1001–1014. [CrossRef]
74. Aljamali, M.; Bior, A.; Sauer, J.; Essenberg, R. RNA interference in ticks: A study using histamine binding protein dsRNA in the female tick Amblyomma americanum. *Insect Mol. Biol.* **2003**, *12*, 299–305. [CrossRef]
75. Manzano-Román, R.; Díaz-Martín, V.; Oleaga, A.; Siles-Lucas, M.; Pérez-Sánchez, R. Subolesin/akirin orthologs from Ornithodoros spp. soft ticks: Cloning, RNAi gene silencing and protective effect of the recombinant proteins. *Vet. Parasitol.* **2012**, *185*, 248–259. [CrossRef] [PubMed]
76. McCusker, P.; Toet, H.; Rathinasamy, V.; Young, N.; Beddoe, T.; Anderson, G.; Dempster, R.; McVeigh, P.; McCammick, E.; Wells, D. Molecular characterisation and vaccine efficacy of two novel developmentally regulated surface tegument proteins of Fasciola hepatica. *Vet. Parasitol.* **2020**, *286*, 109244. [CrossRef]
77. Fernando, D.D.; Marr, E.J.; Zakrzewski, M.; Reynolds, S.L.; Burgess, S.T.; Fischer, K. Gene silencing by RNA interference in Sarcoptes scabiei: A molecular tool to identify novel therapeutic targets. *Parasites Vectors* **2017**, *10*, 289. [CrossRef]
78. Yang, X.; Khan, S.; Zhao, X.; Zhang, J.; Nisar, A.; Feng, X. Suppression of hyaluronidase reduces invasion and establishment of Haemonchus contortus larvae in sheep. *Vet. Res.* **2020**, *51*, 106. [CrossRef]
79. Cagliari, D.; Dias, N.P.; Galdeano, D.M.; Dos Santos, E.Á.; Smagghe, G.; Zotti, M.J. Management of pest insects and plant diseases by non-transformative RNAi. *Front. Plant Sci.* **2019**, *10*, 1319. [CrossRef] [PubMed]
80. De Schutter, K.; Taning, C.N.T.; Van Daele, L.; Van Damme, E.J.; Dubruel, P.; Smagghe, G. RNAi-Based Biocontrol Products: Market Status, Regulatory Aspects, and Risk Assessment. *Front. Insect Sci.* **2022**, *1*, 1–9. [CrossRef]
81. Bachman, P.; Fischer, J.; Song, Z.; Urbanczyk-Wochniak, E.; Watson, G. Environmental Fate and Dissipation of Applied dsRNA in Soil, Aquatic Systems, and Plants. *Front. Plant Sci.* **2020**, *11*, 21. [CrossRef]
82. Fischer, J.R.; Zapata, F.; Dubelman, S.; Mueller, G.M.; Uffman, J.P.; Jiang, C.; Jensen, P.D.; Levine, S.L. Aquatic fate of a double-stranded RNA in a sediment–water system following an over-water application. *Environ. Toxicol. Chem.* **2017**, *36*, 727–734. [CrossRef]
83. Dávalos, A.; Henriques, R.; Latasa, M.J.; Laparra, M.; Coca, M. Literature review of baseline information on non-coding RNA (ncRNA) to support the risk assessment of ncRNA-based genetically modified plants for food and feed. *EFSA Support. Publ.* **2019**, *16*, 1688E. [CrossRef]
84. OECD. *Considerations for the Environmental Risk Assessment of the Application of Sprayed or Externally Applied dsRNA-Based Pesticides*; Organisation for Economic Co-Operation and Development (OECD): Paris, France, 2020; Volume 104.
85. Petrick, J.S.; Brower-Toland, B.; Jackson, A.L.; Kier, L.D. Safety assessment of food and feed from biotechnology-derived crops employing RNA-mediated gene regulation to achieve desired traits: A scientific review. *Regul. Toxicol. Pharmacol.* **2013**, *66*, 167–176. [CrossRef]
86. O'Neill, M.J.; Bourre, L.; Melgar, S.; O'Driscoll, C.M. Intestinal delivery of non-viral gene therapeutics: Physiological barriers and preclinical models. *Drug Discov. Today* **2011**, *16*, 203–218. [CrossRef] [PubMed]
87. Commission, E. Regulation (EC) No 1107/2009 of the European Parliament and of the Council of 21 October 2009 concerning the placing of plant protection products on the market and repealing Council Directives 79/117/EEC and 91/414/EEC. *Off. J. Eur. Union L* **2009**, *309*, 1–50.
88. Heinemann, J.A. Should dsRNA treatments applied in outdoor environments be regulated? *Env. Int.* **2019**, *132*, 104856. [CrossRef] [PubMed]

89. Leahy, J.; Mendelsohn, M.; Kough, J.; Jones, R.; Berckes, N. Biopesticide oversight and registration at the US Environmental Protection Agency. *Biopestic. State Art Future Oppor.* **2014**, *1172*, 3–18.
90. Shaffer, L. Inner Workings: RNA-based pesticides aim to get around resistance problems. *Proc. Natl. Acad. Sci. USA* **2020**, *117*, 32823–32826. [CrossRef]
91. Dietz-Pfeilstetter, A.; Mendelsohn, M.; Gathmann, A.; Klinkenbuß, D. Considerations and regulatory approaches in the USA and in the EU for dsRNA-based externally applied pesticides for plant protection. *Front. Plant Sci.* **2021**, *12*, 974. [CrossRef]
92. Cooper, A.M.; Silver, K.; Zhang, J.; Park, Y.; Zhu, K.Y. Molecular mechanisms influencing efficiency of RNA interference in insects. *Pest Manag. Sci.* **2019**, *75*, 18–28. [CrossRef]
93. Baum, J.; Roberts, J. Progress Towards RNAi-Mediated Insect Pest Management. *Adv. Insect Physiology.* **2014**, *47*, 249–295.

applied biosciences

MDPI

Review

Multifilm Mass Transfer and Time Constants for Mass Transfer in Food Digestion: Application to Gut-on-Chip Models

Timothy A. G. Langrish

School of Chemical and Biomolecular Engineering, Faculty of Engineering, The University of Sydney, Sydney, NSW 2006, Australia; timothy.langrish@sydney.edu.au; Tel.: +61-2-9351-4568

Abstract: This review highlights the involvement of mass transfer in animal food-digestion processes. There may be several mass-transfer steps during the dissolution of food components, starting from the food itself, moving into the digestive juices, then moving through the walls of the gastrointestinal tract. These steps create a sequence of film resistances to mass transfer, where one film resistance often limits the overall mass-transfer process. Mass-transfer rates, mass-transfer coefficients, and the time scales and time constants for different parts of the food-digestion process are all interlinked, and the connections have been explained. In some parts of the food-digestion process, the time constants for the mass-transfer process are similar to the residence times for food digestion, emphasising the importance of mass transfer in these parts of food digestion, such as the duodenum. The mass-transfer and transport behaviour for in vivo human digestive systems and in vitro guts-on-a-chip may be very similar, suggesting that cells on the intestine walls, whether in vitro (guts-on-a-chip) or in vivo, may see similar transport behaviour for both nutrients towards the cells, and waste products away from them.

Keywords: in vitro; food digestion; mass-transfer coefficients; gut-on-a-chip; reaction engineering; time constants

Citation: Langrish, T.A.G. Multifilm Mass Transfer and Time Constants for Mass Transfer in Food Digestion: Application to Gut-on-Chip Models. *Appl. Biosci.* **2022**, *1*, 101–112. https://doi.org/10.3390/applbiosci1020007

Academic Editor: Robert Henry

Received: 17 May 2022
Accepted: 23 June 2022
Published: 24 June 2022

1. Introduction

The transport of nutrients from foods inside the gastrointestinal tract to the bloodstream is an important part of all animal digestion, including that of humans, and nutrients are examples of solutes. This solute transport is an example of mass transfer, because various nutrient (solute) components of the food may be transported at different mass-transfer rates. There appear to be two main general schools of thought about in vitro food digestion: those that emphasise the importance of the chemistry [1,2] and those that emphasise the geometry and operational appearance of the in vitro system [3]. One theory that connects these two points of emphasis is mass-transfer theory.

When discussing mass-transfer theory, the focus here is specifically on the fundamentals of the mass-transfer process, from the basic equation to the implementation in food digestion. The overall equation for the mass-transfer rate of any component [4] is

$$N_A = K\,A\,(C_1 - C_2) \tag{1}$$

Here, N_A is the mass-transfer rate (kg s^{-1}), K is the overall mass-transfer coefficient (m·s^{-1}), A is the interfacial surface area (between phases, m^2), and C_1 and C_2 are the concentrations inside the food and inside the bloodstream, respectively (kg m^{-3}).

An underlying theme of some reviews of food digestion [3] is the suggestion that the "realism" of the in vitro food-digestion system is critical in achieving good agreement between in vitro and in vivo food-digestion outcomes. In this context, "realism" appears to be connected with the geometry, shape and morphology and anatomy of the in vitro system. It is suggested here that the main reason why the morphology and anatomy matter is because they affect the surface area (A) and the mass-transfer coefficient (K).

However, the same values of these parameters (A and K) may be obtained from different geometries and operating conditions. Apart from the surface area (A), which is a fairly obvious parameter affecting the mass-transfer rate, the mass-transfer coefficient, which is not such an obvious parameter as the surface area, is the main parameter affecting the mass-transfer rate between food and the gastric juices and between the gastric juices and the walls of the gastrointestinal tract. The main parameters affecting the concentrations and concentration differences are the state of the biochemical environment, including the chemicals (enzymes, reactants, products) and the microorganisms that are present. It is at least possible to keep the biochemical environment the same, or very similar, in in vivo and in vitro systems.

If there are two systems—for example, a real in vivo system and any in vitro (model) system—they should behave in the same way from a mass-transfer and biochemical-reaction perspective if the mass-transfer coefficient is the same and if all aspects of the biochemical environment (enzymes, solutions, microorganisms) are also the same. These considerations mean that the reasons for disagreement between in vivo and in vitro systems may be seen in the light of differences in surface areas in mass-transfer coefficients and in the biochemical environments. It is possible to suggest that the digestion process is more complicated than this apparently simple picture, but the complications are included in this theory. For example, particle-size reduction is part of the digestion process, and according to the above perspective, the particle size of a food affects the mass-transfer coefficient. Hence, particle-size reduction fits into the mass-transfer perspective, and particle size also affects the residence time of food in the gastrointestinal system [5].

The comparison of in vivo and in vitro (model) systems has been extensively reviewed in the literature [3,6], and interest continues [7,8] in the development and use of in vitro systems. The perspective given in this paper (mass transfer) may be useful to translate results from the static to the dynamic simulations, because the dynamics of gastrointestinal microbiome composition are associated with many factors, including age, nutrition, health status, and drug treatments. A "unit operations" approach to considering food digestion as a multiscale process has been discussed by Bornhorst et al. [9], which mentions mass transfer and discusses it qualitatively.

The aim of this review is to show several ways in which mass-transfer theory can contribute to the understanding of food digestion. In the first section, it is shown that there may be several mass-transfer resistances at any point in a system, and these resistances tend to be additive, as will now be reviewed. In the second section, the time constants for mass transfer will be reviewed, showing how they contribute a time scale to the understanding of various sections of the gastrointestinal (GI) tract. Finally, an example will be given, showing how the calculations of individual mass-transfer coefficients for the time constants can also be used to estimate overall multifilm mass-transfer resistances and coefficients, and a gut-on-a chip system will also be reviewed in this context.

2. Multifilm Mass-Transfer Theory

As shown in Figure 1, this transport process, as described in this paper, includes several mass-transfer resistances, including the resistance inside a food (R_1), outside the food in the external boundary layer (R_2), outside the walls of the intestine in another boundary layer (R_3), inside the walls of the intestine (R_4), and between the intestine walls and the bloodstream (R_5) [4]. The internal mass-transfer resistance inside a food may be zero if the food is a pure substance, which is very rare. There will always be external mass-transfer resistances (R_2 and R_3) because of boundary layers that must form, due to velocity differences between the walls or external surfaces of the food and of the intestine, on one hand, and the bulk of the solutions inside the gastrointestinal tract, on the other hand.

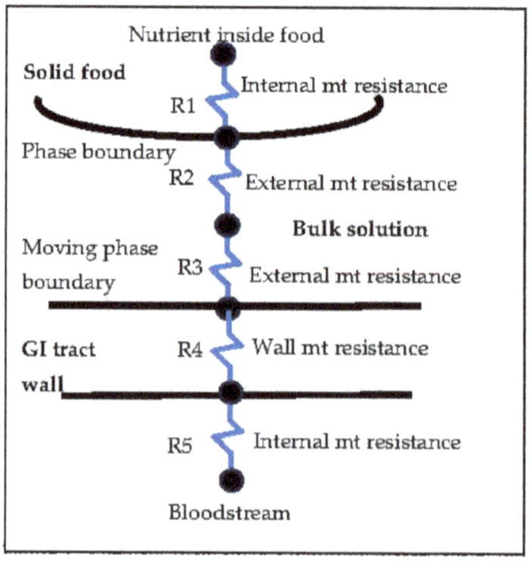

Figure 1. Schematic diagram of a multifilm mass-transfer model for food digestion.

Two-film mass-transfer theory [4] can be extended to this situation, giving the equation

$$\frac{1}{K\,A} = R_1 + R_2 + R_3 + R_4 + R_5 = \frac{1}{k_1\,A_1} + \frac{H_{12}}{k_2\,A_2} + \frac{H_{13}}{k_3\,A_3} + \dots = \frac{1}{k_1\,A_1} + \sum_{i=2}^{n} \frac{H_{1i}}{k_i\,A_i} \tag{2}$$

where the subscript i refers to each of the boundary layers or film mass-transfer resistances shown in Figure 1, R_i is the mass-transfer resistance for each part of the transport process, A_i is the corresponding interfacial surface area, k_i is the film mass-transfer coefficient, K is the overall coefficient, and H_{1i} is the equilibrium or partition coefficient between the food (1) and each location in Figure 1.

A potential objection to the use of two-film mass-transfer theory is that, anecdotally, some might say that it has been developed for steady-state mass transfer and therefore that it cannot be applied to the inherently unsteady-state mass-transfer situation in food digestion. However, this is not true. The development of two-film theory for steady-state mass transfer does not limit its application to steady-state situations, due to the following part of the derivation (e.g., pp. 238–239, [4]). The only requirement in two-film theory for steady state is that the interface region should be thin, so the flux across it must achieve steady state quickly relative to the rates of change for the bulk concentrations. There is nothing in the derivations that limits the theory to overall steady-state situations. In addition, there are some papers in the literature, such as Tharakan et al. [10], where two-film mass-transfer theory has been applied to food digestion with some success.

A particular feature of Equation (2), which has important implications for the overall mass-transfer rate, is the addition of the inverses of various film mass-transfer resistances, which means that the largest film mass-transfer coefficients have the smallest impact on the overall mass-transfer coefficient. The equation also implies that adding another film to the series of mass-transfer resistances will reduce the overall mass-transfer coefficient. Since the overall mass-transfer coefficient is proportional to the rate of change for the bulk concentration in a system with a fixed geometry, adding another film to the series of mass-transfer resistances will also reduce the rate of bulk concentration change.

A dimensionless group, the Biot number (Bi), has been used to compare the internal and external resistances to heat and mass transfer in many unsteady-state situations, such as drying [11,12]. In the case of a food, the internal resistance to mass transfer might be represented by L/D, with L being the effective half-thickness and D being the diffusion coefficient, while the external resistance to mass transfer might be represented by $1/k$, where

k is the external-film mass-transfer coefficient. Then, the Biot number is given by the ratios of the internal to the external resistances, or $Bi = (L/D)/(1/k) = (Lk/D)$. This dimensionless group is useful for indicating where the main resistance to nutrient movement occurs in a system.

Further reviewing the concept of multifilm mass-transfer theory and its application to the unsteady-state situation of food digestion, the fundamental process of diffusion out of a food can also be viewed from the perspective of mass-transfer film theory, in the following way. As highlighted by Hallström et al. [13], the unsteady-state diffusion process proceeds by a diffusive front moving through a material (shown in Figure 2 for a finite slab, but similar in general shape and behaviour for any shape of material) initially penetrating through the material. This penetration period extends until the diffusive front has moved through the material (or halfway, if the penetration occurs from both sides of a slab). The remaining period, the so-called regular regime, involves the concentrations decreasing throughout the material, more quickly at the surface and more slowly in the bulk.

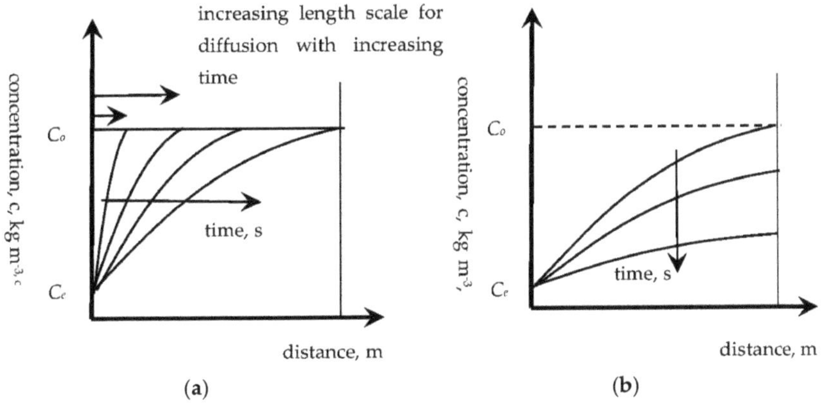

Figure 2. (**a**) penetration period; (**b**) regular regime. Typical concentration profiles as functions of distance for the diffusive movement of a solute through a solid material (typical profiles for a slab, and a cylinder and sphere are generally similar in shape). C_o is the initial concentration, and C_e is the final concentration.

Therefore, the development of a diffusive profile has some similarities with the concept of a receding front, although a receding front typically has a sharper and steeper concentration change and the two transport processes are, of course, different. It is possible to define a distance, L, during the penetration period, over which the main change in concentration occurs, where the distance L varies from zero to the centre, or the centreline, or the halfway point through the solid material. In this case, the overall mass-transfer resistance consists of external and internal resistances, as expressed in the following equation:

$$\frac{1}{K} = \frac{1}{k_1} + \frac{H_{12}}{k_2} \tag{3}$$

Here, k_1 is the external mass-transfer coefficient, H_{12} is the partition or equilibrium coefficient for the solute between the two phases, and k_2 is the internal mass-transfer coefficient, which can be further expressed as the diffusion coefficient (D) divided by the distance over which the diffusion occurs (L). In the early stages, when the solute is just starting to diffuse out of the food, the effective distance over which diffusion occurs (L) is small, so the effective internal mass-transfer coefficient is very high ($k_2 = D/L$), so the effective internal mass-transfer resistance is very low ($H_{12}/k_2 = H_{12}/(D/L)$). This situation means that, in the early stages, the overall mass-transfer coefficient (K) is dominated by the external mass-transfer coefficient (k_1).

This situation also means that as the diffusion of the solute out of the food continues, the effective diffusion distance L increases, increasing the effective internal mass-transfer resistance ($H_{12}/k_2 = H_{12}/(D/L) = L\,H_{12}/D$) and decreasing the effective internal mass-transfer coefficient. This decrease in the internal mass-transfer coefficient decreases the overall mass-transfer coefficient (K) as diffusion increases. It is now shown how this decrease in the overall mass-transfer coefficient affects the overall mass-transfer rate (N_A), as shown in Equation (1), from the perspective of how the bulk concentration (C_b) in the solution surrounding the food changes as a function of time ($d{\cdot}C_b/dt$).

This discussion connects with some tablet dissolution models, particularly the zero-order model, in that zero-order kinetics [14] may be expected if there is a resistive layer of thickness L, with a diffusivity through that layer of D, giving an effective internal mass-transfer coefficient of $k_2 = D/L$, and an effective internal mass-transfer resistance of $H_{12}/k_2 = H_{12}/(D/L)$). A similar type of diffusive resistance across a finite distance may be imagined for the mass transfer of solutes (including reactants and products) across a plant cell wall, as described for the hydrolysis of starch by Li et al. [15].

The external mass-transfer coefficients, outside the foods, have received some attention (R_2 and R_3 in Figure 1) in Langrish et al. [16], who measured these external coefficients outside tablets. They found that the Ranz–Marshall correlation could be used to give a lower estimate of the external coefficients, and this work could be extended to foods, with irregular shapes, using equivalent diameters. The equivalent diameter may be used to estimate the Reynolds number (Re), the fluid properties are then used to estimate the Schmidt number (Sc), and the resulting Sherwood number (Sh) is then used to estimate the external mass-transfer coefficient (k_1). The Reynolds (Re) and Schmidt (Sc) numbers are defined by the following two equations:

$$Re = \frac{\rho\,U\,d}{\mu} \tag{4}$$

$$Sc = \frac{\mu}{\rho\,D} \tag{5}$$

The Ranz–Marshall correlation [17] for mass transfer and the Sherwood number (Sh) are defined as follows:

$$Sh = \frac{k_1\,d}{D} = 2 + 0.6\,Re^{0.6}\,Sc^{0.3} \tag{6}$$

In the above equations, d is the equivalent diameter of the food (m), D is the diffusivity of the solute (nutrient) through the solvent (m$^2{\cdot}$s^{-1}), U is the relative velocity between the food and the solvent (m$\cdot$s^{-1}), and ρ (kg$\cdot$m^{-3}) and μ (kg$\cdot$m$^{-1}\cdot$s^{-1}) are the density and viscosity of the solvent, respectively. The external mass-transfer coefficient, $k_1\cdot$(m$\cdot$s^{-1}), has been discussed previously.

The relationship between the external mass-transfer coefficient and the shear stresses and shear strains in the solid–fluid system for food digestion (as reviewed by Zhong and Langrish [6]) may be estimated using analogies between mass and momentum transfer (fluid mechanics). These analogies allow, for example, momentum-transfer parameters to be estimated from external mass-transfer coefficients. These analogies have a long track record, reviewed in Cussler ([4], pp. 509–511). The fundamental idea is that heat, mass and momentum transfer are all similar processes, limited by transport processes through a boundary layer inside or outside a material (here, a food). The shear stress (τ) is related to the friction factor (f) by the following equation:

$$\tau = f\left(\frac{1}{2}\rho\,u^2\right) \tag{7}$$

The Chilton–Colburn analogy is a well-known way to relate friction factors, heat-transfer coefficients (h) and external mass-transfer coefficients (k):

$$\frac{k}{u} Sc^{2/3} = \frac{h}{\rho \, C_p \, u} Pr^{2/3} = \frac{f}{2} = \frac{\tau}{\rho \, u^2} \tag{8}$$

Here, Pr is the Prandtl number, which is a dimensionless group containing fluid properties for heat transfer, and C_p is the specific heat capacity of the fluid. Equations (7) and (8) allow the shear stresses and the external mass-transfer coefficients to be related to each other and compared, by rearranging Equation (8), as follows:

$$\tau = k \, \rho \, u \, Sc^{2/3} \tag{9}$$

This approach may be used to compare trends in estimated shear stresses with corresponding trends in mass-transfer coefficients and/or dissolution rates.

Hopgood et al. [18] have presented a comparison of two in vitro systems, namely TIMagc (a relatively realistic stomach system, in terms of the geometry) and the industrially common USP-II dissolution apparatus (a formalised beaker and stirrer system)for the dissolution of erosion-limited tablets. Shear rates were predicted in both type of equipment using Computational Fluid Dynamics, and erosion rates were both predicted and measured experimentally. The predicted Reynolds numbers were also compared with in vivo estimates (Re = 0.01–30). The TIMagc apparatus, featuring intermittent flows, showed wall shear rates that were time-dependent, while the rates in the USP-II apparatus were more constant and were directly proportional to the impeller speed. Regarding the Reynolds numbers, they concluded that the conditions (Re) in the TIMagc were mainly in the same range as found with in vivo stomachs, but that the Reynolds numbers at the bottom of the USP-II apparatus were above the range of in vivo conditions for impeller speeds of 25–100 rpm, with the flow regime for these USP-II conditions being turbulent. The overall conclusion was that on the basis of the wall shear rates and range of Reynolds number, the TIMagc apparatus was more likely to successfully approach in vivo dissolution rates for tablets. Mass-transfer coefficients were not compared in this study.

The discussion of mass-transfer coefficients (and hence mass-transfer rates) is also relevant to estimating the time scales over which the mass-transfer processes occur, as will be reviewed in the following section.

3. Time Constants and Time Scale for Food Digestion

As demonstrated by Langrish et al. [16], the time constant may be used to characterise a mass-transfer process, where the time constant for mass transfer (τ, s) may be defined in terms of the mass-transfer coefficient (k, m·s^{-1}), the volume of solution (V, m^3), and the interfacial surface area (A, m^2), $\tau = V/(A\,k)$. The general significance of a time constant is that it characterises the dynamic response of any system [19], such that a period of one time constant means that 63% of a step change in an input will be seen in the output. For periods of two and three time constants, the percentages (of a complete response to a step change in the input) increase to 87% and 95%, respectively.

For each section of the gastrointestinal tract, volumes and surface areas for absorption (including villi, which are small projections which increase the surface areas over the values for a cylinder) may be estimated from the values of the lengths and diameters given by Ritschel [20]. The volumes of each section may be estimated from the volumes of equivalent cylinders with the stated lengths and diameters. In addition, it is possible to estimate mass-transfer coefficients and time constants for each section. These mass-transfer coefficients represent the mass-transfer process from the fluid inside the GI tract to the walls of different parts of this GI tract (R3 in Figure 1).

A sample calculation for the time constant or time scale for the duodenum follows, using a flow velocity of 0.02 m·s^{-1} [3]. Using the absorption area of 0.09 m^2 from Ritschel [20], the length of 0.25 m, and the diameter of 0.05 m from Ritschel [20], with the physical properties of water (density, ρ, 1000 kg·m^{-3}; viscosity, μ, 0.001 kg·m^{-1}·s^{-1}) gives the Reynolds number, according to the equation:

$$Re = \frac{\left(1000 \text{ kg m}^{-3}\right) 0.02 \text{ m s}^{-1} \left(0.05 \text{ m}\right)}{0.001 \text{ kg m}^{-1} \text{ s}^{-1}} = 1000 \tag{10}$$

An appropriate correlation for laminar flow along a tube, for mass transfer to the tube wall, is given in Table 8.3.3 of Cussler [4], with an estimate of the diffusivity being given by a representative value from Table 5.2-1 of Cussler [4], 0.5228 × 10^{-9} m^2·s^{-1}, for sucrose at infinite dilution through water at 25 °C. The correlation gives:

$$\begin{aligned} k &= \left(\frac{D}{d}\right)1.62\left[\frac{d^2U}{L\,D}\right]^{\frac{1}{3}} \\ k &= \left(\frac{0.5228\times10^{-9}\text{ m}^2\text{ s}^{-1}}{0.05\text{ m}}\right)1.62\left[\frac{(0.05\text{ m})^2\ 0.02\text{ m s}^{-1}}{(0.25\text{ m})\ 0.5228\times10^{-9}\text{ m}^2\text{ s}^{-1}}\right]^{\frac{1}{3}} \\ &= 1.23\times10^{-5}\text{ m s}^{-1} \end{aligned} \tag{11}$$

From the length (0.25 m) and the diameter (0.05 m), the volume of this cylindrical region is 0.491 × 10^{-3} m^3. With the previous absorption area of 0.09 m^2, the time constant is then given by the following equation:

$$\tau = \frac{V}{k\,A} = \frac{0.491\times10^{-3}\text{ m}^3}{1.23\times10^{-6}\text{ m s}^{-1}0.09\text{ m}^2}\frac{1\text{ h}}{3600\text{ s}} = 1.23\text{ h} \tag{12}$$

This time constant may be compared with the typical range of residence times in the small intestine stated by Li et al. [3] of 2–5 h, suggesting that the characteristic times for mass transfer (1–3 time constants = 1.2–3.6 h) are a substantial part of the residence time. This comparison further suggests that mass transfer is a significant process in that section (the duodenum) of the gastrointestinal tract. This conclusion is perhaps not too surprising, because the duodenum is well-known [3] to be responsible for a significant part of the absorption processes in the GI tract, and absorption is a mass-transfer process. However, this comparison does highlight and validate the use of mass-transfer time constants for assessing the importance of mass transfer in different sections of the GI tract.

These calculations have been repeated for various sections of the gastrointestinal tract, including calculations for the corresponding Reynolds numbers, mass-transfer coefficients, and time constants, as shown in Table 1. The residence time in the large intestine quoted by Li et al. [3] of 12–24 h may be compared with the time constants from Table 1 of 3.6 h for the cecum and 4.8 h for the colon. Again, the mass-transfer time constants are smaller than the residence time, but the orders of magnitude are similar, pointing to some mass-transfer processes occurring in the large intestine too, such as the absorption of water from foods.

Apart from the mass-transfer process from the fluid solution inside the GI tract to the walls of different parts of this GI tract (R_3 in Figure 1), there is also some relevant research on the mass-transfer process from the outside of the food into the fluid solution inside the GI tract (R_2 in Figure 1). This work was reviewed and extended by Langrish et al. [16], who found that the external mass-transfer coefficients from 13 mm diameter tablets ranged from 0.193–4.48 × 10^{-5} m·s^{-1} in a beaker and stirrer system and 0.222–3.45 × 10^{-5} m·s^{-1} in a copy of the USP dissolution apparatus 2. It is reasonable to suggest that a similar range of external mass-transfer coefficients should be found outside tablets and outside foods of the same sizes and dimensions, since the concept of an external mass-transfer coefficient is that it does not depend on the nature of the internal material. For predicting these external mass-transfer coefficients, the use of dimensional analysis showed that the Ranz–Marshall correlation gave reasonable estimates.

Table 1. Dimensions and absorption areas for various sections of the gastrointestinal tract (from Ritschel [20]), and corresponding Reynolds numbers, mass-transfer coefficients, and time constants, as calculated here. Note that the Reynolds numbers are all laminar, allowing the valid application of Equation (11).

Region	*A*, Absorption Area (m^2)	*V*, Volume (cc)	*L*, Length (cm)	*d*, Diameter (cm)	*Re*, Reynolds Number (-)	*k*, Mass-Transfer Coefficient (m/s)	τ, Time Constant (h)
Duodenum	0.09	491	25	5	1000	1.23×10^{-6}	1.23
Small intestine	120	11,780	600	5	1000	4.26×10^{-7}	0.064
Cecum	0.05	770	20	7	1400	1.18×10^{-6}	3.6
Colon	0.25	2945	150	5	1000	6.77×10^{-7}	4.8
Rectum	0.015	79	16	2.5	500	1.80×10^{-6}	0.81

These external mass-transfer coefficients from tablets or foods are of the order of 10^{-5} m·s^{-1}, which are at least an order of magnitude larger than those listed in Table 1, which are generally around 10^{-6} m·s^{-1}. However, the interfacial areas for the tablets or foods are also smaller than the wall areas for the GI tract. Given these competing considerations, it is not too surprising that the time constants for external mass transfer from the surface of the food into the gastrointestinal fluid may also be significant. Langrish et al. [16] estimated time constants for external mass transfer outside tablets as ranging from 67 s for 0.3 mm diameter tablets to 6750 s for 3 mm objects, or up to 2 h. These time constants cover the same range as the stomach residence times quoted by Zhong and Langrish [6] and Li et al. [3] of 85–120 min and 15 min to 3 h, respectively.

4. Sample Calculation

To give an example of how multifilm mass-transfer theory might be applied to the transfer of food nutrients, consider the dissolution of a 3 mm cube of sugar (sucrose), and its transfer into solution and then through the walls of the gastrointestinal tract, as shown in Figure 3.

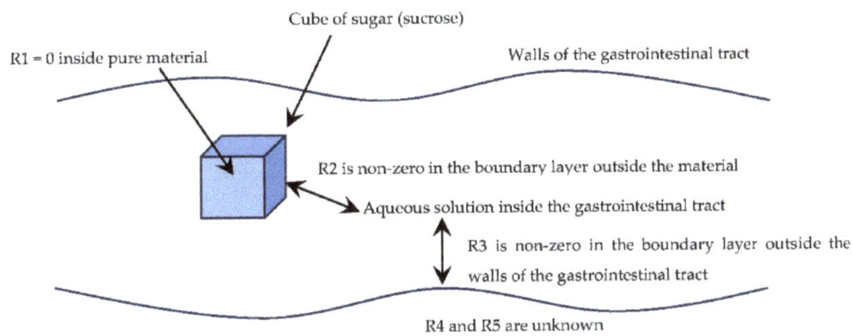

Figure 3. Example of a situation where a cube of sugar (sucrose) dissolves and is transferred through the walls of a sausage casing (references to the mass-transfer resistances, $R_1 \ldots R_5$, follow the nomenclature shown in Figure 1).

In the situation shown in Figure 3, the mass-transfer resistance R_1 is zero, because a pure sugar cube is being considered. Mass-transfer resistance R_2 is certainly not zero, being governed by the external mass-transfer coefficient outside the sugar cube. The mass-transfer resistance R_3 is likewise non-zero, being governed by the external mass-transfer coefficient outside the walls of the gastrointestinal tract. The mass-transfer resistances R_4 and R_5 are unknown, being within the walls of the gastrointestinal tract. Even if these resistances are initially neglected, it is meaningful to compare resistances R_2 and R_3, as follows.

R2, Resistance 2 in Figure 1

In this case, the focus is the external boundary layer outside the tablet, so the approach suggested in Langrish et al. [16] is appropriate. With a flow velocity of 0.02 m·s^{-1} [3], it is necessary to estimate the effective or equivalent diameter of the sugar cube. For a cube, the equivalent diameter, based on the same surface-area-to-volume ratio as a sphere, is the side length of the cube, in this case 0.003 m. The surface area is then 6 × (0.003 m)^2 = 5.4 × 10^{-5} m^2. Using the physical properties of water (density, ρ, 1000 kg·m^{-3}; viscosity, μ, 0.001 kg·m^{-1}·s^{-1}) gives the Reynolds number, according to the equation

$$Re = \frac{\left(1000 \text{ kg m}^{-3}\right) 0.02 \text{ m s}^{-1} \left(0.003 \text{ m}\right)}{0.001 \text{ kg m}^{-1} \text{ s}^{-1}} = 60 \tag{13}$$

An appropriate correlation [17] for flow along a tube, for mass transfer to the tube wall, is given by the Ranz–Marshall equation [16], with the same diffusivity as before, 0.5228 × 10^{-9} m^2·s^{-1}. The correlation gives:

$$\begin{gathered} k = \left(\frac{D}{d}\right)\left[2.0 + 0.6\, Re^{0.5} Sc^{1/3}\right] \\ k = \left(\frac{0.5228\times10^{-9} \text{ m}^2 \text{ s}^{-1}}{0.003 \text{ m}}\right) \\ \left[2.0 + 0.6\,(60)^{0.5}\left(\frac{0.001 \text{ kg m}^{-1} \text{ s}^{-1}}{\left(1000 \text{ kg m}^{-3}\right)0.5228\times10^{-9} \text{ m}^2 \text{ s}^{-1}}\right)^{1/3}\right] \\ k = 1.04 \times 10^{-5} \text{ m s}^{-1} \end{gathered} \tag{1}$$

where Sc is the Schmidt number, as defined in Equation (6). This information may be summarised as k_1 = 1.04 × 10^{-5} m·s^{-1} and A_1 = 5.4 × 10^{-5} m^2.

R3, Resistance 3, Figure 1

This calculation follows from the worked example given above in the section on time constants and time scales, for a sample calculation of the mass-transfer coefficient for the duodenum, between the liquid inside the duodenum and the walls of the gastrointestinal tract, and the same film mass-transfer coefficient applies here (1.23 × 10^{-6} m·s^{-1}). The absorption area is 0.09 m^2, from Ritschel [20]. This information may be summarised as k_2 = 1.23 × 10^{-6} m·s^{-1} and A_2 = 0.09 m^2.

Rtotal: Then the two mass-transfer resistances are as follows:

$$\begin{gathered} \text{R2}: \frac{1}{k_1 A_1} = \frac{1}{1.04\times10^{-5} \text{ m s}^{-1}\left(5.4\times10^{-5} \text{ m}^2\right)} = 1.78 \times 10^9 \text{ s m}^{-3} \\ \text{R3}: \frac{1}{k_2 A_2} = \frac{1}{1.23\times10^{-6} \text{ m s}^{-1}\left(0.09 \text{ m}^2\right)} = 9.04 \times 10^6 \text{ s m}^{-3} \\ \text{Rtotal} = \text{R2} + \text{R3} = 1.78 \times 10^9 + 9.04 \times 10^6 = 1.79 \times 10^9 \text{ s m}^{-3} \end{gathered} \tag{2}$$

Since the overall mass-transfer coefficient is the inverse of the overall mass-transfer resistance, and based on the surface area of the cube, the overall mass-transfer coefficient is then given by the equation

$$K = \frac{1}{\left(5.4 \times 10^{-5} \text{ m}^2\right) 1.79 \times 10^9 \text{ s m}^{-3}} = 1.03 \times 10^{-5} \text{ m s}^{-1} \tag{16}$$

This value is less than 1% different to the individual external mass-transfer coefficient outside the cube, since the external resistance outside the sugar cube dominates the total mass-transfer resistance in this case, due to the small surface area of the cube. The resistance of the intestine wall is complicated by the presence of villi, which add to the interfacial surface area, and the movement of the villi and their pumping action reduces the mass-transfer resistance (R4) of the wall [21,22].

5. Applications to Guts-on-a-Chip

This engineering (mass transfer) perspective on digestion provides strong fundamental support for the development of gut-on-chip models (Fois et al. [23,24]) in a number of ways, both qualitatively and quantitatively. In the qualitative sense, the mass-transfer coefficient and mass-transfer behaviour do not require strict geometrical similarity with in vivo systems. From a mass-transfer perspective, this consideration means that if a gut-on-a-chip has a similar mass-transfer coefficient to an in vivo system, then the mass-transfer behaviour of the in vitro gut-on-a-chip system can readily be made very similar to an in vivo system by ensuring that the chemistry on the two systems is also very similar, and hence that the driving force, C_1–C_2, from Equation (1) is very similar. This situation then begs the question about the quantitative similarity between the mass-transfer coefficients for the in vitro gut-on-a-chip system of Fois et al. [24] and the coefficients estimated in Table 1 for the in vivo human digestive system. The following quantitative analysis uses the geometrical and operating parameters quoted in the paper of Fois et al. [24], and this analysis provides some good news for such a comparison.

Quantitative Analysis of Mass and Momentum Transfer, comparing in vitro guts-on-a-chip with the in vivo human digestive system, Table 1.

From the viewpoint of transport phenomena, the most important single parameter in the work of Fois et al. [24] is the wall shear stress, 0.002 N m^{-2}. This shear stress will now be translated, through an analogy between momentum and mass transfer, into a mass-transfer coefficient for this in vitro gut-on-a-chip system that will be compared with the mass-transfer coefficients estimated in Table 1 for the in vivo human digestive system, particularly for the duodenum and the small intestine.

Considering now the analogy between momentum and mass transfer, the Chilton–Colburn analogy, given in Equation (8), is valid up to Schmidt numbers (Sc) of about 160–200, while the simpler Reynolds analogy ($f/2 = k/u$) is valid for Sc about 1. For higher Sc (less than 3000), the Friend–Metzger analogy [25] should be used, and it is given by the following equation:

$$\frac{k}{u} = \frac{f/2}{1.20 + 11.8\sqrt{f/2}\,(Sc - 1)\,Sc^{-1/3}} \tag{17}$$

From the work of Fois et al. [24], the shear stress (τ) for which their apparatus was designed was 0.02 dyne cm^{-2} (0.002 N·m^{-2}). They also used flow rates of 18 μL·h^{-1} and 29 μL·h^{-1} through a rectangular flow passage with cross-sectional dimensions of 1 mm × 0.15 mm ($A = 1.5 \times 10^{-7}$ m^2), giving average velocities (u) of (3.33–5.37) × 10^{-5} m·s^{-1}. These values of the shear stress and average velocities give values for the group ($f/2 = \tau/(\rho \cdot u^2)$) in the range from 693 to 1800 (dimensionless).

The Schmidt number ($Sc = \mu/(\rho\ D)$) depends on the diffusivity—and hence on the selection—of solute and solvent. Fois et al. [24] used water as the solvent, so the density and viscosity were 1000 kg·m^{-3} and 9.3 × 10^{-4} kg·m^{-1}·s^{-1}, respectively. If the dilute sucrose-water system is selected as a representative one, as discussed previously, the diffusivity (sucrose through water) is 0.5228 × 10^{-9} m^2·s^{-1} [4]. These parameters give a Schmidt number of (9.3 × 10^{-4} kg·m^{-1}·s^{-1}/(1000 kg·m^{-3} × 0.5228 × 10^{-9} m^2·s^{-1}) = 1779).

Once these values for ($f/2$) and Sc are substituted into Equation (17), the resulting predictions for the values of k, irrespective of the velocity u in this range of flow rates, are around 8.2 × 10^{-7} m·s^{-1}, which are very close to the predicted k values for the duodenum and small intestine of 1.23 × 10^{-6} m·s^{-1} and 5.37 × 10^{-7} m·s^{-1} from Table 1, respectively. This similarity should be rather reassuring for developers of in vitro guts-on-a-chip, such as Fois et al. [24], in the following sense. Specifically, there is no reason, providing that the chemistry for the in vitro guts-on-a-chip is maintained similarly to that for in vitro human digestive systems (Table 1), why the mass-transfer and transport behaviour for in vivo human digestive systems should not be well-mimicked by in vitro guts-on-a-chip. Similar mass-transfer behaviour should ensure similarity for the transport of nutrients to, and waste products away from, bowel cancer cells, as studied by Fois et al. (2021). It is noteworthy that no wider claim of reproducing the entire food-digestion process, from the food to the bloodstream, is made for such a gut-on-a chip model, which is reasonable given that pieces of food cannot physically fit inside a 1 mm × 0.15 mm flow passage. This analysis has suggested that such gut-on-a-chip (in vitro) models are likely to reproduce the in vivo process of mass transfer to and from the walls of the intestine (and any cells on these walls) very well.

6. Conclusions

The role of mass transfer in the food-digestion process has been discussed in this work, including the importance of the series of film mass-transfer resistances in determining the limitations to the overall mass-transfer rate. The interaction between mass transfer and the time scales for food digestion have also been reviewed, showing that the time constants for many parts of the digestion process are very similar to the residence times in these parts of the gastrointestinal tract. There is no reason why the mass-transfer and transport behaviour for in vivo human digestive systems should not be well-mimicked by in vitro guts-on-a-chip, implying that the transport of nutrients to, and waste products away from, cells on the intestine walls should also be very similar in both devices.

Funding: This research received no external funding.

Institutional Review Board Statement: This study did not require ethical approval.

Informed Consent Statement: Not applicable.

Data Availability Statement: Not applicable.

Conflicts of Interest: The author declares no conflict of interest.

References

1. Minekus, K.; Alminger, M.; Alvito, P.; Ballance, S.; Bohn, T.; Bourlieu, C.; Carrière, F.; Boutrou, R.; Corredig, M.; Dupont, D.; et al. A standardised static in vitro digestion method suitable for food–An international consensus. *Food Funct.* **2014**, *5*, 1113–1124. [CrossRef]
2. Bohn, T.; Carriere, F.; Day, L.; Deglaire, A.; Egger, L.; Freitas, D.; Golding, M.; Le Feunteun, S.; Macierzanka, A.; Menard, O.; et al. Correlation between in vitro and in vivo data on food digestion. What can we predict with static in vitro digestion models? *Crit. Rev. Food Sci. Nutr.* **2018**, *58*, 2239–2261. [CrossRef]
3. Li, C.; Yu, W.; Wu, P.; Chen, X.D. Current in-vitro digestion systems for understanding food digestion in human upper gastrointestinal tract. *Trends Food Sci. Technol.* **2020**, *96*, 114–126. [CrossRef]
4. Cussler, E.L. *Diffusion: Mass Transfer in Fluid Systems*, 3rd ed.; Cambridge University Press: New York, NY, USA, 2009; pp. 211–241, 338–346, 427–466, 503–528. ISBN 9780521871211.
5. Keppler, S.; O'Meara, S.; Bakalis, S.; Fryer, P.J.; Bornhorst, G.M. Characterization of individual particle movement during in vitro digestion in the Human Gastric Simulator (HGS). *J. Food Eng.* **2020**, *264*, 109674. [CrossRef]
6. Zhong, C.; Langrish, T.A.G. A comparison of different physical stomach models and an analysis of shear stresses and strains in these systems. *Food Res. Int.* **2020**, *135*, 109296. [CrossRef]
7. Miralles, B.; Sanchon, J.; Sanchez-Rivera, L.; Martinez-Maqueda, D.; LeGouar, Y.; Dupont, D.; Amigo, L.; Recio, I. Digestion of micellar casein in duodenum cannulated pigs. Correlation between in vitro simulated gastric digestion and in vivo data. *Food Chem.* **2021**, *343*, 128424. [CrossRef] [PubMed]
8. Sasaki, T. Influence of xanthan gum and gluten on in vitro digestibility of rice bread. *Int. J. Food Sci. Tech.* **2022**, *57*, 2376–2383. [CrossRef]
9. Bornhorst, G.M.; Gouseti, O.; Wickham, M.S.J.; Bakalis, S. Engineering digestion: Multiscale processes of food digestion. *J. Food Sci.* **2016**, *81*, R534–R543. [CrossRef]
10. Tharakan, A.; Norton, I.T.; Fryer, P.J.; Bakalis, S. Mass transfer and nutrient absorption in a simulated model of small intestine. *J. Food Sci.* **2010**, *75*, E339–E346. [CrossRef]
11. Pordage, L.J.; Langrish, T.A.G. Simulation of the effect of air velocity in the drying of hardwood timber. *Dry. Technol.* **1997**, *17*, 237–255. [CrossRef]
12. Patel, K.C.; Chen, X.D. Surface-center temperature differences within milk droplets during convective drying and drying-based Biot number analysis. *AIChE J.* **2008**, *54*, 3273–3290. [CrossRef]
13. Hallström, B.; Gekas, V.; Sjöholm, I.; Romulus, A.M. Mass Transfer in Foods. In *Handbook of Food Engineering*; Heldman, D.R., Lund, D.B., Eds.; Chapter 7; CRC Press: Boca Raton, FL, USA, 2006; ISBN 1420014374.
14. Leena, M.M.; Antoniraj, M.G.; Moses, J.A.; Anandharamakrishnan, C. Three fluid nozzle spray drying for co-encapsulation and controlled release of curcumin and resveratrol. *J. Drug Deliv. Sci. Technol.* **2020**, *57*, 101678. [CrossRef]

15. Li, H.; Dhital, S.; Gidley, M.J.; Gilbert, R.G. A more general approach to fitting digestion kinetics of starch in food. *Carbohydr. Polym.* **2019**, *225*, 115244. [CrossRef]
16. Langrish, T.A.G.; Zhong, C.; Sun, L. Probing differences in mass-transfer coefficients in beaker and stirrer digestion systems and the USP dissolution apparatus 2 using benzoic acid tablets. *Processes* **2021**, *9*, 2168. [CrossRef]
17. Ranz, W.E.; Marshall, W.R. Evaporation from drops. *Chem. Eng. Progress* **1952**, *48*, 141–146.
18. Hopgood, M.; Reynolds, G.; Barker, R. Using computational fluid dynamics to compare shear rate and turbulence in the TIM-automated gastric compartment with USP apparatus II. *J. Pharm. Sci.* **2018**, *107*, 1911–1919. [CrossRef]
19. Stephanopoulos, G. *Chemical Process Control: An Introduction to Theory and Practice*; Chapter 10; Prentice Hall: Englewood Cliffs, NJ, USA, 1984; ISBN 0-13-128629-3.
20. Ritschel, W.A. Targeting in the gastrointestinal tract: New approaches. *Methods Find. Exp. Clin. Pharmacol.* **1991**, *13*, 313–336. [PubMed]
21. Lentle, R.G.; Janessen, P.W.M.; DeLoubens, C.; Lim, Y.F.; Hulls, C.; Chambers, P. Mucosal manifolds augment mixing at the wall of the distal ileum of the brushtail possum. *Neurogastroenterol. Motil.* **2013**, *25*, 881-e700. [CrossRef]
22. Lim, Y.F.; DeLoubens, C.; Love, R.J.; Lentle, R.G.; Janessen, P.W.M. Flow and mixing by small intestine villi. *Food Funct.* **2016**, *6*, 1787–1795. [CrossRef]
23. Fois, C.A.M.; Le, T.Y.L.; Schindeler, A.; Naficy, S.; McClure, D.D.; Read, M.N.; Valtchev, P.; Khademhosseini, A.; Dehghani, F. Models of the gut for analyzing the impact of food and drugs. *Adv. Healthc. Mater.* **2019**, *8*, 1900968. [CrossRef]
24. Fois, C.A.M.; Schindeler, A.; Valtchev, P.; Dehghani, F. Dynamic flow and shear stress as key parameters for intestinal cells morphology and polarization in an organ-on-chip model. *Biomed. Microdevices* **2021**, *23*, 55. [CrossRef] [PubMed]
25. Friend, W.L.; Metzger, A.B. Turbulent heat transfer inside tubes and the analogy among heat, mass and momentum transfer. *AIChE J.* **1958**, *4*, 393–402. [CrossRef]

applied biosciences

MDPI

Review

Progress in Plant Genome Sequencing

Robert J. Henry [1,2]

[1] Queensland Alliance for Agriculture and Food Innovation, University of Queensland, Brisbane, QLD 4072, Australia; robert.henry@uq.edu.au
[2] ARC Centre of Excellence for Plant Success in Nature and Agriculture, University of Queensland, Brisbane, QLD 4072, Australia

Abstract: The genome sequence of any organism is key to understanding the biology and utility of that organism. Plants have diverse, complex and sometimes very large nuclear genomes, mitochondrial genomes and much smaller and more highly conserved chloroplast genomes. Plant genome sequences underpin our understanding of plant biology and serve as a key platform for the genetic selection and improvement of crop plants to achieve food security. The development of technology that can capture large volumes of sequence data at low costs and with high accuracy has driven the acceleration of plant genome sequencing advancements. More recently, the development of long read sequencing technology has been a key advance for supporting the accurate sequencing and assembly of chromosome-level plant genomes. This review explored the progress in the sequencing and assembly of plant genomes and the outcomes of plant genome sequencing to date. The outcomes support the conservation of biodiversity, adaptations to climate change and improvements in the sustainability of agriculture, which support food and nutritional security.

Keywords: DNA sequencing; plant genome; long read sequencing; chromosome assembly

Citation: Henry, R.J. Progress in Plant Genome Sequencing. *Appl. Biosci.* **2022**, *1*, 113–128. https://doi.org/10.3390/applbiosci1020008

Academic Editor: Adriana Basile

Received: 15 May 2022
Accepted: 1 July 2022
Published: 4 July 2022

1. Introduction

Advances in the analysis of DNA sequences have been a key driver of enhanced biological understanding and the application of biological knowledge [1]. DNA sequencing in the 20th century was largely based on Sanger sequencing, which limited both the quality (accuracy) and volume of data that could be generated relative to the next generation sequencing that we have today [2]. The introduction and rapid development of next generation sequencing has resulted in an acceleration in the development of plant genome sequencing, especially over the last decade [3]. This technology has evolved rapidly, resulting in continuous major changes to the strategies that are used to sequence and assemble genomes. For example, when only short-read sequences were available, physical mapping was a key strategy. Large fragments of the genomes were cloned in bacterial artificial chromosomes (BACs) [4]. The BACs were then sequenced and the genomes were assembled by covering the genetic maps with BAC tiles [5]. The availability of accurate long read sequencing has made these approaches largely redundant [6]. A review in 2018 [7] reported that 236 angiosperm genome sequences had been reported. Since then, many more genomes have been sequenced and the quality of the genome sequences has increased significantly. The NCBI database (https://www.ncbi.nlm.nih.gov/genome/browse#!/overview/flowering%20plants; accessed 6 June 2022) includes 831 flowering plant genomes, with 373 at the chromosome level. The de novo assembly of long read sequences allows very large contigs to be assembled, sometimes representing a complete plant chromosome [8].

2. Diversity of Plant Genomes

Plant genomes vary enormously in size, even within closely related groups of plants [9]. The nuclear genomes of flowering plants (angiosperms) vary more than 1000-fold, from less than 100 kb to more than 100 Gb [10]. The genomes of gymnosperms are generally large and complex and represent an even greater challenge for genome sequencing [11]. The large (10 Gb) genome of *Ginkgo biloba* has recently been reported [12], which provides the first reference genome for gymnosperms. Genomes also vary greatly in terms of their content of repetitive sequences, the level of gene duplication, their ploidy and their heterozygosity, providing a range of challenges and degrees of difficulty within genome sequencing and assembly.

3. Applications of Plant Genome Sequencing

3.1. Model Genomes

The challenge of sequencing plant genomes using early technologies made it necessary to focus on sequencing model genomes that could be used to study related, but more complex, species. The first plant to have a sequenced genome was *Arabidopsis thaliana* [13], which was chosen because it is a small plant with a rapid generation time and a very small genome, thereby making it an ideal model plant for research use. The first crop plant with a sequenced genome [14] was rice (*Oryza sativa*), which was chosen because it is a major food crop plant with a relatively small genome. This became a model for cereal and grass genomes. Similarly, *Brachypodium distachyon* was sequenced [15] as a model grass genome, which is especially relevant for the wheat genome. Recent advances in genome sequencing technology have greatly reduced the need for models as it is now possible to sequence most species easily.

3.2. Crop Plant Genomes

The sequencing of the genomes of crop species has become a key enabling tool for plant improvement. Most major crops now have reference genome sequences [16] and as the technology becomes more powerful and the costs reduce, genomes are also being generated for many other minor crops. This usually involves the production of a reference genome sequence for a species and the re-sequencing of many individuals to define allelic variations within that species. Current efforts recognize that a single reference genome cannot always serve the needs of plant breeders, so pan-genomes that capture the variations in many diverse genomes within the gene pool are being produced as breeding platforms.

3.3. Sequencing Plant Biodiversity

Many diverse plant genomes have now been sequenced with an increasing coverage of the major groups, especially among flowering plants. The coverage of plant orders is high and the genomes from many plant families have now been reported; however, coverage at the genus level is still very low for most plant groups. Systematic efforts to obtain plant genome sequences may take a top-down approach to sequencing a member of each plant family, then each genus and, finally, each species would become available as resources. Ultimately, the re-sequencing of the diversity within each species is of value. A knowledge of the diversity within plant populations is a fundamental tool that can guide the effective conservation of the diversity within species.

3.4. Sequencing Rare and Threatened Species

Targeted efforts are now being made to sequence rare and threatened species of plants as a tool to aid conservation, both in situ [17] and ex situ [18]. This is more urgent among critically endangered species, for which a genome sequence may be all we can retain as the species are lost to extinction. Efforts to sequence biodiversity often focus on rare species as the highest priority.

The critically endangered wild crop relative *Macadamia jansenii* has been used to compare plant genome sequencing and assembly methods [19]. This has allowed for the comparison of sequencing platforms and bioinformatics tools for genome assembly using a common sample. The generation of a chromosome-level genome sequence for a plant involves the preparation of a DNA sample, the sequencing of that DNA, the assembly of the sequence reads into contigs and, finally, the assembly of the sequence contigs into a chromosome-level assembly (Figure 1).

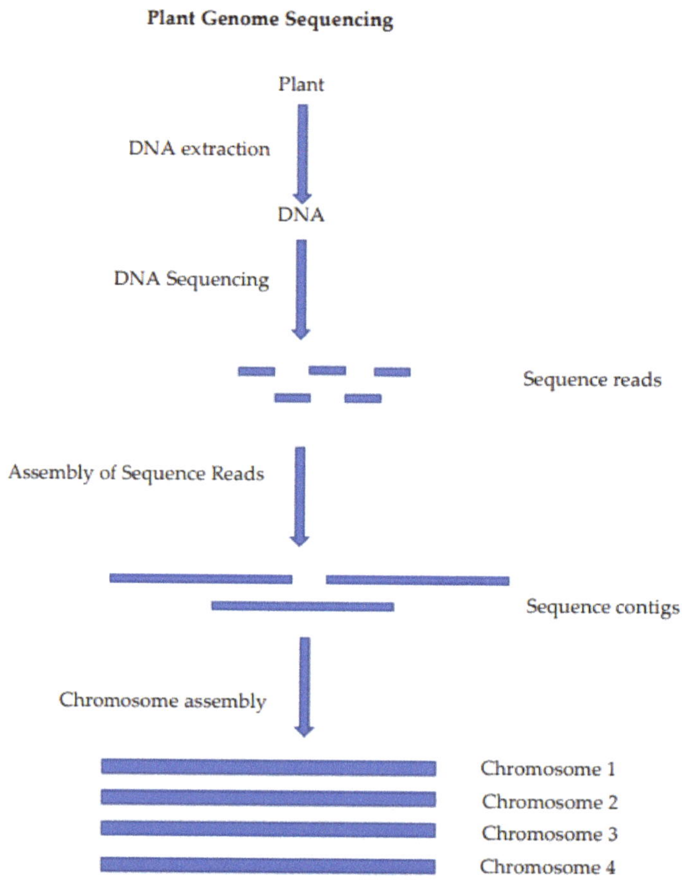

Figure 1. Steps in the sequencing and assembly of a plant genome: DNA extraction is used to produce a DNA sample that is suitable for sequencing, the sequencing of the DNA produces long read sequences, the reads are self-assembled into contigs (often at or near chromosome length) and these contigs are then assembled at the chromosome level using chromatin mapping or genetic mapping.

4. Sequencing Technology

4.1. DNA Isolation

The starting point for the sequencing of plant genomes is obtaining a sample of DNA to sequence. The ease of obtaining a DNA sample of suitable quality for sequencing varies greatly between species. Plants contain many secondary metabolites, proteins and polysaccharides that may interfere with DNA extraction and become a source of contaminants that often reduce the efficiency of DNA extraction. Current technologies require a minimum amount of DNA and the DNA must be pure (free from contaminants that may inhibit sequencing) to facilitate efficient sequencing and the generation of large volumes of data. The amount of DNA required for long read sequencing has been greater (usually µg quantities) than that needed for short read sequencing (https://dnatech.genomecenter.ucdavis.edu/pacbio-library-prep-sequencing; accessed on 6 June 2022). For long read sequencing, the DNA must be intact (not degraded) so that long sequences are present in the sample and can be sequenced to extract long reads.

Simple methods that were suitable for DNA extraction in the past for purposes such as PCR analysis [20] are no longer adequate, so the development of species- [21] or tissue-specific methods that can support next generation sequencing are often required [22]. Some species are especially difficult and require the isolation of the nucleus [23] first as a source of DNA that may be free from contaminants from other parts of the plant cell. The isolation of nuclei in plants is challenging for some of the same reasons that DNA isolation is difficult. The disruption of the plant cell wall requires forces that may damage organelles, such as nuclei and shear DNA.

4.2. Sort Read Sequences

The first set of next generation sequencing technologies provided lager volumes of short DNA sequences. The accuracy of these short sequences and the volume of data have since increased dramatically. The length of these sequences started at around 30 bp and has rapidly advanced to 100–150 bp. Paired-end sequencing has extended this technology to allow for the production of sequences of around 400 bp, but most applications currently deliver sequences of around 150 bp. Illumina sequencing platforms are the dominant technology used for short read sequencing. This technology conducts sequencing by synthesis in a very large number of parallel reactions. The incorporation of nucleotides is monitored as the DNA is copied. Other techniques (e.g., solid [24] and 454 sequencing [25]) have been replaced by new technologies because they generally offered lower accuracy or data volumes, resulting in relatively higher costs. Ion Torrent sequencing is used for the rapid determination of a sequence of large numbers of small sequences, such as amplicon sequencing and 16 S metagenomic sequencing. In plants, this has been used for chloroplast sequencing [26].

4.3. Long Read Sequences

The assembly of plant genomes with large numbers of repetitive sequences is not possible with only short read sequences. Therefore, technology that allows much longer sequences to be generated has been key to simplifying genome assembly. The length of these sequences and their accuracy have improved greatly since the technology was first introduced.

4.3.1. PacBio

Pacific Biosciences (PacBio) has developed a long-read sequencing platform that provides accurate long read sequencing. The single-molecule real-time (SMRT) sequencing involves monitoring the incorporation of fluorescent-labeled nucleotides [27]. Recently, single long reads (also known as continuous long reads (CLR)) have largely been replaced by HiFi reads, which provide a consensus sequence based on sequencing a long fragment of DNA (approximately 15,000 bp) multiple times by first circularizing the DNA and reading around the circle many times [28]. The repeated sequencing of the same molecule allows a highly accurate sequence to be generated as the circular consensus sequence (CCS) is read. The quality of the genomes that are generated by the assembly of these reads into contigs has been improved by the application of optimized assembly tools, such as those provided by hifiasm.

4.3.2. ONT

Oxford Nanopore Technologies (ONT) provides a long read sequencing technology that delivers accurate sequence data quickly. The sequence is determined by measuring the changes in electrical currents as the DNA is passed through a pore. The ONT platform generates very long reads and has the advantage of very low instrumentation costs. This platform has continued to improve and deliver very long read sequences with increasing accuracy [29]. ONT sequencing has been widely applied to very rapid sequencing, such as that required for diagnostics [30], due to the advantage of having portable instruments.

The chromosome-level assemblies of plant genomes can be achieved in combination with methods, such as optical mapping [31].

4.3.3. Other

Several technology providers have developed pseudo-long reads that are created by linking short reads. These techniques may produce long reads at lower costs, but the long reads that are generated often do not match the accuracy of the current long read methods [19]. These technologies have been developed by Universal Sequencing Technology [32], MGI [33] and 10× genomics [34]. Despite the great contribution that long read sequencing technology has made to the efficient production of high-quality plant genomes, the emergence of further advances in long read sequencing technologies remains one of the key areas that may contribute to future advances.

4.3.4. Advances

Genome sequencing and assembly requires an adequate depth of sequencing. The size of contigs that can be assembled as long read sequence data has been shown to increase in an almost linear way [8]. The size of the assembled genomes reduces slightly with more contiguous assemblies, probably due to the joining of homologous contig ends, as does the completeness [8]. Improved software has also enabled improvements in the assembly of long read sequences [35]. The use of hifiasm has been shown to allow the haplotype-resolved assembly of the large (30 Gb) genome of the Californian redwood (*Sequoia sempervirens*) [36]. These advances are illustrated by the quality of the early plant genomes relative to those that are being generated by the latest technology. The first rice genome, which was reported in 2002, was highly fragmented while current technology delivers sequence contigs that are often full-length chromosomes [35].

4.4. Chromosome-Level Assembly

The ultimate aim of genome sequencing is to obtain a complete genome sequence of each chromosome, from telomere to telomere. This relies on evidence from beyond the DNA sequence data. Physical and genetic mapping methods have been used to achieve the chromosome-level assembly of contigs that were generated from sequencing data [37]. Recently, the advances in sequencing technology have made it possible to generate many full-length chromosomes from the sequence data alone [35]. The complete assembly of sequence contigs into whole chromosomes has been widely achieved using genetic mapping data, chromatin mapping (Hi-C) or optical mapping. Hi-C [38] involves the mapping of chromatin by crosslinking the DNA in the intact chromatin, digesting the DNA and then sequencing (short reads) the DNA fragments at the ends of the crosslinks. These are used to position the sequence contigs along the chromosome. Optical mapping (Bionano) can also be used to locate sequences along the DNA sequence and to scaffold the sequence contigs [39]. Many projects have combined these technologies to support the generation of high-quality genomes. Recent advances in long read sequencing have enabled the generation of long contigs of highly accurate sequences, reducing reliance on these techniques for high-level assembly. They remain essential for the de novo assembly of most chromosome-level genomes. High-quality sequence contigs in combination with genetic mapping data, Bionano optical data or Hi-C chromatin mapping have generally succeeded in achieving chromosome-level assemblies of plant genomes. A report on more than 100 chromosome-level assemblies in 2021 [40] found only a 73% coverage of the pseudomolecules that represent the chromosomes. The combination of long read sequencing and the use of these tools has resulted in the recent reporting of many high-quality chromosome-level genome sequences (Table 1), with the quality improving greatly along with the most recent technology.

Table 1. Some recent chromosome-level assemblies of plant genomes.

Species		Technique *	Reference
Amorphophallus	*Amorphophallas konjac*	Hi-C	[41]
Apple	*Malus domestica*	Genetic Map	[42]
Avocado	*Persea americana*	Genetic Map	[43]
Banana	*Musa balbisiana*	Hi-C	[44]
Camphor	*Cinnamomum camphora*	Hi-C	[45]
Carrot	*Daucus carota*	Genetic Map	[46]
Chinese Skullcap	*Scutellaria baicalensis*	Hi-C	[47]
Crucihimalaya	*Crucihimalaya lasicocarpa*	Hi-C	[48]
Cucumber	*Cucumis metuliferus*	Hi-C	[49]
Eucalypt	*Corymbia citriodora*	Genetic Map	[37]
Field Pennycresss	*Thlaspi arvense*	Genetic Map/Hi-C/Bionano	[50]
Ginger	*Zingiber officinale*	Hi-C	[51]
Ginkgo	*Ginkgo biloba*	Hi-C	[52]
Jojoba	*Simmondsia chinensis*	Hi-C	[53]
Macadamia	*Macadamia jansenii*	Hi-C	[54]
	Macadamia integrifolia	Genetic Map	[55]
Paper Mulberry	*Broussonetis papyrifera*	Hi-C	[56]
Peach	*Prunus persica*	Hi-C	[57]
Peanut	*Arachis hypogaea*	Hi-C	[58]
Speranskia	*Speranskia yunnanensis*	Hi-C	[59]
Taxus	*Taxus chinensis*	Hi-C	[60]
Tea	*Camellia sinensus*	Hi-C	[61]
Water Caltrop	*Trapa* spp.	Hi-C	[62]

* The technique used to achieve the chromosome-level assembly of the sequence contigs.

4.5. Haplotype-Resolved Genomes

Most published plant genomes are collapsed representations of the diploid genome as a single sequence, with a random inclusion of one of the two alleles at each heterozygous position. Only recently has it become possible to assemble each haplotype separately [63]. This has been the result of advances in both sequencing technology and sequence assembly tools. Current technology suggests that most genomes can now be sequenced at the haplotype level, thereby replacing the reporting of collapsed genomes with the sequences of the two haplotypes.

4.6. Pan-Genomes

The sequencing of plant genomes has shown that significant differences may be found within a plant species, which means that more than one reference genome is required to represent the species. The sequencing of plant genomes has also demonstrated that many genes are variably present in different individuals within a species. These presence/absence differences have led to the construction of pan-genomes, which represent the complete set of genes found within a population. A genome that includes all of the variations within a group of plants is known as a pan-genome. The pan-genome concept is a powerful tool for plant breeders for the analysis of gene pools [64]. Pan-genomes can be generated at different levels to represent the diversity that is found within, for example, domesticated gene pools, species or genera.

4.7. Transcriptomes

Transcriptome sequencing is an important tool for the analysis of the expressed regions of a genome. This is key to understanding gene functions and the determination of the genetic basis of important plant traits [65–70]. Transcriptome sequencing complements genome sequencing in genome characterization. Transcripts provide physical evidence that the sequence is formed of the expressed and complementary predictions of genes, based on the sequence alone. The comparison of the transcriptomes of different genotypes from different tissues or cell types at different stages of development and under different environmental conditions allows for the discovery of the genes that control plant traits and has become a key approach in plant biology and the discovery of genes for selection in plant breeding. Single-cell transcriptomics has become a powerful tool for understanding gene expression at the cell and tissue level but has had limited application in plants [71], partly due to the difficulty in isolating specific plant cells without disrupting expression.

4.7.1. RNAseq

The quantitative analysis of the levels of expression of genes in any specific cell, tissue, organ, genotype or development stage is widely determined by RNA sequencing (RNAseq) [72,73]. RNAseq has largely replaced earlier array-based or gene by gene analysis tools as it provided a more unbiased analysis of the whole transcriptome.

An analysis of the gene expression in the highly polyploid sugarcane genome revealed that while the different alleles of most genes are expressed in direct proportion to their abundance in the genome, some genes show highly biased patterns of expression [74]. In hexaploid wheat, subgenome-specific responses to diseases have also been reported [75].

4.7.2. Long Read Transcriptomes

Long read sequencing is a method that has been applied to the analysis of plant transcriptomes, which reveals the diversity of full-length transcripts and defines the variations in splicing and intron retention in gene expression [76]. The long read sequencing of transcriptomes avoids the challenge of the assembly of many closely related transcripts from short reads. Unique 3′ and 5′ sequences may be separated by common intervening sequences, which creates the risk of incorrectly combining the ends of the transcripts when using short reads.

Some examples of the application of long read sequencing to the analysis of plant transcriptomes of increasing complexity can be found for polyploid species in Table 2.

Table 2. The long read sequencing of polyploid transcriptomes.

Species	Ploidy	Reference
Coffee	4X	[76]
Wheat	6X	[77]
Strawberry	8X	[78]
Sugarcane	12X	[79]

4.8. *Organelle Genome Sequencing*

Plant cells usually contain a single nucleus and many organelles, probably hundreds of mitochondria and thousands of chloroplasts. Sequencing the organelle genomes is complicated by the transfer of genes between these genomes. The nuclear genome often contains many insertions of large and small sequences of organellar genomes. Many early methods struggled to distinguish organellar gene sequences from those of copies that were inserted into the nuclear genome because they relied on PCR amplification or organelle separation [80]. Nuclear inserts may represent versions of organellar genomes that were transferred in the past and that have diverged since insertion.

4.8.1. Chloroplast Genomes

The chloroplast genomes of plants are highly conserved sequences of 100–150 Kb, containing around 100 genes [81]. The structure of most chloroplasts is similar, with four components including inverted repeats that separate large and small single-copy regions. Chloroplasts have been widely used in plant identification due to their presence in all green plants and the high copy numbers in the cell that simplify the detection of chloroplast sequences. Early approaches that relied on chloroplast isolation or PCR amplification were plagued by confusion due to the copies of chloroplast sequences in the nuclear and mitochondrial genomes. Recent approaches [82] rely on the higher abundance of chloroplast genome sequence reads in short read sequence data to clearly distinguish the correct sequence of the relevant chloroplast [83]. The development of software tools now allows for the efficient extraction of accurate whole chloroplast genome sequences from even low (nuclear) coverage sequencing datasets. The annotation of chloroplast genomes that were generated in this way has resulted in the identification of around 100 genes with increasingly well-defined functions [84].

The sequencing of the maternal (e.g., chloroplast) and nuclear genomes of plants has frequently revealed discordant phylogenies [85–87], suggesting widespread reticulate evolution in plant populations (Table 3). Chloroplast transfers between species during rare events results in "chloroplast capture" by closely related species.

Table 3. Discordant phylogenies for chloroplast and nuclear genome sequences.

Species	Reference
Apple Tribe	[88]
Eucalypts	[87]
Osmorhiza spp.	[89]
Pedicularis spp.	[90]
Sorghum	[85]
Rice (*Oryza*)	[86]

4.8.2. Plant Mitochondrial Genomes

The mitochondrial genomes of plants [91] are much larger and less conserved than the chloroplast genomes and as a result, they have been much less studied than chloroplasts. The mitochondrial genome, as with the nuclear genome, may include sequences that were derived from the chloroplast that have been inserted into the genome at various times throughout its evolutionary history. Due to the relatively higher number of mitochondrial genomes in cells, these sequences are even more likely to be confused with chloroplast genome sequences than chloroplast sequences that were inserted into nuclear genomes.

5. Biological Understanding

Sequencing plant genomes provides an enhanced understanding of the biology of plant species. This knowledge informs the better conservation of biodiversity and sustainable use in agriculture and food production. Plant genomes may often explain the response of plants to the environment and may assist in improving the management of crops.

5.1. Whole Genome Duplications

Genome sequencing and assembly defines the presence of duplicated regions of the genome that are often the result of whole genome duplications during the evolutionary history of the species. In many species, evidence can be found for more than one duplication event. The analysis of these events can aid in the determination of evolutionary relationships. The selection of key genomes for sequencing allows evolutionary relationships to be defined. For example, the sequencing of the *Aristolochia fimbriata* genome revealed that this plant lacked the whole genome duplication that was reported in other magnoliid plants, placing it at a basal position in the angiosperm phylogenies [92].

5.2. Polyploid Challenges

The sequencing and assembly of a polyploid plant genome is complicated by the presence of many similar sequences that can be difficult to assemble. Sugarcane is an important global crop, but the high degree of ploidy (12X) in this species has resulted in it being one of the last major crop species to have a sequenced genome. Instead, a monoploid equivalent (based on BAC clones from sugarcane that cover the sorghum genome) has been widely used in sugarcane genomes due to the lack of a polyploid genome [5].

While a genome for the diploid Robusta coffee was reported in 2014 [93], the sequencing of the tetraploid Arabica coffee has been more difficult due to a genome that was based on a doubled diploid currently being used to characterize the origins of Arabica coffee (Arabica Genome Sequencing Consortium).

5.3. *Genomics of Plants with Diverse Reproductive Biology*

While most flowering plants are hermaphrodites with both male and female reproductive organs, separate male and female plants are found in dioecious species. There are dioecious plants in many plant families and they represent around 6% of all plant species. The differences between the genomes of male and female dioecious plants have not been fully characterized. The recent sequencing of the male and female genomes of jojoba revealed large differences between the sex chromosomes, with the presence of many sex-specific genes [53]. The chromosome-level assembly of male and female genomes has provided a perspective on the basis of sexual dimorphism that was not possible with more limited genomic information. The sequencing of the genomes of more dioecious plants may define the diversity of dimorphisms that have evolved into the many separate linages of dioecious members. The genomes of other plants with unusual reproductive biology, such as apomictic plants, have not yet been studied but may explain the adaptive value of these modes of reproduction. An analysis of transcriptomes defined the conserved genes for organogenesis that are associated with reproduction in flowering plants [94].

5.4. *Evolutionary Insights*

The comparison of the genome sequences of plant species is the key basis that we have for defining evolutionary relationships. The phylogenetic comparison of plant genomes has often been based on the analysis of one or a few genes. The availability of whole genome sequence data has allowed the use of much larger numbers of genes. A common approach has been to compare the sequences of many conserved single-copy genes. This approach has been used to define relationships between species in the rice (*Oryza*) [86,95] and sorghum (*Sorghum*) [85] genera.

Genome sequencing provides an opportunity to better understand the process of domestication, through which human selection has resulted in plants that are better suited to survival in agricultural environments rather than though natural selection in nature. The sequencing of the coffee genome revealed that human attraction to caffeine might have resulted in the domestication of plants for coffee, tea and chocolate production from diverse species in which caffeine had evolved separately [93].

5.5. *Maternal Genome Inheritance*

In most plants, the cytoplasmic organelles (chloroplasts and mitochondria) are maternally inherited. In some rare cases, other patterns of inheritance are observed; for example, in cucumber (*Cucumis sativus*), the chloroplasts are inherited maternally while the mitochondria are paternally inherited [96]. Although some plant groups display paternal inheritance, the maternal inheritance of organellar DNA and the paternal contribution to nuclear DNA can result in discordant patterns of evolution. Reticulate evolution, in which there is a gene flow between different linages, may result in the transfer (or "capture") of organellar genomes from closely related species. Many examples have been reported via chloroplast genome analyses that suggest phylogenies that differ from those indicated by the nuclear genomes [87]. The potential for the nuclear, chloroplast and mitochondrial genomes of plants to have evolved along separate paths makes it import to focus phylogenetic analysis on nuclear genomes. Improved methods for applying phylogenetic analysis to whole nuclear genomes are needed. Because nuclear genomes show so many variations at the sequence level, the current approaches rely on aligning a small subset of highly conserved genes from nuclear genomes that can be reliably aligned. As whole genome sequence data become more widely available, it may be possible to develop tools that take advantage of the larger volumes of data to better determine relationships through phylogenetic analysis.

5.6. Importance of Genome Size

Plant genomes vary widely in size, with more than a 1000-fold variation among flowering plants. Even within a single plant family, genome sizes may vary more than 100-fold. The biological significance of these variations remains poorly understood. Differences may be due to gene content, genome duplications, polyploidy or differences in repetitive DNA content. Improvements in genome sequencing technology are likely to allow larger numbers of diverse genomes to be sequenced to facilitate our understanding of genome size diversity.

6. Enabling Plant Breeding

The availability of plant genomes facilitates the breeding and selection of plants, which is essential to support ongoing food security. The need for an accelerated genetic improvement of plants has been made more urgent by climate change, which is demanding new plant genotypes that are adapted to new and more difficult environments. Climate change is altering the physical environment, with higher temperatures and greater water stress, and it is also changing the biological environment, with a wider range of pests and diseases [97].

Plant genomics is critical for the management of plant genetic resources in seed banks [18] and the conservation of the wild crop relatives that provide the genetic resources that are required for sustainable food production.

Plant genomes also enable an understanding of plant biology and the molecular and genetic basis of plant traits.

The analysis of plant genome sequences has facilitated the rapid identification of the key genes that determine the traits of great importance in food crops. This has allowed for more efficient crop breeding by simplifying the selection of these critical traits. The completion of the rice genome allowed for the discovery of the identity of the recessive gene for fragrance in rice [98], which is a trait that can double the value of the rice and can now be detected with a perfect marker [99]. The cooking temperature (gelatinization temperature) of rice was also shown to be determined by a major gene [100]. In wheat, transcriptome analysis has revealed the genetic basis of flour yield [101] and loaf volume for bread [102].

6.1. Molecular Markers and Plant Selection

Sequencing has changed the approaches to marker development and applications in plant breeding. Traditional molecular markers [103], which were linked to traits by being close to the genes that control the phenotype, are now able to be replaced by sequence-based markers [104] for the differences in genomes that are actually responsible for the traits under selection. This greatly increases the reliability of the selection. As the costs for whole genome sequencing reduce, this becomes an alternative to the analysis of large numbers of markers. In a sense, genome sequences are the ultimate genetic marker tool as they capture all of the variations within the genome that may explain the phenotype. They also avoid decisions on which genetic markers to select for analysis needing to be made in advance by capturing all of the possibilities.

6.2. Genetic Manipulation

Genome sequencing is the ultimate method for the characterization of genetically modified plants, revealing the exact changes in genomes that have been produced. This may be a requirement for release into the environment in some jurisdictions. The sequencing of transgenic genotypes defines both the exact point of insertion and the sequence of the added genes, but it also defines the number of copies that have been added and may reveal any other genetic changes that might have been a result of cell culture [105].

6.3. Editing Plant Genomes

The growing application of genome editing is being aided by the use of genome sequencing [106] to support the better targeting of gene editing and to ensure the avoidance of off-target effects. The routine sequencing of the genomes of transgenic- or gene-edited plants can also be used to confirm that the intended changes have been made and that no other unintended changes have occurred. Often, only a single nucleotide needs to be edited or many different loci need to be targeted simultaneously to make the required change in a phenotype; however, in each case, it may be necessary to sequence the genome to confirm that other changes have not occurred.

6.4. Biotechnology Applications (Food, Medicinal and Industrial Crops)

Genome sequencing is a key tool that enables the rapid production of plant varieties with higher nutritional value, enhanced levels of bioactivity, improved biomass composition [107] or the expression of high-value molecules. The availability of sequencing technology allows for the identification of novel genotypes with the desired traits and supports the manipulation of plant genomes to produce plants with novel traits. Genome sequencing can be used to confirm the results of gene editing or any other changes resulting from plant transformation, mutagenesis or conventional crossbreeding. Sequencing may be critical for ensuring compliance with regulations that govern plant genetic manipulation. This may become a major application for plant genome sequencing as gene editing becomes more routine. This application for plant genome sequencing may require the development of standard protocols that can be applied routinely, both in research and in industry, for quality assurance and to protect intellectual property.

7. IP Issues

An issue of growing importance is the ownership and control of genome sequence data [108]. Modern biological science has been built upon the widespread dissemination of sequence data by providing public access to large sequence databases. The Convention on Biological Diversity empowered countries to claim ownership of their biological resources and the more recent Nagoya protocol requires the consideration of access and benefit sharing. These rights may extend to DNA sequence data that were derived from genetic resources. These issues are especially difficult for historical collections because prior informed consent cannot be obtained for the sample collection [109]. International efforts to resolve these issues are urgently required to balance the rights of traditional owners with the need for open access to sequence data to advance biological science. The protection of plant varieties through the use of plant breeders' rights (PBRs) may be supported by the use of genome sequence data to confirm the identity of genotypes. Establishing the distinctness of genotypes to secure PBRs may also become more dependent on genome sequence data. Genome sequence data may be critical for determining whether a new variety was essentially derived from an earlier variety, which is a question that becomes more important as genome editing becomes more widely used.

8. Future Prospects

Many visionary projects aim to ultimately sequence all of the planet's biodiversity as a long-term goal. The sequencing of rare and endangered species can be considered as a priority in this process. Advances in technology have made it difficult to determine when to start such a project as the costs continue to drop and the quality of the data continues to improve. The achievement of plant genome sequences for all species on the planet could accelerate as sequencing technology advances and data storage and handing become more effective.

The difficulty in obtaining a suitable sample of DNA from a plant is one of the remaining challenges in the widespread application of the sequencing of plant genomes. The development of more general methods that can be applied to a wide range of plant samples would represent a major advance, unless DNA sequencing methods become more robust and can cope with poorer quality DNA preparations.

The technology that is now available for plant genome sequencing and assembly make this an increasingly cost-effective strategy for improving our understanding of the biology of all plant species and a key tool for the conservation of plant biodiversity and the use of plants in agriculture and food production. The sequencing of all plant species is a long-term goal that may become key to effectively supporting life on Earth through the improved management of plants in wild populations and their selection and genetic enhancement for use in agriculture and food production. Threats to food security from human conflicts and pandemics [110] have created more interest in food supply from local sources. Plant genome sequencing provides a platform for innovation in plant breeding to deliver a diverse and balanced diet regionally. Adaptations to climate change require the development of plant varieties that can support the adaptation and relocation of agriculture and the development of plant varieties for production in vertical farming [111]. Genomics is a key tool for tackling climate change [112] and for capturing a wider range of diversity from wild crop relatives [113] and other plants to support food production [114].

Funding: This research received no external funding.

Institutional Review Board Statement: Not applicable.

Informed Consent Statement: Not applicable.

Data Availability Statement: Not applicable.

Acknowledgments: The author acknowledges the support of the ARC Centre of Excellence for Plant Success in Nature and Agriculture.

Conflicts of Interest: The author declares no conflict of interest.

References

1. Henry, R.J. Applied Biosciences: Application of Biological Science and Technology. *Appl. Biosci.* **2022**, *1*, 38–39. [CrossRef]
2. Shendure, J.; Balasubramanian, S.; Church, G.M.; Gilbert, W.; Rogers, J.; Schloss, J.; Waterston, R.H. DNA sequencing at 40: Past, present and future. *Nature* **2017**, *550*, 345–353. [CrossRef] [PubMed]
3. Marks, R.A.; Hotaling, S.; Frandsen, P.B.; VanBuren, R. Representation and participation across 20 years of plant genome sequencing. *Nat. Plants* **2021**, *7*, 1571–1578. [CrossRef]
4. Yüksel, B.; Paterson, A.H. Construction and characterization of a peanut HindIII BAC library. *Theor. Appl. Genet.* **2005**, *111*, 630–639. [CrossRef]
5. Garsmeur, O.; Droc, G.; Antonise, R.; Grimwood, J.; Potier, B.; Aitken, K.; Jenkins, J.; Martin, G.; Charron, C.; Hervouet, C.; et al. A mosaic monoploid reference sequence for the highly complex genome of sugarcane. *Nat. Commun.* **2018**, *9*, 2638. [CrossRef] [PubMed]
6. Pucker, B.; Irisarri, I.; de Vries, J.; Xu, B. Plant genome sequence assembly in the era of long reads: Progress, challenges and future directions. *Quant. Plant Biol.* **2021**, *3*, 1–14. [CrossRef]
7. Chen, F.; Dong, W.; Zhang, J.; Guo, X.; Chen, J.; Wang, Z.; Lin, Z.; Tang, H.; Zhang, L. The Sequenced Angiosperm Genomes and Genome Databases. *Front. Plant Sci.* **2018**, *9*, 418. [CrossRef]
8. Sharma, P.A.O.; Alsubaie, B.; Al-Mssallem, I.; Nath, O.; Mitter, N.; Margarido, G.R.A.; Topp, B.; Murigneux, V.; Masouleh, A.K.; Furtado, A.; et al. Improvements in The Sequencing and Assembly of Plant Genomes. *Gigabyte* **2021**. [CrossRef]
9. Wendel, J.F.; Jackson, S.A.; Meyers, B.C.; Wing, R.A. Evolution of plant genome architecture. *Genome Biol.* **2016**, *17*, 37. [CrossRef]
10. Pellicer, J.; Hidalgo, O.; Dodsworth, S.; Leitch, I.J. Genome Size Diversity and Its Impact on the Evolution of Land Plants. *Genes* **2018**, *9*, 88. [CrossRef]
11. Uddenberg, D.; Akhter, S.; Ramachandran, P.; Sundström, J.F.; Carlsbecker, A. Sequenced genomes and rapidly emerging technologies pave the way for conifer evolutionary developmental biology. *Front. Plant Sci.* **2015**, *6*, 970. [CrossRef] [PubMed]
12. Liu, H.; Wang, X.; Wang, G.; Cui, P.; Wu, S.; Ai, C.; Hu, N.; Li, A.; He, B.; Shao, X.; et al. The nearly complete genome of Ginkgo biloba illuminates gymnosperm evolution. *Nat. Plants* **2021**, *7*, 748–756. [CrossRef] [PubMed]
13. Kaul, S.; Koo, H.L.; Jenkins, J.; Rizzo, M.; Rooney, T.; Tallon, L.J.; Feldblyum, T.; Nierman, W.; Benito, M.I.; Lin, X.Y.; et al. Analysis of The Genome Sequence of The Flowering Plant *Arabidopsis Thaliana*. *Nature* **2000**, *408*, 796–815.
14. Jackson, S.A. Rice: The First Crop Genome. *Rice* **2016**, *9*, 14. [CrossRef] [PubMed]
15. Vogel, J.P.; Garvin, D.F.; Mockler, T.C.; Schmutz, J.; Rokhsar, D.; Bevan, M.W.; Barry, K.; Lucas, S.; Harmon-Smith, M.; Lail, K.; et al. Genome Sequencing and Analysis of the Model Grass Brachypodium Distachyon. *Nature* **2010**, *463*, 763–768.
16. Kersey, P.J. Plant genome sequences: Past, present, future. *Curr. Opin. Plant Biol.* **2018**, *48*, 1–8. [CrossRef]

17. Wambugu, P.W.; Henry, R.; Browne, L. Supporting in situ conservation of the genetic diversity of crop wild relatives using genomic technologies. *Mol. Ecol.* **2022**, *31*, 2207–2222. [CrossRef]
18. Wambugu, P.W.; Ndjiondjop, M.-N.; Henry, R.J. Role of genomics in promoting the utilization of plant genetic resources in genebanks. *Brief. Funct. Genom.* **2018**, *17*, 198–206. [CrossRef]
19. Murigneux, V.; Rai, S.K.; Furtado, A.; Bruxner, T.J.C.; Tian, W.; Harliwong, I.; Wei, H.; Yang, B.; Ye, Q.; Anderson, E.; et al. Comparison of long-read methods for sequencing and assembly of a plant genome. *GigaScience* **2020**, *9*, giaa146. [CrossRef]
20. Graham, G.C.; Mayers, P.; Henry, R.J. A simplified method for the preparation of fungal genomic DNA for PCR and RAPD analysis. *BioTechniques* **1994**, *16*, 48–50.
21. Nath, O.; Fletcher, S.J.; Hayward, A.; Shaw, L.M.; Agarwal, R.; Furtado, A.; Henry, R.J.; Mitter, N. A Comprehensive High-Quality DNA and RNA Extraction Protocol for a Range of Cultivars and Tissue Types of the Woody Crop Avocado. *Plants* **2022**, *11*, 242. [CrossRef] [PubMed]
22. Healey, A.; Furtado, A.; Cooper, T.; Henry, R.J. Protocol: A simple method for extracting next-generation sequencing quality genomic DNA from recalcitrant plant species. *Plant Methods* **2014**, *10*, 21. [CrossRef] [PubMed]
23. Guilfoyle, T.J. Chapter 8 Isolation and Characterization of Plant Nuclei. *Method Cell Biol.* **1995**, *50*, 101–112. [CrossRef]
24. Garrido-Cardenas, J.A.; Garcia-Maroto, F.; Alvarez-Bermejo, J.A.; Manzano-Agugliaro, F. DNA Sequencing Sensors: An Overview. *Sensors* **2017**, *17*, 588. [CrossRef] [PubMed]
25. Bundock, P.C.; Eliott, F.G.; Ablett, G.; Benson, A.D.; Casu, R.E.; Aitken, K.S.; Henry, R.J. Targeted single nucleotide polymorphism (SNP) discovery in a highly polyploid plant species using 454 sequencing. *Plant Biotechnol. J.* **2009**, *7*, 347–354. [CrossRef] [PubMed]
26. Brozynska, M.; Furtado, A.; Henry, R.J. Direct Chloroplast Sequencing: Comparison of Sequencing Platforms and Analysis Tools for Whole Chloroplast Barcoding. *PLoS ONE* **2014**, *9*, e110387. [CrossRef]
27. Rhoads, A.; Au, K.F. PacBio Sequencing and Its Applications. *Genom. Proteom. Bioinf.* **2015**, *13*, 278–289. [CrossRef]
28. Hon, T.; Mars, K.; Young, G.; Tsai, Y.-C.; Karalius, J.W.; Landolin, J.M.; Maurer, N.; Kudrna, D.; Hardigan, M.A.; Steiner, C.C.; et al. Highly accurate long-read HiFi sequencing data for five complex genomes. *Sci. Data* **2020**, *7*, 399. [CrossRef]
29. Wang, Y.H.; Zhao, Y.; Bollas, A.; Wang, Y.R.; Au, K.F. Nanopore Sequencing Technology, Bioinformatics and Applications. *Nat. Biotechnol.* **2021**, *39*, 1348–1365. [CrossRef]
30. Wang, J.; Bhakta, N.; Miller, V.A.; Revsine, M.; Litzow, M.; Paietta, E.; Roberts, K.; Gu, Z.; Mullighan, C.; Jones, C.; et al. Nanopore RNA Sequencing as A Diagnostic Tool for Acute Leukemia in Low Resource Settings. *Pediatr. Blood Cancer* **2021**, *68*, S107.
31. Belser, C.; Istace, B.; Denis, E.; Dubarry, M.; Baurens, F.-C.; Falentin, C.; Genete, M.; Berrabah, W.; Chèvre, A.-M.; Delourme, R.; et al. Chromosome-scale assemblies of plant genomes using nanopore long reads and optical maps. *Nat. Plants* **2018**, *4*, 879–887. [CrossRef] [PubMed]
32. Chen, Z.; Pham, L.; Wu, T.-C.; Mo, G.; Xia, Y.; Chang, P.L.; Porter, D.; Phan, T.; Che, H.; Tran, H.; et al. Ultralow-input single-tube linked-read library method enables short-read second-generation sequencing systems to routinely generate highly accurate and economical long-range sequencing information. *Genome Res.* **2020**, *30*, 898–909. [CrossRef] [PubMed]
33. Wang, O.; Chin, R.; Cheng, X.; Wu, M.K.Y.; Mao, Q.; Tang, J.; Sun, Y.; Anderson, E.; Lam, H.K.; Chen, D.; et al. Efficient and unique cobarcoding of second-generation sequencing reads from long DNA molecules enabling cost-effective and accurate sequencing, haplotyping, and de novo assembly. *Genome Res.* **2019**, *29*, 798–808. [CrossRef] [PubMed]
34. Singh, M.; Al-Eryani, G.; Carswell, S.; Ferguson, J.M.; Blackburn, J.; Barton, K.; Roden, D.; Luciani, F.; Phan, T.G.; Junankar, S.; et al. High-throughput targeted long-read single cell sequencing reveals the clonal and transcriptional landscape of lymphocytes. *Nat. Commun.* **2019**, *10*, 3120. [CrossRef] [PubMed]
35. Sharma, P.; Masouleh, A.K.; Topp, B.; Furtado, A.; Henry, R.J. *De novo* chromosome level assembly of a plant genome from long read sequence data. *Plant J.* **2021**, *109*, 727–736. [CrossRef]
36. Cheng, H.; Concepcion, G.T.; Feng, X.; Zhang, H.; Li, H. Haplotype-resolved de novo assembly using phased assembly graphs with hifiasm. *Nat. Methods* **2021**, *18*, 170–175. [CrossRef]
37. Healey, A.L.; Shepherd, M.; King, G.J.; Butler, J.B.; Freeman, J.S.; Lee, D.J.; Potts, B.M.; Silva-Junior, O.B.; Baten, A.; Jenkins, J.; et al. Pests, diseases, and aridity have shaped the genome of *Corymbia citriodora*. *Commun. Biol.* **2021**, *4*, 537. [CrossRef]
38. Harewood, L.; Kishore, K.; Eldridge, M.; Wingett, S.; Pearson, D.; Schoenfelder, S.; Collins, V.P.; Fraser, P. Hi-C as a tool for precise detection and characterisation of chromosomal rearrangements and copy number variation in human tumours. *Genome Biol.* **2017**, *18*, 125. [CrossRef]
39. Chen, P.; Jing, X.; Ren, J.; Cao, H.; Hao, P.; Li, X. Modelling BioNano optical data and simulation study of genome map assembly. *Bioinformatics* **2018**, *34*, 3966–3974. [CrossRef]
40. Shirasawa, K.; Harada, D.; Hirakawa, H.; Isobe, S.; Kole, C. Chromosome-level de novo genome assemblies of over 100 plant species. *Breed. Sci.* **2021**, *71*, 117–124. [CrossRef]
41. Gao, Y.; Zhang, Y.; Feng, C.; Chu, H.; Feng, C.; Wang, H.; Wu, L.; Yin, S.; Liu, C.; Chen, H.; et al. A chromosome-level genome assembly of *Amorphophallus konjac* provides insights into konjac glucomannan biosynthesis. *Comput. Struct. Biotechnol. J.* **2022**, *20*, 1002–1011. [CrossRef] [PubMed]
42. Sun, X.; Jiao, C.; Schwaninger, H.; Chao, C.T.; Ma, Y.; Duan, N.; Khan, A.; Ban, S.; Xu, K.; Cheng, L.; et al. Phased diploid genome assemblies and pan-genomes provide insights into the genetic history of apple domestication. *Nat. Genet.* **2020**, *52*, 1423–1432. [CrossRef] [PubMed]

43. Rendón-Anaya, M.; Ibarra-Laclette, E.; Méndez-Bravo, A.; Lan, T.; Zheng, C.; Carretero-Paulet, L.; Perez-Torres, C.A.; Chacón-López, A.; Hernandez-Guzmán, G.; Chang, T.-H.; et al. The avocado genome informs deep angiosperm phylogeny, highlights introgressive hybridization, and reveals pathogen-influenced gene space adaptation. *Proc. Natl. Acad. Sci. USA* **2019**, *116*, 17081–17089. [CrossRef] [PubMed]
44. Wang, Z.; Miao, H.; Liu, J.; Xu, B.; Yao, X.; Xu, C.; Zhao, S.; Fang, X.; Jia, C.; Wang, J.; et al. Musa balbisiana genome reveals subgenome evolution and functional divergence. *Nat. Plants* **2019**, *5*, 810–821. [CrossRef] [PubMed]
45. Shen, T.; Qi, H.; Luan, X.; Xu, W.; Yu, F.; Zhong, Y.; Xu, M. The chromosome-level genome sequence of the camphor tree provides insights into Lauraceae evolution and terpene biosynthesis. *Plant Biotechnol. J.* **2021**, *20*, 244–246. [CrossRef] [PubMed]
46. Iorizzo, M.; Ellison, S.; Senalik, D.; Zeng, P.; Satapoomin, P.; Huang, J.; Bowman, M.; Iovene, M.; Sanseverino, W.; Cavagnaro, P.; et al. A high-quality carrot genome assembly provides new insights into carotenoid accumulation and asterid genome evolution. *Nat. Genet.* **2016**, *48*, 657–666. [CrossRef] [PubMed]
47. Zhao, Q.; Yang, J.; Cui, M.-Y.; Liu, J.; Fang, Y.; Yan, M.; Qiu, W.; Shang, H.; Xu, Z.; Yidiresi, R.; et al. The Reference Genome Sequence of *Scutellaria baicalensis* Provides Insights into the Evolution of Wogonin Biosynthesis. *Mol. Plant* **2019**, *12*, 935–950. [CrossRef]
48. Feng, L.; Lin, H.; Kang, M.; Ren, Y.; Yu, X.; Xu, Z.; Wang, S.; Li, T.; Yang, W.; Hu, Q. A chromosome-level genome assembly of an alpine plant *Crucihimalaya lasiocarpa* provides insights into high-altitude adaptation. *DNA Res.* **2022**, *29*, dsac004. [CrossRef]
49. Ling, J.; Xie, X.X.; Gu, X.F.; Zhao, J.L.; Ping, X.X.; Li, Y.; Yang, Y.H.; Mao, Z.C.; Xie, B.Y. High-quality chromosome-level genomes of *Cucumis metuliferus* and *Cucumis melo* provide insight into *Cucumis* genome evolution. *Plant J.* **2021**, *107*, 136–148. [CrossRef]
50. Nunn, A.; Rodríguez-Arévalo, I.; Tandukar, Z.; Frels, K.; Contreras-Garrido, A.; Carbonell-Bejerano, P.; Zhang, P.; Cruz, D.R.; Jandrasits, K.; Lanz, C.; et al. Chromosome-level *Thlaspi arvense* genome provides new tools for translational research and for a newly domesticated cash cover crop of the cooler climates. *Plant Biotechnol. J.* **2022**, *20*, 944–963. [CrossRef]
51. Li, H.L.; Wu, L.; Dong, Z.M.; Jiang, Y.S.; Jiang, S.J.; Xing, H.T.; Li, Q.; Liu, G.C.; Tian, S.M.; Wu, Z.Y.; et al. Haplotype-Resolved Genome of Diploid Ginger (*Zingiber Officinale*) and Its Unique Gingerol Biosynthetic Pathway. *Hortic. Res.* **2021**, *8*, 189. [CrossRef] [PubMed]
52. Shan, H.; Kong, H. The genome of Ginkgo biloba refined. *Nat. Plants* **2021**, *7*, 714–715. [CrossRef] [PubMed]
53. Al-Dossary, O.; Alsubaie, B.; Kharabian-Masouleh, A.; Al-Mssallem, I.; Furtado, A.; Henry, R.J. The jojoba genome reveals wide divergence of the sex chromosomes in a dioecious plant. *Plant J.* **2021**, *108*, 1283–1294. [CrossRef] [PubMed]
54. Sharma, P.; Murigneux, V.; Haimovitz, J.; Nock, C.J.; Tian, W.; Masouleh, A.K.; Topp, B.; Alam, M.; Furtado, A.; Henry, R.J. The genome of the endangered *Macadamia jansenii* displays little diversity but represents an important genetic resource for plant breeding. *Plant Direct* **2021**, *5*, e364. [CrossRef] [PubMed]
55. Nock, C.J.; Baten, A.; Mauleon, R.; Langdon, K.S.; Topp, B.; Hardner, C.; Furtado, A.; Henry, R.J.; King, G.J. Chromosome-Scale Assembly and Annotation of the Macadamia Genome (*Macadamia integrifolia* HAES 741). *G3-Genes Genomes Genet.* **2020**, *10*, 3497–3504. [CrossRef]
56. Peng, X.; Liu, H.; Chen, P.; Tang, F.; Hu, Y.; Wang, F.; Pi, Z.; Zhao, M.; Chen, N.; Chen, H.; et al. A Chromosome-Scale Genome Assembly of Paper Mulberry (*Broussonetia papyrifera*) Provides New Insights into Its Forage and Papermaking Usage. *Mol. Plant* **2019**, *12*, 661–677. [CrossRef] [PubMed]
57. Lian, X.D.; Zhang, H.P.; Jiang, C.; Gao, F.; Yan, L.; Zheng, X.B.; Cheng, J.; Wang, W.; Wang, X.B.; Ye, X.; et al. De Novo Chromosome-Level Genome of A Semi-Dwarf Cultivar of Prunus Persica Identifies The Aquaporin Pptip2 as Responsible for Temperature-Sensitive Semi-Dwarf Trait And Ppb3-1 for Flower Type And Size. *Plant Biotechnol. J.* **2022**, *20*, 886–902. [CrossRef]
58. Zhuang, W.; Chen, H.; Yang, M.; Wang, J.; Pandey, M.K.; Zhang, C.; Chang, W.-C.; Zhang, L.; Zhang, X.; Tang, R.; et al. The genome of cultivated peanut provides insight into legume karyotypes, polyploid evolution and crop domestication. *Nat. Genet.* **2019**, *51*, 865–876. [CrossRef]
59. Yuan, G.; Tan, S.; Wang, D.; Yang, Y.; Tian, B. Chromosome-Level Genome Assembly of the Rare and Endangered Tropical Plant *Speranskia yunnanensis* (Euphorbiaceae). *Front. Genet.* **2022**, *12*, 755564. [CrossRef]
60. Xiong, X.; Gou, J.; Liao, Q.; Li, Y.; Zhou, Q.; Bi, G.; Li, C.; Du, R.; Wang, X.; Sun, T.; et al. The Taxus genome provides insights into paclitaxel biosynthesis. *Nat. Plants* **2021**, *7*, 1026–1036. [CrossRef]
61. Zhang, X.; Chen, S.; Shi, L.; Gong, D.; Zhang, S.; Zhao, Q.; Zhan, D.; Vasseur, L.; Wang, Y.; Yu, J.; et al. Haplotype-resolved genome assembly provides insights into evolutionary history of the tea plant Camellia sinensis. *Nat. Genet.* **2021**, *53*, 1250–1259. [CrossRef] [PubMed]
62. Lu, R.; Chen, Y.; Zhang, X.; Feng, Y.; Comes, H.P.; Li, Z.; Zheng, Z.; Yuan, Y.; Wang, L.; Huang, Z.; et al. Genome sequencing and transcriptome analyses provide insights into the origin and domestication of water caltrop (*Trapa* spp., Lythraceae). *Plant Biotechnol. J.* **2022**, *20*, 761–776. [CrossRef] [PubMed]
63. Low, W.Y.; Tearle, R.; Liu, R.; Koren, S.; Rhie, A.; Bickhart, D.; Rosen, B.D.; Kronenberg, Z.N.; Kingan, S.B.; Tseng, E.; et al. Haplotype-resolved genomes provide insights into structural variation and gene content in Angus and Brahman cattle. *Nat. Commun.* **2020**, *11*, 2071. [CrossRef] [PubMed]
64. Tao, Y.; Zhao, X.; Mace, E.; Henry, R.; Jordan, D. Exploring and Exploiting Pan-genomics for Crop Improvement. *Mol. Plant* **2018**, *12*, 156–169. [CrossRef] [PubMed]
65. Cheng, B.; Furtado, A.; Henry, R.J. The coffee bean transcriptome explains the accumulation of the major bean components through ripening. *Sci. Rep.* **2018**, *8*, 11414. [CrossRef]

66. Gillies, S.A.; Futardo, A.; Henry, R.J. Gene expression in the developing aleurone and starchy endosperm of wheat. *Plant Biotechnol. J.* **2012**, *10*, 668–679. [CrossRef]
67. Kasirajan, L.; Hoang, N.V.; Furtado, A.; Botha, F.C.; Henry, R.J. Transcriptome analysis highlights key differentially expressed genes involved in cellulose and lignin biosynthesis of sugarcane genotypes varying in fiber content. *Sci. Rep.* **2018**, *8*, 11612. [CrossRef]
68. Nirmal, R.C.; Furtado, A.; Wrigley, C.; Henry, R.J. Influence of Gene Expression on Hardness in Wheat. *PLoS ONE* **2016**, *11*, e0164746. [CrossRef]
69. Rangan, P.; Furtado, A.; Henry, R. Differential response of wheat genotypes to heat stress during grain filling. *Exp. Agric.* **2019**, *55*, 818–827. [CrossRef]
70. Rangan, P.; Furtado, A.; Henry, R.J. The transcriptome of the developing grain: A resource for understanding seed development and the molecular control of the functional and nutritional properties of wheat. *BMC Genom.* **2017**, *18*, 766. [CrossRef]
71. Cuperus, J.T. Single-cell genomics in plants: Current state, future directions, and hurdles to overcome. *Plant Physiol.* **2021**, *188*, 749–755. [CrossRef] [PubMed]
72. Hoang, N.V.; Furtado, A.; Thirugnanasambandam, P.P.; Botha, F.C.; Henry, R.J. De novo assembly and characterizing of the culm-derived meta-transcriptome from the polyploid sugarcane genome based on coding transcripts. *Heliyon* **2018**, *4*, e00583. [CrossRef] [PubMed]
73. Tirumala, R.K.; Subbaiyan, G.K.; Singh, A.K.; Furtado, A.; Henry, R.J. RNA-Seq to Understand Transcriptomes and Application in Improving Crop Quality. In *Comprehensive Foodomics*, 1st ed.; Cifuentes, A., Ed.; Elsevier: Amsterdam, The Netherlands, 2021; Volume 1, pp. 472–485. [CrossRef]
74. Margarido, G.R.A.; Correr, F.H.; Furtado, A.; Botha, F.C.; Henry, R.J. Limited allele-specific gene expression in highly polyploid sugarcane. *Genome Res.* **2022**, *32*, 297–308. [CrossRef] [PubMed]
75. Powell, J.J.; Fitzgerald, T.L.; Stiller, J.; Berkman, P.J.; Gardiner, D.M.; Manners, J.M.; Henry, R.J.; Kazan, K. The defence-associated transcriptome of hexaploid wheat displays homoeolog expression and induction bias. *Plant Biotechnol. J.* **2016**, *15*, 533–543. [CrossRef] [PubMed]
76. Cheng, B.; Furtado, A.; Henry, R.J. Long-read sequencing of the coffee bean transcriptome reveals the diversity of full-length transcripts. *GigaScience* **2017**, *6*, gix086. [CrossRef] [PubMed]
77. Zhou, S.; Zhang, J.; Han, H.; Zhang, J.; Ma, H.; Zhang, Z.; Lu, Y.; Liu, W.; Yang, X.; Li, X.; et al. Full-length transcriptome sequences of Agropyron cristatum facilitate the prediction of putative genes for thousand-grain weight in a wheat-A. cristatum translocation line. *BMC Genom.* **2019**, *20*, 1025. [CrossRef] [PubMed]
78. Yuan, H.Z.; Yu, H.M.; Huang, T.; Shen, X.J.; Xia, J.; Pang, F.H.; Wang, J.; Zhao, M.Z. The Complexity of The *Fragaria X Ananassa* (Octoploid) Transcriptome by Single-Molecule Long-Read Sequencing. *Hortic. Res.* **2019**, *6*, 46. [CrossRef] [PubMed]
79. Hoang, N.V.; Furtado, A.; Mason, P.J.; Marquardt, A.; Kasirajan, L.; Thirugnanasambandam, P.P.; Botha, F.C.; Henry, R.J. A survey of the complex transcriptome from the highly polyploid sugarcane genome using full-length isoform sequencing and de novo assembly from short read sequencing. *BMC Genom.* **2017**, *18*, 395. [CrossRef]
80. Hoang, N.V.; Furtado, A.; McQualter, R.B.; Henry, R.J. Next generation sequencing of total DNA from sugarcane provides no evidence for chloroplast heteroplasmy. *New Negat. Plant Sci.* **2015**, *1–2*, 33–45. [CrossRef]
81. Dobrogojski, J.; Adamiec, M.; Luciński, R. The chloroplast genome: A review. *Acta Physiol. Plant.* **2020**, *42*, 98. [CrossRef]
82. Henry, R.; Rice, N.; Edwards, M.; Nock, C. Next-Generation Technologies to Determine Plastid Genome Sequences. In *Chloroplast Biotechnology*; Maliga, P., Ed.; Humana Press: Totova, NJ, USA, 2014; Volume 1132, pp. 39–46.
83. Nock, C.J.; Waters, D.L.; Edwards, M.A.; Bowen, S.G.; Rice, N.; Cordeiro, G.M.; Henry, R.J. Chloroplast genome sequences from total DNA for plant identification. *Plant Biotechnol. J.* **2010**, *9*, 328–333. [CrossRef] [PubMed]
84. Guyeux, C.; Charr, J.-C.; Tran, H.T.M.; Furtado, A.; Henry, R.J.; Crouzillat, D.; Guyot, R.; Hamon, P. Evaluation of chloroplast genome annotation tools and application to analysis of the evolution of coffee species. *PLoS ONE* **2019**, *14*, e0216347. [CrossRef] [PubMed]
85. Ananda, G.; Norton, S.; Blomstedt, C.; Furtado, A.; Møller, B.; Gleadow, R.; Henry, R. Phylogenetic relationships in the *Sorghum* genus based on sequencing of the chloroplast and nuclear genes. *Plant Genome* **2021**, *14*, e20123. [CrossRef] [PubMed]
86. Brozynska, M.; Copetti, D.; Furtado, A.; Wing, R.A.; Crayn, D.; Fox, G.; Ishikawa, R.; Henry, R.J. Sequencing of Australian wild rice genomes reveals ancestral relationships with domesticated rice. *Plant Biotechnol. J.* **2016**, *15*, 765–774. [CrossRef]
87. Healey, A.; Lee, D.J.; Furtado, A.; Henry, R.J. Evidence of inter-sectional chloroplast capture in *Corymbia* among sections *Torellianae* and *Maculatae*. *Aust. J. Bot.* **2018**, *66*, 369. [CrossRef]
88. Hodel, R.G.J.; Zimmer, E.A.; Liu, B.-B.; Wen, J. Synthesis of Nuclear and Chloroplast Data Combined With Network Analyses Supports the Polyploid Origin of the Apple Tribe and the Hybrid Origin of the Maleae—Gillenieae Clade. *Front. Plant Sci.* **2022**, *12*, 820997. [CrossRef] [PubMed]
89. Yoo, K.; Lowry, P.P.; Wen, J. Discordance of chloroplast and nuclear ribosomal DNA data in *Osmorhiza* (Apiaceae). *Am. J. Bot.* **2002**, *89*, 966–971. [CrossRef]
90. Yu, W.-B.; Huang, P.-H.; Li, D.-Z.; Wang, H. Incongruence between Nuclear and Chloroplast DNA Phylogenies in *Pedicularis* Section *Cyathophora* (Orobanchaceae). *PLoS ONE* **2013**, *8*, e74828. [CrossRef]
91. Gualberto, J.M.; Mileshina, D.; Wallet, C.; Niazi, A.K.; Weber-Lotfi, F.; Dietrich, A. The plant mitochondrial genome: Dynamics and maintenance. *Biochimie* **2014**, *100*, 107–120. [CrossRef]

92. Qin, L.; Hu, Y.; Wang, J.; Wang, X.; Zhao, R.; Shan, H.; Li, K.; Xu, P.; Wu, H.; Yan, X.; et al. Insights into angiosperm evolution, floral development and chemical biosynthesis from the Aristolochia fimbriata genome. *Nat. Plants* **2021**, *7*, 1239–1253. [CrossRef]
93. Denoeud, F.; Carretero-Paulet, L.; Dereeper, A.; Droc, G.; Guyot, R.; Pietrella, M.; Zheng, C.; Alberti, A.; Anthony, F.; Aprea, G.; et al. The coffee genome provides insight into the convergent evolution of caffeine biosynthesis. *Science* **2014**, *345*, 1181–1184. [CrossRef] [PubMed]
94. Julca, I.; Ferrari, C.; Flores-Tornero, M.; Proost, S.; Lindner, A.-C.; Hackenberg, D.; Steinbachová, L.; Michaelidis, C.; Gomes Pereira, S.; Misra, C.S.; et al. Comparative transcriptomic analysis reveals conserved programmes underpinning organogenesis and reproduction in land plants. *Nat. Plants* **2021**, *7*, 1143–1159. [CrossRef] [PubMed]
95. Stein, J.C.; Yu, Y.; Copetti, D.; Zwickl, D.J.; Zhang, L.; Zhang, C.; Chougule, K.; Gao, N.; Iwata, A.; Goicoechea, J.L.; et al. Publisher Correction: Genomes of 13 domesticated and wild rice relatives highlight genetic conservation, turnover and innovation across the genus Oryza. *Nat. Genet.* **2018**, *50*, 285–296. [CrossRef] [PubMed]
96. Park, H.-S.; Lee, W.K.; Lee, S.-C.; Lee, H.O.; Joh, H.J.; Park, J.Y.; Kim, S.; Song, K.; Yang, T.-J. Inheritance of chloroplast and mitochondrial genomes in cucumber revealed by four reciprocal F1 hybrid combinations. *Sci. Rep.* **2021**, *11*, 2506. [CrossRef] [PubMed]
97. Henry, R.J. Genomics strategies for germplasm characterization and the development of climate resilient crops. *Front. Plant Sci.* **2014**, *5*, 68. [CrossRef]
98. Bradbury, L.M.T.; Fitzgerald, T.L.; Henry, R.J.; Jin, Q.; Waters, D.L.E. The gene for fragrance in rice. *Plant Biotechnol. J.* **2005**, *3*, 363–370. [CrossRef]
99. Bradbury, L.M.T.; Henry, R.J.; Jin, Q.; Reinke, R.F.; Waters, D. A Perfect Marker for Fragrance Genotyping in Rice. *Mol. Breed.* **2005**, *16*, 279–283. [CrossRef]
100. Waters, D.L.E.; Henry, R.J.; Reinke, R.F.; Fitzgerald, M.A. Gelatinization temperature of rice explained by polymorphisms in starch synthase. *Plant Biotechnol. J.* **2006**, *4*, 115–122. [CrossRef]
101. Nirmal, R.C.; Furtado, A.; Rangan, P.; Henry, R.J. Fasciclin-like arabinogalactan protein gene expression is associated with yield of flour in the milling of wheat. *Sci. Rep.* **2017**, *7*, 12539. [CrossRef]
102. Furtado, A.; Bundock, P.; Banks, P.M.; Fox, G.; Yin, X.; Henry, R.J. A novel highly differentially expressed gene in wheat endosperm associated with bread quality. *Sci. Rep.* **2015**, *5*, 10446. [CrossRef]
103. Henry, R.J. *Plant Genotyping: The DNA Fingerprinting of Plants*; CABI Publishing: Oxford, UK, 2001; p. 325.
104. Henry, R.J.; Edwards, M.; Waters, D.L.E.; Krishnan, S.G.; Bundock, P.; Sexton, T.R.; Kharabian-Masouleh, A.; Nock, C.; Pattemore, J. Application of large-scale sequencing to marker discovery in plants. *J. Biosci.* **2012**, *37*, 829–841. [CrossRef] [PubMed]
105. Krishna, H.; Alizadeh, M.; Singh, D.; Singh, U.; Chauhan, N.; Eftekhari, M.; Sadh, R.K. Somaclonal variations and their applications in horticultural crops improvement. *3 Biotech* **2016**, *6*, 54. [CrossRef] [PubMed]
106. Henry, R.J. Genomics and Gene-Editing Technologies Accelerating Grain Product Innovation. *Cereal Foods World* **2019**, *64*. [CrossRef]
107. Henry, R.J. Advances in DNA sequencing enabling more rapid development of improved biomass and biofuel conversion technologies. *Biofuels* **2012**, *3*, 507–509. [CrossRef]
108. Sherman, B.; Henry, R.J. Access to biodiversity for food production: Reconciling open access digital sequence information with access and benefit sharing. *Mol. Plant* **2021**, *14*, 701–704. [CrossRef]
109. Sherman, B.; Henry, R.J. The Nagoya Protocol and historical collections of plants. *Nat. Plants* **2020**, *6*, 430–432. [CrossRef]
110. Henry, R. Innovations in Agriculture and Food Supply in Response to the COVID-19 Pandemic. *Mol. Plant* **2020**, *13*, 1095–1097. [CrossRef]
111. Henry, R.J. Innovations in plant genetics adapting agriculture to climate change. *Curr. Opin. Plant Biol.* **2019**, *56*, 168–173. [CrossRef]
112. Abberton, M.; Batley, J.; Bentley, A.; Bryant, J.; Cai, H.; Cockram, J.; De Oliveira, A.C.; Cseke, L.J.; Dempewolf, H.; De Pace, C.; et al. Global agricultural intensification during climate change: A role for genomics. *Plant Biotechnol. J.* **2015**, *14*, 1095–1098. [CrossRef]
113. Henry, R.J. Sequencing of wild crop relatives to support the conservation and utilization of plant genetic resources. *Plant Genet. Resour.* **2014**, *12*, S9–S11. [CrossRef]
114. McCouch, S.; Navabi, Z.K.; Abberton, M.; Anglin, N.L.; Barbieri, R.L.; Baum, M.; Bett, K.; Booker, H.; Brown, G.L.; Bryan, G.J.; et al. Mobilizing Crop Biodiversity. *Mol. Plant* **2020**, *13*, 1341–1344. [CrossRef] [PubMed]

applied biosciences

MDPI

Article

Enhancing the Thermal and Kinetic Stability of Ketol-Acid Reductoisomerase, a Central Catalyst of a Cell-Free Enzyme Cascade for the Manufacture of Platform Chemicals

You Lv [1,†], Shan Zheng [1,†], Adi Goldenzweig [2], Fengjiang Liu [3], Yan Gao [3], Xiuna Yang [3], Ajit Kandale [1], Ross P. McGeary [1], Simon Williams [1], Bostjan Kobe [1,4], Mark A. Schembri [1,4], Michael J. Landsberg [1], Bin Wu [5], Thomas B. Brück [6], Volker Sieber [7], Mikael Boden [1], Zihe Rao [3], Sarel J. Fleishman [2], Gerhard Schenk [1,4,8] and Luke W. Guddat [1,*]

1 School of Chemistry and Molecular Biosciences, The University of Queensland, Brisbane, QLD 4072, Australia
2 Department of Biomolecular Sciences, Weizmann Institute of Science, Rehovot 7600001, Israel
3 Shanghai Institute for Advanced Immunochemical Studies, School of Life Science and Technology, ShanghaiTech University, Shanghai 201210, China
4 Australian Infectious Diseases Research Centre, The University of Queensland, Brisbane, QLD 4072, Australia
5 College of Biotechnology & Pharmaceutical Engineering, Nanjing Tech University, Nanjing 210037, China
6 Werner Siemens Chair of Synthetic Biotechnology, Department of Chemistry, Technical University of Munich, 85748 Garching, Germany
7 Chair of Chemistry of Biogenic Resources, Technical University of Munich, TUM Campus Straubing, 94315 Straubing, Germany
8 Australian Institute of Bioengineering and Nanotechnology, The University of Queensland St. Lucia, Brisbane, QLD 4072, Australia
* Correspondence: luke.guddat@uq.edu.au
† These authors contributed equally to this work.

Citation: Lv, Y.; Zheng, S.; Goldenzweig, A.; Liu, F.; Gao, Y.; Yang, X.; Kandale, A.; McGeary, R.P.; Williams, S.; Kobe, B.; et al. Enhancing the Thermal and Kinetic Stability of Ketol-Acid Reductoisomerase, a Central Catalyst of a Cell-Free Enzyme Cascade for the Manufacture of Platform Chemicals. *Appl. Biosci.* **2022**, *1*, 163–178. https://doi.org/10.3390/applbiosci1020011

Academic Editor: Hervé Quiquampoix

Received: 1 June 2022
Accepted: 29 July 2022
Published: 10 August 2022

Abstract: The branched-chain amino acids (BCAAs) leucine, isoleucine and valine are synthesized via a common biosynthetic pathway. Ketol-acid reductoisomerase (KARI) is the second enzyme in this pathway. In addition to its role in BCAA biosynthesis, KARI catalyzes two rate-limiting steps that are key components of a cell-free biofuel biosynthesis route. For industrial applications, reaction temperature and enzyme stability are key factors that affect process robustness and product yield. Here, we have solved the cryo-EM structure (2.94 Å resolution) of a homododecameric Class I KARI (from *Campylobacter jejuni*) and demonstrated how a triad of amino acid side chains plays a crucial role in promoting the oligomerization of this enzyme. Importantly, both its thermal and solvent stability are greatly enhanced in the dodecameric state when compared to its dimeric counterpart (apparent melting temperatures (Tm) of 83.1 °C and 51.5 °C, respectively). We also employed protein design (PROSS) for a tetrameric Class II KARI (from *Escherichia coli*) to generate a variant with improved thermal and solvent stabilities. In total, 34 mutations were introduced, which did not affect the oligomeric state of this enzyme but resulted in a fully functional catalyst with a significantly elevated Tm (58.5 °C vs. 47.9 °C for the native version).

Keywords: cell-free enzyme cascades; biomanufacturing; enzyme stability; protein design; PROSS; branched chain amino acid biosynthesis; ketol-acid reductoisomerase

1. Introduction

The branched-chain amino acids' (BCAAs) pathway for the biosynthesis of leucine, isoleucine and valine is present in bacteria, fungi and plants, but not in animals [1]. Consequently, this biosynthesis route has long been recognized as a target for herbicides and antimicrobial agents [2–6]. More recently, enzymes from this pathway attracted increasing interest as components of designed enzymatic cascades to convert renewable raw material into high-value products [7]. For example, in *Escherichia coli*, the BCAA pathway was re-engineered to produce higher alcohols, including isobutanol, 1-butanol, 2-methyl-1-butanol,

3-methyl-1-butanol and 2-phenylethanol from glucose [7]. These products are alternatives to current fossil fuels because they exhibit a higher energy density, lower hygroscopicity, lower vapor pressure and are compatible with existing fossil fuel infrastructures [8].

Recently, ab initio designed, cell-free biosynthesis routes were reported that facilitate the conversion of renewable sugar streams into the biofuel building blocks ethanol or isobutanol [9]. This approach is partly reliant on the use of the BCAA pathway enzymes, acetohydroxyacid synthase (AHAS), ketol-acid reductoisomerase (KARI) and dihydroxy-acid dehydratase (DHAD) [10–15]. Compared to conventional fermentation approaches, cell-free strategies have several advantages. The cell-free production route enhances the resilience of the process towards accumulating reaction products, which are toxic to cellular production systems. Furthermore, reactions can be performed at higher temperatures, which can improve the rate and efficiency of production leading to higher product yields.

KARI is a bifunctional enzyme that converts 2-acetolactate into 2,3-dihydroxy-3-isovalerate (precursor of valine and leucine) or 2-aceto-2-hydroxybutyrate into 2,3-dihydroxy-3-methylvalerate (precursor of isoleucine) via a two-step reaction within a single active site [16–19]. The initial Mg^{2+}-dependent alkyl group rearrangement is followed by a NAD(P)H-dependent reduction (Figure 1) [16,18].

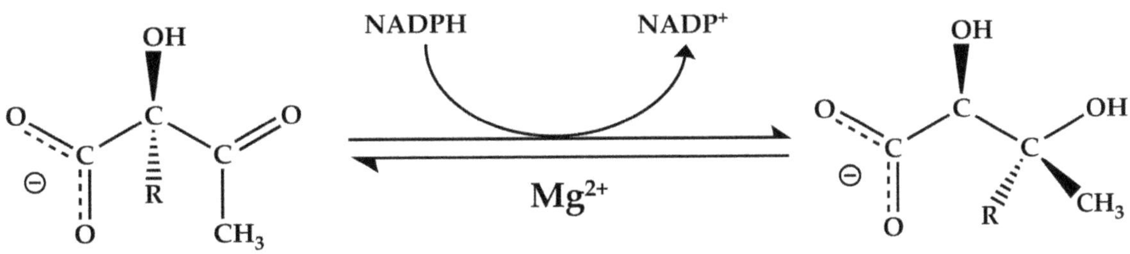

R=CH_3 (2-acetolactate)
R=CH_2CH_3 (2-aceto-2-hydroxybutyrate)

R= CH_3 (2,3-dihydroxy-3-isovalerate)
R=CH_2CH_3 (2,3-dihydroxy-3-methylvalerate)

Figure 1. Reaction catalyzed by KARI. NADPH is the commonly used cofactor, but some KARIs can use NADH as the preferred reductant. KARI catalyzes the concerted isomerization and reduction in the native substrate, 2-acetolactate.

KARIs can be divided into two classes according to the length of the polypeptide. Class I KARIs are ~330 amino acids in length, containing an N-terminal domain with a Rossmann-fold and a C-terminal domain that is largely α-helical [19]. Class I KARIs require the C-terminal domain from an adjacent subunit to complete the active site, forming a dimer (e.g., KARI from *Mycobacterium tuberculosis* (*Mt*KARI) [20] or *Staphylococcus aureus* (*Sa*KARI) [21]) or dodecamer (e.g., *Pseudomonas aeruginosa* KARI (*Pa*KARI) [22] and *Sulfolobus solfataricus* KARI (*Ss*KARI) [23]); the dodecamer assembly requires six interlocking dimeric units [22,24]; see below. Class II KARIs, exemplified by enzymes from *E. coli* (*Ec*KARI) [5] and plants [25], are ~500 amino acids in length and contain an additional C-terminal domain that appears to have arisen as the result of a gene duplication event. With this extension, Class II KARIs can form a complete active site within a single subunit, a feature that could be favorable for cell-free chemical production strategies, as this may simplify attachment of the enzyme to a solid support.

The advantages of using enzymes as catalysts include their extremely high efficiency, exquisite substrate specificity, stereo-selectivity and potentially low environmental impact. Such properties have been widely exploited in applications that include biosensors, bioreactors and in enzymatic fuel cells [26–29]. However, compared with chemical catalysts, a major drawback of using enzymes in industry is their short half-life and low thermal stability [30]. Another major limitation inherent to KARIs is their low turnover number, which is the lowest amongst the enzymes in the BCAA pathway [9]. However, previous studies demonstrated that KARIs are amenable to modifications which may enhance their potential for industrial applications. For instance, the cofactor preference of *Slackia exigua* KARI was switched from NADPH to NADH (a more stable hydride donor) through targeted mutations [31]. More recently, we employed ancestral sequence reconstruction (ASR) to design an ancient KARI variant that both increased thermal stability and enhanced reactivity towards the substrate, 2-acetolactate [10].

In this study, we focused our attention on (i) investigating factors that contribute to the thermal stability of KARI and (ii) designing a variant with enhanced stability using the protein design algorithm, Protein Repair One Stop Shop (PROSS) [32]. Specifically, we demonstrate that the emergence of higher oligomeric forms of KARI may be a response for the need to have enhanced thermal stability. The cumulative data provide detailed molecular insight into factors that are essential to engineer KARI variants suitable for industrial applications.

2. Materials and Methods

2.1. Expression and Purification of the Enzymes and Their Mutants

Here, *Cj*KARI is defined as the wild-type KARI from *Campylobacter jejuni*, and *Cj*KARI_Dm is the K290A/L294A double mutant. The *Cj*KARI and *Cj*KARI_Dm genes were synthesized by Epoch Life Science Inc. The DNA fragments were inserted between the NdeI and XhoI sites and cloned into a pET-21a(+) vector. The final constructs contain the gene sequences of the enzyme and a -LEHHHHHH- sequence at the C-terminus for purification. The plasmids containing the enzyme were then transformed into *E. coli* BL21 (DE3) cells. These cells were then grown in LB medium in the presence of 100 mg/L ampicillin at 37 °C until the OD_{600} reached ~0.6. The expression of protein was induced by the addition of 1 mM isopropyl β-D-1-thiogalactopyranoside (IPTG), followed by growth for 10 hours at 20 °C. Cells were centrifuged and the pellet resuspended in ice-cold lysis buffer (20 mM Tris-HCl pH 8.0, 20 mM imidazole, 250 mM $MgCl_2$, 500 mM NaCl and 10% glycerol). The cells were homogenized by ultrasonication, followed by further centrifugation. The supernatant was collected for protein purification using a QIAGEN Ni-NTA-agarose resin column and by gel filtration chromatography with a Superdex 200 Hiload 26/60 column (GE Healthcare) in gel filtration buffer (20 mM Tris-HCl, pH 8.0, 250 mM $MgCl_2$, 500 mM NaCl and 10% glycerol). The protein concentration was determined by measurement of A_{280}. The molar extinction coefficient, 20,400 cm^{-1} M^{-1}, was calculated using the method of Gill and von Hippel [33] by considering the theoretical molecular weight for the monomer and adding the two additional residues from the construct (-LE-) and the hexa-histidine tag (38032 Da for *Cj*KARI and 37,933 Da for *Cj*KARI_Dm). Purity was assessed by 12% SDS-PAGE. For long-term storage, the enzyme was kept at −70 °C in gel filtration buffer. The preparation of *Mt*KARI and *Ec*KARI (including its mutant) followed previously described protocols [5,20].

2.2. Preparation of the CjKARI for Cryo-EM Studies

Crosslinking was performed with the enzyme in 50 mM HEPES, pH 8.0, by incubating with 0.05% glutaraldehyde for 15 min at 25 °C. The reaction was terminated by adding ~65 mM Tris-HCl, pH 8.2. The crosslinked enzyme was repurified at 4 °C by gel filtration using a Superose 6 Increase 10/300 GL column (GE Healthcare). The buffer consisted of 50 mM NaCl and 20 mM Tris-HCl, pH 8.0. The concentration of the enzyme was ~1.1 mg/mL.

2.3. Cryo-Electron Microscopy Studies

The enzyme samples were applied to ultrathin carbon film supported by a holey carbon film on Quantifoil R1.2/1.3 300 mesh copper grids. Prior to sample application, the grids were glow-discharged using a Gatan Solarus 950 Advanced Plasma cleaning system (Gatan, Inc., Pleasanton, CA, USA). Dithiothreitol (DTT; 5 mM) was added to the protein, and 3 μL of the prepared sample were applied onto the grids. The grids were blotted for 3.5 s and then plunge-frozen in liquid ethane cooled with liquid nitrogen using a FEI Vitrobot Mark IV (FEI company, Hillsboro, OR, USA) operated at 10 °C and 100% humidity. A Titan

Krios transmission electron microscope was used to image the grids under 300 kV, while maintaining the specimen at liquid nitrogen temperatures. Automated data acquisition was performed using SerialEM. Images for these samples were recorded using a Gatan, Inc. K3 direct detection camera at 29,000-fold magnification operated in super-resolution counting mode with a physical pixel size of 0.41 Å and binned to a pixel size of 0.82 Å. The defocus range was −1.2–−1.8 μm, and each exposure had an accumulated dose of ~50 $e^-/Å^2$ and a total of 40 frames per micrograph.

2.4. Cryo-EM Image Processing, Model Building, Refinement and Validation

A number of micrograph movies were collected, initial and final particles picked, and all the information related to data collection and refinement is summarized in Table 1. All processing steps were performed using CryoSPARC [34]. All the original images were aligned, and local motion correction was conducted by patching motion correction. Suitable particles were automatically selected through reference-free 2D classification and used for ab initio 3D reconstruction to generate 3D initial models as a reference for heterogeneous refinement. After heterogeneous refinement, final particles were selected for non-uniform refinement (Nu-refinement) to generate the final cryo-EM map. The final resolution was estimated according to a global-standard Fourier shell correlation cutoff of 0.5 [35]. Local resolution maps were also generated by CryoSPARC. The cryo-EM maps of *Cj*KARI were used for model building. The maps were segmented into submaps where each contains a dimer (i.e., two monomers) using UCSF Chimera [36]. Then, all the residues in the dimer model were manually fitted into the 3D map density using Coot [37]. The fitted structure was further refined with real space refinement in PHENIX [38] to avoid over-fitting. All the figures were generated with UCSF Chimera and Pymol.

Table 1. Cryo-EM data statistics.

Data Collection and Processing	
PDB Entry	8CY8
EMDB Entry	EMD-27070
EM equipment	FEI Titan Krios
Camera	K3
Frames per micrograph	40
Magnification	29,000
Voltage (keV)	300
Electron exposure (e-/Å^2)	50
Defocus range (μm)	−1.2 to −1.8
Pixel size (Å)	0.82
Symmetry	T
Micrographs (no.)	2247
Initial particle images (no.)	2,472,436
Final particle images (no.)	74,899
Map resolution (Å)	2.94
FSC threshold	0.143
Map local resolution range (Å)	2.5–4.2
Local resolution FSC	0.5
Refinement	
Model composition	
Non-hydrogen atoms	28,800
Protein residues	3744
Metal	0
Cofactor	0
Inhibitor	0
Ligands	12

Table 1. *Cont.*

Data Collection and Processing	
R.m.s. deviations	
Bond length (Å)	0.010
Bond angles (°)	0.772
Validation	
MolProbity score	1.13
Clash score	3.38
Poor rotamers (%)	0.00
Ramachandran plot (%)	
Favored (%)	99.37
Allowed (%)	0.63
Outliers (%)	0.00
B factors (Å^2)	
Protein (min/max/mean)	10.51/116.90/53.22
Ligand (min/max/mean)	33.80/36.87/35.02

2.5. Oligomeric States Analysis

Size-exclusion chromatography-multi-angle laser light scattering (SEC-MALS) was performed using an inline Superdex Increase 200 5/150 GL SEC column (GE Healthcare) combined with a Dawn Heleos II 18-angle light-scattering detector coupled with an Optilab TrEX refractive index detector (Wyatt Technology, Santa Barbara, CA, USA). A total of 100 µg of purified *Cj*KARI and *Cj*KARI_Dm were eluted at 0.25 ml/min in 10 mM Tris-HCl, pH 8.0, 250 mM $MgCl_2$ and 500 mM NaCl at room temperature. The molecular mass calculations were performed using Astra6.1 software (Wyatt Technology). The input of the refractive increment (dn/dc values) was fixed at 0.186 mL/g, with an assumption that dn/dc is invariable for unmodified proteins. The molecular mass was determined across the protein elution peak.

2.6. Enzyme Kinetics and Thermal Stability Studies

To remove glycerol from the stored enzyme samples, the enzymes were loaded onto a PD-10 column using a buffer containing 20 mM Tris-HCl, pH 8.0, 250 mM $MgCl_2$ and 500 mM NaCl. The enzyme concentrations were measured using Direct Detect. The activity of the enzymes was monitored by measuring the conversion of NADPH to $NADP^+$ at 340 nm using 2-acetolactate or 3-hydroxypyruvate as the substrate. In a standard assay, the enzyme activity was measured at 25 °C 100 mM Tris-HCl pH 8.0, 4 mM $MgCl_2$ and 0.2 mM of NADPH. The catalytic parameters (k_{cat}, K_m) were calculated using the Michaelis–Menten equation in Prism 6.0. In order to determine the thermal stability of the KARI variants, samples were aliquoted into 100 µLs volumes and placed into PCR tubes at a final concentration of 1 mg/mL. The samples were incubated on a heating block at temperatures ranging from 30 °C to 95 °C for 10 min. The samples were then assayed under standard conditions (see above). The activity of the samples incubated at 30 °C for 10 min was considered to represent 100% activity. The thermal stability profile of the enzyme was plotted using Prism 6.0. In order to assess the effect of isobutanol on KARI activity, 10% isobutanol (v/v) was added to a protein sample, which was then incubated at 25 °C. The activity was assayed under standard conditions at various time points.

2.7. PROSS Design of EcKARI

The PROSS algorithm was applied as previously described [32] with the sequence of *Ec*KARI as the query for the sequence homologue search and its structure (PDB entry: 1YRL) [19] as the template structure. The final model used for expression was selected by manual examination of each mutation position of the design models. The redesigned *Ec*KARI is defined as *Ec*KARI_De.

3. Results and Discussion

Recombinantly expressed KARIs from *C. jejuni* (*Cj*KARI) and *E. coli* (*Ec*KARI) were selected as representatives for a dodecameric Class I and a tetrameric Class II KARI, respectively, to probe and optimize their thermal stability.

3.1. Overall Structure of the Class I CjKARI and the Relevance of the Oligomeric State on Enzyme Stability

The structure of the apo form of *Cj*KARI was solved by cryo-electron microscopy (cryo-EM) at 2.94 Å resolution, showing the homododecameric structure of the enzyme (Figure 2; Table 1). The backbone of the polypeptide could be traced throughout except for two regions, 50–59 and 142–147. The former represents the region where the adenine of NADPH is expected to bind. Based on other KARI structures, this region is known to be flexible [20–25]. The latter is a loop in the C-terminal domain and is located on the surface of the protein.

The cryo-EM map resolved the location of the majority of the side chains including Y258, L294 and K290 (Figure 3). Similar to *Pa*KARI and *Ss*KARI, the dodecamer of *Cj*KARI represents an assembly of six identical dimers with the active sites located between an N-domain and two interlocking C-domains. The C-domain is deeply knotted [39] and harbors the catalytic pocket that accommodates two essential metal ions (Mg^{2+} in most naturally occurring KARIs). The interactions that hold the *Cj*KARI dodecamer together involve conserved residues from the C-domains (Figure 3).

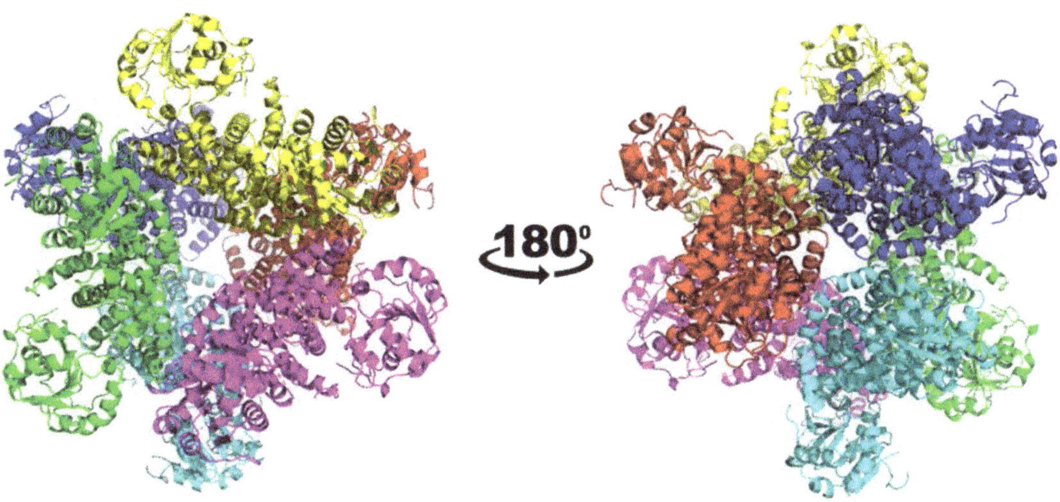

Figure 2. Overall structure of *Cj*KARI in two orientations. The cryo-EM structure of *Cj*KARI was solved to a resolution of 2.94 Å and illustrates how the dodecamer is composed of six dimeric units.

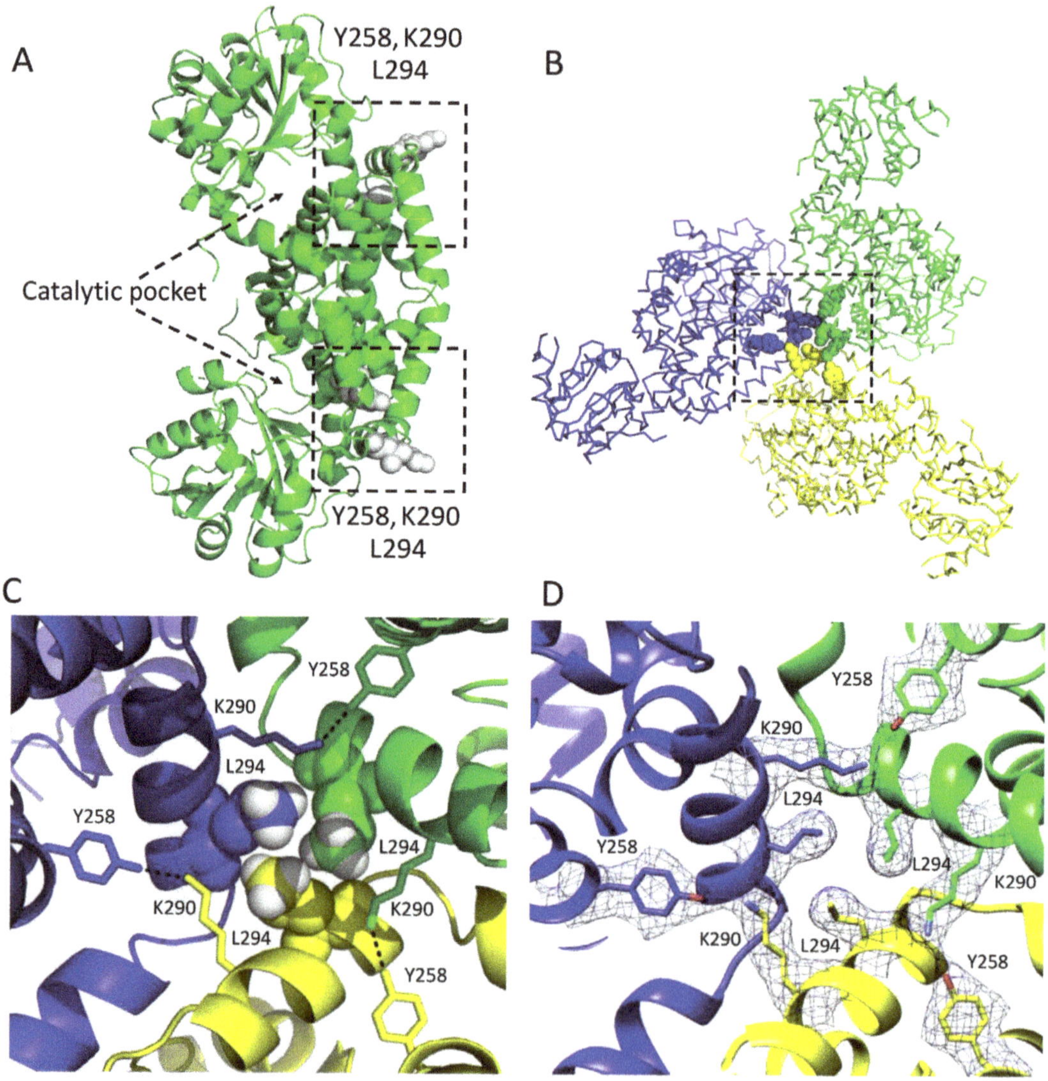

Figure 3. Subunit interactions in a dodecameric KARI. (**A**) A dimeric unit showing the location of the regions (boxed) involved in the formation of the dodecamer in *Cj*KARI. Residues colored white in the box are the contact residues that are key to stabilization of the three dimers in this image. (**B**) The assembly of three of the six dimers in the dodecamer. An equivalent arrangement occurs with the other three dimers of the dodecamer. (**C**) Zoom showing the interactions of three contact residues Y258, K290 and L294. The hydrogen atoms (attached as riding models) of the distal carbon that make contact with the neighboring subunits are shown as white CPK models. Y258 and K290 from the different subunits form hydrogen bonds (3.0 Å). (**D**) Cryo-EM map of the residues Y258, K290 and L294.

In particular, the side chains of Y258, K290 and L294 (Figure 3A) appear to be critically important for interactions that maintain the dodecameric structure. These residues are highly conserved among the dodecameric KARIs. Specifically, Y258 from one subunit forms a hydrogen bond with K290 from a neighboring subunit, while the L294 residues from three subunits form a hydrophobic cluster (Figure 3B–D). We, therefore, predicted that mutations of K290 and L294 to alanine would be sufficient to alter the oligomeric state from a dodecamer to a dimer. Hence, we generated the K290A/L294A double mutant of *Cj*KARI (*Cj*KARI_Dm) by site-directed mutagenesis and expressed and purified this enzyme using the same protocol as for its wild-type counterpart.

Using size-exclusion chromatography (SEC), coupled with multi-angle laser light scattering (MALS), the oligomerization status of *Cj*KARI and *Cj*KARI_Dm was investigated. As expected, *Cj*KARI has a molecular mass of ~450 kDa consistent with a dodecamer (the theoretical molecular mass of the dodecamer is 456 kDa). In contrast, *Cj*KARI_Dm elutes later and has a refractive index corresponding to a molecular mass of ~74.5 kDa, consistent with the formation of a dimer (the theoretical molecular mass of a dimer is 75.8 kDa). Thus, the two mutations, as predicted, disrupt formation of the dodecamer (Figure 4).

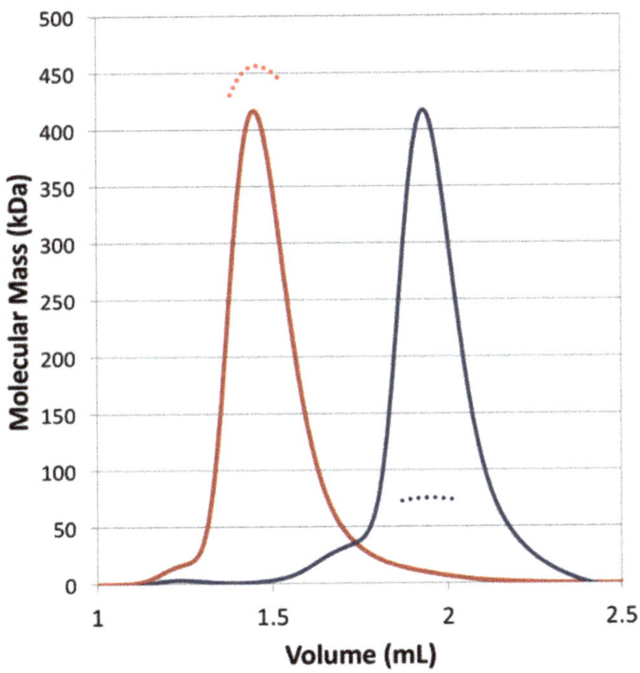

Figure 4. SEC-MALS of *Cj*KARI and *Cj*KARI_Dm. The chromatogram (refractive index) and the measured molecular mass for *Cj*KARI are represented by a red solid line and a red dashed line, respectively. For *Cj*KARI_Dm, these are represented by grey lines. *Cj*KARI elutes with a MW of ~450 kDa, consistent with it being a dodecamer. *Cj*KARI_Dm elutes with a MW of 74.5 kDa, consistent with it being a dimer.

Kinetic assays show that *Cj*KARI_Dm has a k_{cat} of $0.47 \pm 0.02\ s^{-1}$ (K_m: 1.98 ± 0.19 mM) for the reaction with the native substrate, 2-acetolactate, and a k_{cat} of $3.30 \pm 0.15\ s^{-1}$ (K_m: 4.35 ± 0.40 mM) when 3-hydroxypyruvate (Figure 1) is used as substrate. In comparison, the wild-type enzyme has k_{cat} values of $0.85 \pm 0.05\ s^{-1}$ (K_m: 881 ± 50 μM) and $2.20 \pm 0.05\ s^{-1}$ (K_m: 1.70 ± 0.10 mM) for 2-acetolactate and 3-hydroxypyruvate, respectively. Thus, the catalytic efficiency (i.e., k_{cat}/K_m) towards the two substrates is four- and two-fold greater, respectively, for the native enzyme. However, while the effect of the mutations on the catalytic parameters is relatively modest, their effect on the apparent melting temperature (Tm) is more dramatic (83.1 °C vs. 51.5 °C for *Cj*KARI vs. *Cj*KARI_Dm; Figure 5A). *Mt*KARI, which is dimeric in its native form [20], has a similar thermal stability as *Cj*KARI_Dm with a Tm of 51.8 °C.

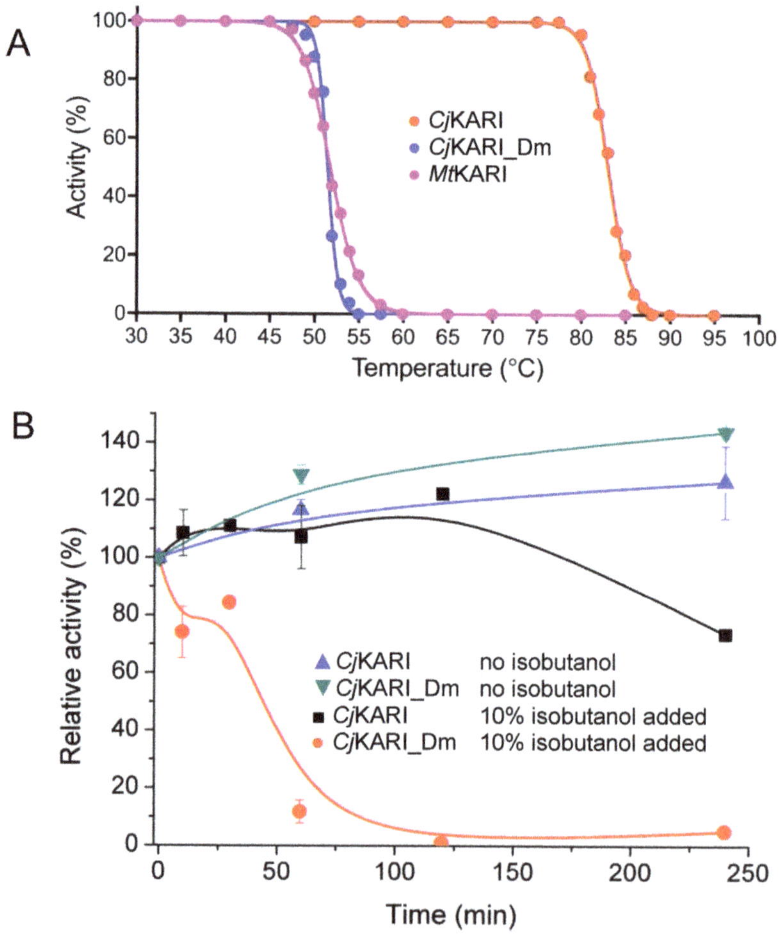

Figure 5. Activities of dimeric and dodecameric KARIs. (**A**) The normalized activities of *Cj*KARI, *Cj*KARI_Dm and *Mt*KARI after 10 min incubation at temperatures ranging from 30 °C to 95 °C. (**B**) Stability of *Cj*KARI and *Cj*KARI_Dm in the presence or absence of 10% isobutanol.

Cell-based isobutanol production systems have a maximal total volumetric yield as low as 1–2% before the growth of the cell culture is affected, due to end-product toxicity effects [7,40,41]. We thus tested the activities of wild-type *Cj*KARI and *Cj*KARI_Dm in the presence of isobutanol at a concentration of 10% (v/v) (Figure 5B). After four hours of incubation at 25 °C, the dodecamer retained 80% of its activity, whereas the dimeric enzyme was inactive. Thus, not only does the dodecameric enzyme have a higher apparent thermal stability compared to its dimeric counterpart, it also is less susceptible to loss of activity in the presence of isobutanol.

Studies to modify the quaternary structures of dimeric Class II KARIs (i.e., plant KARIs) have been reported [42]. In an attempt to correlate the oligomeric state of such

enzymes with their function, a loop that crosses over from one subunit to the other was removed via mutagenesis. This led to the dissociation of the dimer into monomers, with a concomitant 3.8-fold reduction in the specific activity at 30 °C. Furthermore, the denaturation of the monomeric enzyme occurred at 35 °C, whereas the wild-type enzyme was still fully active at temperatures > 50 °C [42]. Thus, the higher oligomeric states of both classes of KARI appear to be critical for maintaining stability and activity.

In addition to *Cj*KARI, the structures of a number of dodecameric class I KARIs have been determined. These include *Pa*KARI, *Alicyclobacillus acidocaldarius* KARI (*Aa*KARI), *Azotobacter vinelandii* KARI (*Av*KARI) and *Ss*KARI [22–24,43]. Note that *Ss*KARI from the thermophile *S. solfataricus* is thermostable but catalytically inefficient [9] for industrial applications as it catalyzes reactions two to three orders of magnitude slower than *Cj*KARI. A comparison of these dodecameric structures reveals similar dimeric unit assemblies as *Cj*KARI (Figure 3). Thus, these dimer-dimer interactions can be predicted by sequence analyses, which can then be used as a guide to distinguish potentially highly thermostable Class I dodecameric KARIs from their dimeric counterparts. As the dimeric units that assemble in dimeric class I KARIs are also found in the dodecamer, new dodecameric enzymes can be assembled from dimers using a protein-protein interface design.

3.2. Improving Enzyme Stability through Molecular Design

Above, we demonstrated how changes to the oligomeric structure contribute to enhanced stability of a Class I KARI. We also explored the possibility to increase the thermal stability of a Class II KARI using a computational analysis with the program Protein Repair One Stop Shop (PROSS) [32], which combines phylogenetic analysis with Rosetta atomistic design calculations to increase thermal stability without impairing activity. This computational approach effectively bypasses laborious iterative design/experiment cycles typically used in directed evolution. We used PROSS to predict a set of mutations in *Ec*KARI, which were further qualified through a detailed evolutionary and structural analysis. We collected 634 full-length reviewed proteins of the gene *ilvC* in UniProt, aligned them using MAFFT [44] and generated a phylogenetic tree using RAxML [45] (where the root was placed to represent the hypothesized internal domain duplication event, separating Class I from Class II KARIs). We inferred the most probable amino acid states at all ancestral branch points to recover the joint substitutions that best explain the composition of existent KARI sequences. The crystallographic coordinates for chain A of *Ec*KARI (PDB code: 1YRL) were used as a guide structure, and seven structural models were generated using Rosetta. PROSS predicted a model *Ec*KARI_De with 34 changes. These changes were examined by (a) matching each target amino acid to inferred substitutions between the Class II ancestral branch point and *Ec*KARI, (b) mapping and rationalizing each mutation relative to chain A of *Ec*KARI and (c) experimentally characterizing *Ec*KARI_De.

From the 34 predicted substitutions in *Ec*KARI_De, 24 matched changes were suggested by our evolutionary analysis as the most likely to have occurred sometime between the class II ancestor and *Ec*KARI [10]. This indicates that a majority of (predicted) thermal stability-enhancing mutations are also recovered by ASR (and hence not reliant on access to a structure). Only eight substitutions corresponded to changes that are assigned near-zero probability to have occurred in nature (K69P/EQ where E or Q are the most likely ancestral states, not P; A73D/AK; D124S/KQ; K184R/QHK; M188W/L; S314K/RS; E384P/E; G456N/Q). We note that these eight mutations suggested by PROSS are available broadly in the KARI family alignment [10] but not favored within the *Ec*KARI branch or across close homologs. Exceptional cases are E384P, where P is associated almost exclusively with class I KARI, and G456N where N is generally rare across the KARI family.

A comparison of the crystal structure of wild-type *Ec*KARI and the *Ec*_KARI_De Rosetta/PROSS model indicates that the protein backbones are similar (RMSD of 0.276 Å after superposition of all Cα-atoms). However, the orientation of the side chains of several mutated residues varies to create new interactions (Table 2). These can be categorized into four types of change that increase structural stability by (i) introducing polarity to the surface of the enzyme, (ii) contributing to the core packing, (iii) strengthening turn structures by inclusion of a proline side-chain and (iv) adding new hydrogen bonds or salt bridges. Examples of substitutions are illustrated in Figure 6. To enhance the polarity of the surface, thirteen changes were made to the enzyme. For eight examples, this could be achieved by replacing a small hydrophobic residue (e.g., glycine or alanine) with a bulkier polar or charged residue (Figure 6A,B; Table 2). For the A189E change, it was predicted that a change in rotamer conformation of the nearby Y34 could occur, creating additional hydrophobic interactions to stabilize the enzyme (Figure 6B). To improve the core packing, three changes were made. In Figure 6C, the effect of the M188W mutation is depicted, which, as well as introducing a bulkier hydrophobic group, induces a conformational change in the nearby valine side chain to enhance interactions further. Three changes were also made to facilitate the introduction of stabilizing proline side chains. In two cases, charged amino acids were substituted by proline, and in the third example an alanine was replaced by proline. The K69P mutation is illustrated in Figure 6D; mutations to Pro may lower the entropy loss upon folding, thereby improving folding free energy [46]. Conformational changes are also predicted to occur to the side chains of R68 and E70. Nine changes were made that are predicted to result in an increase in the number of hydrogen bonds or salt bridges in the structure. These are illustrated in Figure 6E–L. Three other changes appeared to show no apparent effect that may lead to improved stabilization, but similarly these changes also did not seem to be detrimental to stability.

*Ec*KARI and *Ec*KARI_De were expressed and purified to homogeneity. The catalytic efficiencies of wild-type *Ec*KARI and *Ec*KARI_De are similar despite the introduction of 34 mutations in the latter. With 2-acetolactate as the substrate, *Ec*KARI_De has a k_{cat} of 0.75 ± 0.01 s^{-1} (K_m of 490 ± 25 μM), while for the reaction with 3-hydroxypyruvate a k_{cat} of 8.1 ± 0.2 s^{-1} (K_m of 3.4 ± 0.2 mM) was determined. The corresponding values for *Ec*KARI are: $k_{cat} = 2.15 \pm 0.02$ s^{-1} and $K_m = 230 \pm 8$ μM (for 2-acetolactate), and $k_{cat} = 7.9 \pm 0.2$ s^{-1} and $K_m = 3.5 \pm 0.2$ mM (for 3-hydroxypyruvate). Thus, while the catalytic efficiency for the full reaction is approximately six-fold greater in the wild-type enzyme, for the reduction-only reaction, the two enzymes are virtually identical. The structure of *Ec*KARI was determined in the apo-form [19] and with both cofactors (i.e., Mg^{2+} and NADPH) [5]. The major difference in the two complexes is that the active site is closed in the apo-form but is open when the cofactors are bound. Therefore, there are several ligand-induced conformational changes that occur to prepare the enzyme for substrate binding, during catalysis and upon release of the products [5]. This inherent flexibility is therefore of prime importance to allow catalysis to commence. As none of the mutations in the designed mutant is close to the active- or the NADPH-binding site, the reduction in catalytic efficiency may be due to the fact that the enzyme has lost some structural flexibility important for preparing the active site for catalysis.

Table 2. Differences in amino acid composition between wild-type *Ec*KARI and redesigned *Ec*KARI_De.

Polar or Charged Surface	Core Packing	Proline at Turn Region	Hydrogen Bonds/Ion Pairs	Molecular Force Unclear
G24D S33N Q36K A73D A77Q R116N D124S V139E A189E F256Y A297R G315T K326N E346D G456N	A32C E141M M188W L283I S425A	K69P E383P A469P	K81N K184R L234I A287E S314K T341N V411R Y426H E438T	T245V E251A

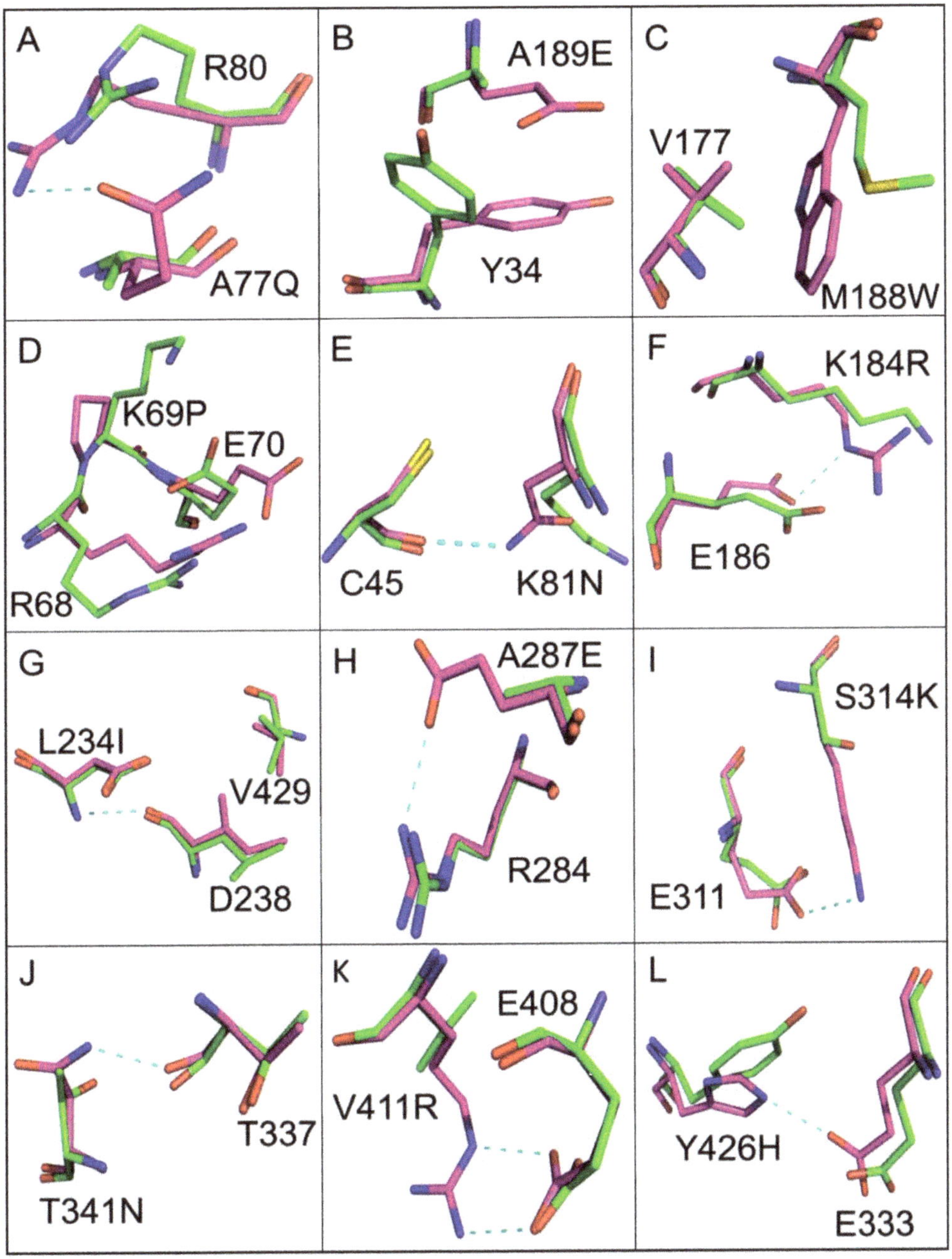

Figure 6. The redesign of *Ec*KARI. Superposition of the crystal structure of *Ec*KARI (shown in green) and the PROSS-Rosetta model of *Ec*KARI_De (shown in magenta). The carbon atoms of the residues of the wild-type enzyme are shown in green, and carbon atoms of the designed enzyme are shown in magenta. Panels (**A–L**) show the individual site mutations when *Ec*KARI and *Ec*KARI_De are compared. Dashed cyan lines represent the new hydrogen bond interactions formed as a result of the change. In B and C new hydrophobic contacts are formed. In D there are multiple conformational changes due to the K69P change.

The gel-filtration elution time of *Ec*KARI_De was the same as for the wild-type protein, showing that the tetrameric state of *Ec*KARI_De is maintained, but the Tm of *Ec*KARI_De increased by over 10 °C from 47.9 °C to 58.5 °C (Figure 7A). Next, we tested how a prolonged incubation with 10% (*v*/*v*) isobutanol affects the activity of both *Ec*KARI and *Ec*KARI_De (Figure 7B). While not as impressive as wild-type *Cj*KARI (Figure 5B), the rate of activity loss was significantly reduced for the designed protein when compared to *Ec*KARI. *Ec*KARI_De retained ~75% activity after 1 hour of incubation with 10% isobutanol, while the wild-type enzyme had less than 40% of its initial value. Even after four hours of incubation, the mutant retained nearly 20% of its initial activity, while *Ec*KARI was inactive.

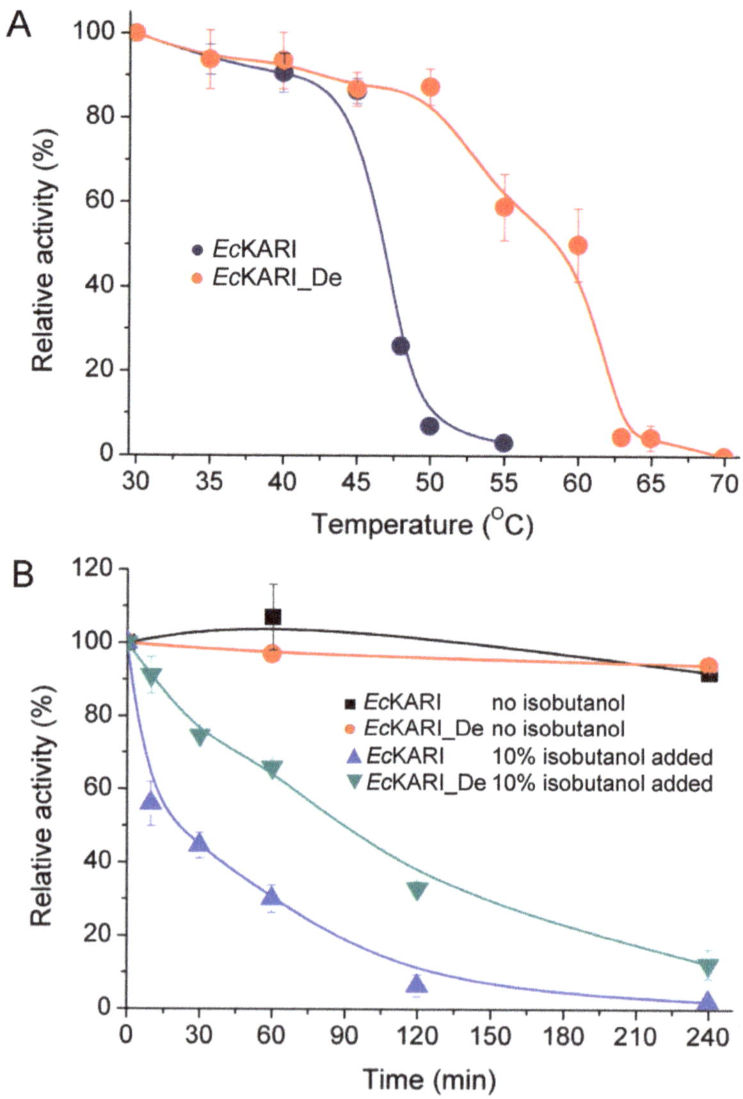

Figure 7. Activity of *Ec*KARI and *Ec*KARI_De. (**A**) The normalized activity of *Ec*KARI and *Ec*KARI_De after 10 min incubation at temperatures ranging from 30 °C to 70 °C. (**B**) Stability of *Ec*KARI and *Ec*KARI_De in the presence or absence of 10% isobutanol.

4. Conclusions

KARI is an established target for biocidal agents (herbicides and antimicrobials) [2,3,6,21,47–50], but this enzyme also plays a central role in synthetic biomanufacturing cascades [7,9,51]. This study demonstrated that the oligomeric status is critical to the thermostability of KARI. Here, we demonstrated that while *Cj*KARI can function as a dimer, its apparent Tm value increases by ~30 °C upon dodecamer formation. Furthermore, we could provide evidence that the *Ec*KARI thermostability and resistance to the high energy biofuel isobutanol can be enhanced by rational protein design. Enzyme engineering has been shown to benefit from guidance from both evolution and structure; here, we rationalize a range of changes suggested by PROSS by reference to substitutions that trace back to ancestral variants and/or by reference to structure. This is the first PROSS application that has enhanced the thermostability of an *E. coli* protein, impressively in the presence of a denaturant (i.e., isobutanol). Thus, an integrated approach of structure and phylogenetic analysis holds great promise for enhancing the properties of enzymes in a wide range of applications.

Author Contributions: Conceptualization, Y.L., V.S., S.J.F., G.S. and L.W.G.; methodology, Y.L., S.Z., F.L., Y.G., A.K., X.Y., S.W. and B.W.; software, A.G., M.B. and S.J.F.; formal analysis, Y.L., S.Z. and B.W.; investigation, Y.L., S.Z., A.G., F.L., Y.G., X.Y., A.K., S.W. and B.W.; data curation, A.G., M.J.L., M.B., Z.R., S.J.F. and L.W.G.; writing—original draft preparation, Y.L., S.Z., G.S. and L.W.G.; writing—review and editing, T.B.B., M.A.S., M.J.L., V.S., M.B., G.S. and L.W.G.; visualization, S..Z., F.L., Y.G. and M.J.L.; supervision, R.P.M., B.K., M.A.S., Z.R., S.J.F.,M.J.L., G.S. and L.W.G.; project administration, G.S. and L.W.G.; funding acquisition, R.P.M., V.S., G.S. and L.W.G. All authors have read and agreed to the published version of the manuscript.

Funding: This work was funded by a Discovery Grant from the Australian Research Council (ARC; DP210101802) to G.S., L.W.G. and V.S. Research in the Fleishman laboratory was funded by a Starter's Grant from the European Research Council (335439) and by the Israel Science Foundation through its Center for Research Excellence in Structural Cell Biology (1775/12). V.S. and G.S. would like to thank the Federal Ministry of Education and Research (BMBF), Germany for funding. T.B. gratefully acknowledges funding by the Werner Siemens foundation to establish the new field of Synthetic Biotechnology at the Technical University of Munich, Germany. B.K. is an ARC Laureate Fellow (FL180100109).

Institutional Review Board Statement: Not applicable.

Informed Consent Statement: Not applicable.

Data Availability Statement: The coordinates for the structure of CjKARI have been deposited to the Protein Databank (PDB) and the Electron Microscopy Data Bank (EMDB).

Acknowledgments: The authors are grateful to Thierry Lonhienne for his careful proof-reading of the manuscript.

Conflicts of Interest: The authors declare no conflict of interest.

References

1. Duggleby, R.G.; Pang, S.S. Acetohydroxyacid synthase. *J. Biochem. Mol. Biol.* **2000**, *33*, 1–36.
2. Grandoni, J.A.; Marta, P.T.; Schloss, J.V. Inhibitors of branched-chain amino acid biosynthesis as potential antituberculosis agents. *J. Antimicrob. Chemother.* **1998**, *4*, 475–482. [CrossRef]
3. Schulz, A.; Sponemann, P.; Kocher, H.; Wengenmayer, F. The herbicidally active experimental compound Hoe 704 is a potent inhibitor of the enzyme acetolactate reductoisomerase. *FEBS Lett.* **1988**, *238*, 375–378. [CrossRef]
4. Epelbaum, S.; LaRossa, R.A.; VanDyk, T.K.; Elkayam, T.; Chipman, D.M.; Barak, Z. Branched-chain amino acid biosynthesis in Salmonella typhimurium: A quantitative analysis. *J. Bacteriol.* **1998**, *180*, 4056–4067. [CrossRef]
5. Wong, S.H.; Lonhienne, T.G.; Winzor, D.J.; Schenk, G.; Guddat, L.W. Bacterial and plant ketol-acid reductoisomerases have different mechanisms of induced fit during the catalytic cycle. *J. Mol. Biol.* **2012**, *424*, 168–179. [CrossRef] [PubMed]
6. Pue, N.; Guddat, L.W. Acetohydroxyacid synthase: A target for antimicrobial drug discovery. *Curr. Pharm. Des.* **2014**, *20*, 740–753. [CrossRef] [PubMed]
7. Atsumi, S.; Hanai, T.; Liao, J.C. Non-fermentative pathways for synthesis of branched-chain higher alcohols as biofuels. *Nature* **2008**, *451*, 86–89. [CrossRef]

8. Wang, B.W.; Shi, A.Q.; Tu, R.; Zhang, X.L.; Wang, Q.H.; Bai, F.W. Branched-chain higher alcohols. *Adv. Biochem. Eng. Biotechnol.* **2012**, *128*, 101–118.
9. Guterl, J.K.; Garbe, D.; Carsten, J.; Steffler, F.; Sommer, B.; Reisse, S.; Philipp, A.; Haack, M.; Ruhmann, B.; Koltermann, A.; et al. Cell-free metabolic engineering: Production of chemicals by minimized reaction cascades. *ChemSusChem* **2012**, *5*, 2165–2172. [CrossRef]
10. Gumulya, Y.; Baek, J.-M.; Wun, S.-J.; Thomson, R.E.S.; Harris, K.L.; Hunter, D.J.B.; Behrendorff, J.B.Y.H.; Kulig, J.; Zheng, S.; Wu, X.; et al. Engineering highly functional thermostable proteins using ancestral sequence reconstruction. *Nature Catal.* **2018**, *1*, 878–888. [CrossRef]
11. Lonhienne, T.; Low, Y.S.; Garcia, M.D.; Croll, T.; Gao, Y.; Wang, Q.; Brillault, L.; Williams, C.M.; Fraser, J.A.; McGeary, R.P.; et al. Structures of fungal and plant acetohydroxyacid synthases. *Nature* **2020**, *586*, 317–321. [CrossRef]
12. Flint, D.H.; Emptage, M.H.; Finnegan, M.G.; Fu, W.; Johnson, M.K. The role and properties of the iron-sulfur cluster in Escherichia coli dihydroxy-acid dehydratase. *J. Biol. Chem.* **1993**, *268*, 14732–14742. [CrossRef]
13. Sutiono, S.; Teshima, M.; Beer, B.; Schenk, G.; Sieber, V. Enabling the direct enzymatic dehydration of D-glycerate to pyruvate as the key step in synthetic enzyme cascades used in the cell-free production of fine chemicals. *ACS Catal.* **2020**, *10*, 3110–3118. [CrossRef]
14. Melse, O.; Sutiono, S.; Haslbeck, M.; Schenk, G.; Antes, I.; Sieber, V. Structure-guided modulation of the catalytic properties of dihydroxyacid dehydratases. *ChemBioChem* **2022**, *23*, e202200088. [CrossRef]
15. Bayaraa, T.; Gaete, J.; Sutiono, S.; Kurz, J.; Lonhienne, T.; Harmer, J.R.; Bernhardt, P.V.; Sieber, V.; Guddat, L.W.; Schenk, G. Dihydroxy-acid dehydratases from pathogenic bacteria: Emerging drug targets to combat antibiotic resistance. *Chem. Eur. J.* **2022**, *28*, e202200927. [CrossRef] [PubMed]
16. Tadrowski, S.; Pedroso, M.M.; Sieber, V.; Larrabee, J.A.; Guddat, L.W.; Schenk, G. Metal ions play an essential catalytic role in the mechanism of ketol-acid reductoisomerase. *Chemistry* **2016**, *22*, 7427–7436. [CrossRef] [PubMed]
17. Dumas, R.; Biou, V.; Halgand, F.; Douce, R.; Duggleby, R.G. Enzymology, structure, and dynamics of acetohydroxy acid isomeroreductase. *Acc. Chem. Res.* **2001**, *34*, 399–408. [CrossRef] [PubMed]
18. Tyagi, R.; Lee, Y.T.; Guddat, L.W.; Duggleby, R.G. Probing the mechanism of the bifunctional enzyme ketol-acid reductoisomerase by site-directed mutagenesis of the active site. *FEBS J.* **2005**, *272*, 593–602. [CrossRef]
19. Tyagi, R.; Duquerroy, S.; Navaza, J.; Guddat, L.W.; Duggleby, R.G. The crystal structure of a bacterial class II ketol-acid reductolsomerase: Domain conservation and evolution. *Protein Sci.* **2005**, *14*, 3089–3100. [CrossRef]
20. Lv, Y.; Kandale, A.; Wun, S.J.; McGeary, R.P.; Williams, S.J.; Kobe, B.; Sieber, V.; Schembri, M.A.; Schenk, G.; Guddat, L.W. Crystal structure of *Mycobacterium tuberculosis* ketol-acid reductoisomerase at 1.0 Å resolution-a potential target for anti-tuberculosis drug discovery. *FEBS J.* **2016**, *283*, 1184–1196. [CrossRef]
21. Patel, K.; Teran, D.; Zheng, S.; Kandale, A.; Garcia, M.; Lv, Y.; Schembri, M.A.; McGeary, R.P.; Schenk, G.; Guddat, L.W. Crystal structures of *Staphylococcus aureus* ketol-acid reductoisomerase in complex with two transition state analogs that have biocidal activity. *Chem. Eur. J.* **2017**, *23*, 18289–18295. [CrossRef] [PubMed]
22. Ahn, H.J.; Eom, S.J.; Yoon, H.J.; Lee, B.I.; Cho, H.J.; Suh, S.W. Crystal structure of class I acetohydroxy acid isomeroreductase from *Pseudomonas aeruginosa*. *J. Mol. Biol.* **2003**, *328*, 505–515. [CrossRef]
23. Chen, C.Y.; Chang, Y.C.; Lin, B.L.; Lin, K.F.; Huang, C.H.; Hsieh, D.L.; Ko, T.P.; Tsai, M.D. Use of Cryo-EM to uncover structural bases of pH effect and cofactor bispecificity of ketol-acid reductoisomerase. *J. Am. Chem. Soc.* **2019**, *141*, 6136–6140. [CrossRef] [PubMed]
24. Cahn, J.K.; Brinkmann-Chen, S.; Spatzal, T.; Wiig, J.A.; Buller, A.R.; Einsle, O.; Hu, Y.; Ribbe, M.W.; Arnold, F.H. Cofactor specificity motifs and the induced fit mechanism in class I ketol-acid reductoisomerases. *Biochem. J.* **2015**, *468*, 475–484. [CrossRef] [PubMed]
25. Leung, E.W.W.; Guddat, L.W. Conformational changes in a plant ketol-acid reductoisomerase upon Mg^{2+} and NADPH binding as revealed by two crystal structures. *J. Mol. Biol.* **2009**, *389*, 167–182. [CrossRef] [PubMed]
26. Kirk, O.; Borchert, T.V.; Fuglsang, C.C. Industrial enzyme applications. *Curr. Opin. Biotechnol.* **2002**, *13*, 345–351. [CrossRef]
27. Van Beilen, J.B.; Li, Z. Enzyme technology: An overview. *Curr. Opin. Biotechnol.* **2002**, *13*, 338–344. [CrossRef]
28. Ahuja, S.K.; Ferreira, G.M.; Moreira, A.R. Utilization of enzymes for environmental applications. *Crit. Rev. Biotechnol.* **2004**, *24*, 125–154. [CrossRef] [PubMed]
29. Sheldon, R.A. Engineering a more sustainable world through catalysis and green chemistry. *J. R. Soc. Interface* **2016**, *13*. [CrossRef]
30. Bornscheuer, U.T. Trends and challenges in enzyme technology. *Adv. Biochem. Eng. Biotechnol.* **2005**, *100*, 181–203.
31. Brinkmann-Chen, S.; Flock, T.; Cahn, J.K.; Snow, C.D.; Brustad, E.M.; McIntosh, J.A.; Meinhold, P.; Zhang, L.; Arnold, F.H. General approach to reversing ketol-acid reductoisomerase cofactor dependence from NADPH to NADH. *Proc. Natl. Acad. Sci. USA* **2013**, *110*, 10946–10951. [CrossRef]
32. Goldenzweig, A.; Goldsmith, M.; Hill, S.E.; Gertman, O.; Laurino, P.; Ashani, Y.; Dym, O.; Unger, T.; Albeck, S.; Prilusky, J.; et al. Automated structure- and sequence-based design of proteins for high bacterial expression and stability. *Mol. Cell* **2016**, *63*, 337–346. [CrossRef]
33. Gill, S.C.; von Hippel, P.H. Calculation of protein extinction coefficients from amino acid sequence data. *Anal. Biochem.* **1989**, *182*, 319–326. [CrossRef]

34. Punjani, A.; Rubinstein, J.L.; Fleet, D.J.; Brubaker, M.A. cryoSPARC: Algorithms for rapid unsupervised cryo-EM structure determination. *Nat. Methods* **2017**, *14*, 290–296. [CrossRef] [PubMed]
35. Grigorieff, N. Frealign: An exploratory tool for single-particle cryo-EM. *Methods Enzym.* **2016**, *579*, 191–226.
36. Pintilie, G.D.; Zhang, J.; Goddard, T.D.; Chiu, W.; Gossard, D.C. Quantitative analysis of cryo-EM density map segmentation by watershed and scale-space filtering, and fitting of structures by alignment to regions. *J. Struct. Biol.* **2010**, *170*, 427–438. [CrossRef]
37. Emsley, P.; Lohkamp, B.; Scott, W.G.; Cowtan, K. Features and development of Coot. *Acta Crystallogr. Sect. D Biol. Crystallogr.* **2010**, *66*, 486–501. [CrossRef]
38. Adams, P.D.; Afonine, P.V.; Bunkoczi, G.; Chen, V.B.; Davis, I.W.; Echols, N.; Headd, J.J.; Hung, L.-W.; Kapral, G.J.; Grosse-Kunstleve, R.W.; et al. PHENIX: A comprehensive Python-based system for macromolecular structure solution. *Acta Crystallogr. Sect. D Biol. Crystallogr.* **2010**, *66*, 213–221. [CrossRef]
39. Taylor, W.R. A deeply knotted protein structure and how it might fold. *Nature* **2000**, *406*, 916–919. [CrossRef]
40. Atsumi, S.; Wu, T.Y.; Machado, I.M.; Huang, W.C.; Chen, P.Y.; Pellegrini, M.; Liao, J.C. Evolution, genomic analysis, and reconstruction of isobutanol tolerance in *Escherichia coli*. *Mol. Syst. Biol.* **2010**, *6*, 449. [CrossRef]
41. Brynildsen, M.P.; Liao, J.C. An integrated network approach identifies the isobutanol response network of *Escherichia coli*. *Mol. Syst. Biol.* **2009**, *5*, 277. [CrossRef] [PubMed]
42. Wessel, P.M.; Biou, V.; Douce, R.; Dumas, R. A loop deletion in the plant acetohydroxy acid isomeroreductase homodimer generates an active monomer with reduced stability and altered magnesium affinity. *Biochemistry* **1998**, *37*, 12753–12760. [CrossRef] [PubMed]
43. Brinkmann-Chen, S.; Cahn, J.K.B.; Arnold, F.H. Uncovering rare NADH-preferring ketol-acid reductoisomerases. *Metab. Eng.* **2014**, *26*, 17–22. [CrossRef] [PubMed]
44. Yamada, K.D.; Tomii, K.; Katoh, K. Application of the MAFFT sequence alignment program to large data-reexamination of the usefulness of chained guide trees. *Bioinformatics* **2016**, *32*, 3246–3251. [CrossRef]
45. Stamatakis, A. RAxML version 8: A tool for phylogenetic analysis and post-analysis of large phylogenies. *Bioinformatics* **2014**, *30*, 1312–1313. [CrossRef] [PubMed]
46. Matthews, B.W.; Nicholson, H.; Becktel, W.J. Enhanced protein thermostability from site-directed mutations that decrease the entropy of unfolding. *Proc. Natl. Acad. Sci. USA* **1987**, *84*, 6663–6667. [CrossRef]
47. Bayaraa, T.; Kurz, J.L.; Patel, K.M.; Hussein, W.M.; West, N.P.; Schenk, G.; McGeary, R.P.; Guddat, L.W. Discovery, synthesis and evaluation of a novel ketol-acid reductoisomerase inhibitor. *Chem. Eur. J.* **2020**, *26*, 8958–8968. [CrossRef]
48. Wun, S.J.; Johnson, L.A.; McGeary, R.P.; Schenk, G.; Guddat, L.W. Inhibition studies of ketol-acid reductoisomerases from pathogenic microorganisms. *Arch. Biochem. Biophys.* **2020**, *692*, 108516. [CrossRef]
49. Lin, X.; Kurz, J.L.; Patel, K.M.; Wun, S.J.; Hussein, W.M.; Lonhienne, T.; West, N.P.; McGeary, R.P.; Schenk, G.; Guddat, L.W. Discovery of a pyrimidine-dione derivative with potent inhibitory activity against *Mycobacterium tuberculosis* ketol-acid reductoisomerase. *Chem. Eur. J.* **2020**, *27*, 3130–3141. [CrossRef]
50. Kandale, A.; Patel, K.; Waleed, H.; Wun, S.J.; Zheng, S.; Tan, L.; West, N.; Schenk, G.; Guddat, L.W.; McGeary, R.P. Analogs of the herbicide, N-hydroxy-N-isopropyloxamate (IpOHA), inhibit *Mycobacterium tuberculosis* ketol-acid reductoisomerase and their prodrugs are promising anti-TB drug leads. *J. Med. Chem.* **2021**, *64*, 1670–1684. [CrossRef]
51. Valera, A.; Wang, S.; Carr, R.; Trembleau, L.; Deng, H. Characterization of a class II ketol-acid reductoisomerase from *Mycobacterium tuberculosis*. *RSC Adv.* **2022**, *12*, 10540–10544. [CrossRef] [PubMed]

applied biosciences

MDPI

Review

Bioencapsulation of Microbial Inoculants: Mechanisms, Formulation Types and Application Techniques

Blanca Rojas-Sánchez [1], Paulina Guzmán-Guzmán [1], Luzmaria R. Morales-Cedeño [1], Ma. del Carmen Orozco-Mosqueda [2,*], Blanca C. Saucedo-Martínez [1], Juan M. Sánchez-Yáñez [1], Ayomide Emmanuel Fadiji [3], Olubukola Oluranti Babalola [3], Bernard R. Glick [4] and Gustavo Santoyo [1,*]

1 Instituto de Investigaciones Químico-Biológicas, Universidad Michoacana de San Nicolás de Hidalgo, Morelia 58030, Michoacán, Mexico
2 Departamento de Bioquímica e Ingeniería Ambiental, Tecnológico Nacional de México en Celaya, Celaya 38110, Guanajuato, Mexico
3 Food Security and Safety Focus Area Niche, Faculty of Natural and Agricultural Sciences, North-West University, Mmabatho 2735, South Africa
4 Department of Biology, University of Waterloo, Waterloo, ON N2L 3G1, Canada
* Correspondence: carmen.orozco@itcelaya.edu.mx (M.d.C.O.-M.); gustavo.santoyo@umich.mx (G.S.)

Abstract: The excessive use of agrochemicals in the field to increase production and counteract the negative effects caused by biotic and abiotic factors has led to a deterioration in soil fertility, plus an increment in negative impacts on the environment and human health. Therefore, the application of beneficial microorganisms as bioinoculants is an eco-friendly alternative to agrochemicals. Plant growth-promoting bacteria and fungi have been effective in promoting plant growth and production, as well as reducing the action of pathogens in multiple crops. However, successful application of such beneficial microorganisms in the agricultural field has faced several difficulties, such as survival, colonization efficiency and short periods of shelf storage. Therefore, it is essential to explore novel ways to encapsulate, formulate and apply bioinoculants. To obtain the expected quality in bioencapsulated products, it is essential to determine the type of polymer, capsule size, encapsulation technique and use the correct chemical and physical cofactors involved in the production process. Thus, this review highlights the various formulation types and application techniques, as well as discussing the multiple advantages of using microbial encapsulates to have better results in agricultural production.

Keywords: bioinoculants; beneficial fungi; microbial bioencapsulation; PGPR; pellet; agriculture

Citation: Rojas-Sánchez, B.; Guzmán-Guzmán, P.; Morales-Cedeño, L.R.; Orozco-Mosqueda, M.d.C.; Saucedo-Martínez, B.C.; Sánchez-Yáñez, J.M.; Fadiji, A.E.; Babalola, O.O.; Glick, B.R.; Santoyo, G. Bioencapsulation of Microbial Inoculants: Mechanisms, Formulation Types and Application Techniques. *Appl. Biosci.* **2022**, *1*, 198–220. https://doi.org/10.3390/applbiosci1020013

Academic Editor: Robert Henry

Received: 8 June 2022
Accepted: 29 August 2022
Published: 1 September 2022

1. Introduction

The continuous increase in the world population is accompanied by a high demand for agricultural products that must be satisfied in quantity and quality. In recent decades, and particularly since the advent of the green revolution, excessive use of chemical fertilizers has taken place to maximize production and improve the quality of crops and try to increase the productivity of nutritionally poor soils [1]. However, the benefits initially observed by the use of agrochemicals on crop productivity have been overshadowed by studies that show the adverse effects of excessive use of these products on the environment [2].

Among the damages caused to the soil are the deterioration in its structure and texture and a reduction in the populations of microflora and microfauna, which together trigger a nutritional imbalance within the soil. In addition, chemical inputs are the main source of contamination in soils used for agricultural production; they contribute significantly to the contamination of water and the atmosphere, triggering diseases in living beings [2].

In addition to the above, it has been shown that the use of chemical fertilizers by plants is inefficient, since they only take advantage of ~50% or less of the chemical doses applied, regardless of the nitrogen source with which they are formulated [3], causing an accumulation of the products used in the soil.

The damages that have been caused over time by the excessive use of agrochemicals require the development and implementation of technologies that have a minimal impact on the environment and that are designed in such a way that they can maintain and preserve the productivity of the soil. In addition, agricultural products intended for human consumption that are free of chemicals gives added value to these products [4].

One of the technological options to decrease the amount of chemicals in the soil environment is to use beneficial microorganisms that can promote plant growth and reduce synthetic fertilizers without negatively affecting crop productivity [5,6]. The use of these microorganisms, also known as biofertilizers, is one of the most important contributions of biotechnology and microbiology to modern agriculture, and it is an alternative for the reduction in production costs and the environmental impact caused by the excessive use of agrochemicals [7,8].

Therefore, the inoculation of microorganisms that promote plant growth in crops is a practical alternative to agrochemicals and can be applied directly to the soil, sprayed on plants or used as a coating for seeds [9].

2. The Role of Beneficial Microorganisms as Inoculants

The interaction of plants with microbial communities results from co-evolution over millions of years, contributing to the adaptation of plants on earth [10]. The interactions between microorganisms and plants occurs mainly in the portion of the soil that is in close contact with the plant root, known as the rhizosphere. This zone is defined as the volume of soil associated with and influenced by plant roots [11], constituting a favorable environment for the development of microorganisms in quantities that are much higher than those found in the rest of the soil. These high microorganism concentrations are a consequence of the fact that plants provide the necessary nutrients for the development of these microorganisms, which in turn provide the plants with substances that promote their growth, establishing a mutualistic relationship between both organisms [12].

The interactions between plants and beneficial microorganisms have been the subject of various scientific investigations, since this relationship provides a viable alternative for sustainable plant development and the conservation of the environment [4]. Some microorganisms that promote plant growth include mycorrhizal fungi, beneficial fungi or promoters of plant growth and certain rhizobacteria [13–15].

Mycorrhizal fungi are a group of root biotrophs that exchange mutual benefits with approximately 80% of plants and include arbuscular mycorrhizae and ectomycorrhizae from multiple fungal clades, such as *Glomeromycota*, *Ascomycota* and *Basidiomycota* [16]. Among the benefits of this mutualism is the supply of soil nutrients to plants in exchange for carbon from the host plants. This relationship results in an increase in the absorption of nutrients, the production of bioactive compounds and an increase in the production of fruits and tubers. In addition, this relationship has been highly effective as a nematicide, in addition to increasing the uptake of water to plants in certain environmental conditions [16,17].

Beneficial fungi or plant growth-promoting fungi (PGPF) have taken on great importance since it has been proven that they promote plant growth and, in turn, control numerous foliar and root pathogens by activating induced systemic resistance (ISR) in the host plant through various signaling pathways [8,18]. This group of organisms includes species of genera, such as *Trichoderma*, *Aspergillus* and *Phoma* [19,20].

Plant growth-promoting bacteria (PGPB) are a set of bacteria that inhabit the rhizosphere. Through different mechanisms, they promote plant growth and provide them with tolerance to both biotic and abiotic stress conditions [21]. Within this group we find species belonging to the genera *Azospirillum*, *Azotobacter*, *Bacillus*, *Burkholderia*, *Enterobacter*, *Klebsiella* and *Pseudomonas*, as well as some endophytic species, such as *Axoarcus*, *Gluconacetobacter* and *Herbaspirillum* [22–24].

The role of inoculants, such as PGPF and PGPB, on plants is to improve plant growth, production and resistance against several phytopathogens. These microorganisms are used as different types of formulations, prepared accordingly to the desired function or effect of the microorganism to be used. These formulations contain live or latent microorganisms (bacteria or fungi, alone or in combination) and, depending on the mechanism they use to promote plant growth (direct, indirect or both), are classified into one of three categories, i.e., biofertilizers, biostimulants or biopesticides [4,25]. The mechanisms of action of beneficial microorganisms are discussed in the next section.

Biofertilizers are formulations with one or several microorganisms that provide and improve the bioavailability of nutrients when applied to crops, biostimulants include microorganisms that promote plant growth directly through the production of hormones and biopesticides include microorganisms that are used to control phytopathogenic agents [26–28].

3. Mechanisms of Action of Beneficial Microorganisms

Certain microorganisms can promote plant growth through direct mechanisms or indirect mechanisms, although some mechanisms can work both directly and indirectly [29]. These mechanisms of action are briefly described below. We recommend the reader to other excellent recent reviews in this area [30,31].

3.1. Direct Mechanisms of Action

Direct mechanisms refer to the promotion of plant growth in two ways: microorganisms make it easier for plants to acquire the nutrients they need or they help to modulate the levels of plant hormones involved in the development and growth of plants [32–34].

With the increased use of chemicals in agriculture, much of the nutrients, such as soluble inorganic phosphorus used as a chemical fertilizer, becomes immobilized soon after its application, making it unavailable to plants [35]. Naturally, in the soil, insoluble phosphorus is found as apatite or in some organic forms, such as inositol phosphate (phytate), phosphomonoesters and phosphodiesters [36], forms which plants cannot directly assimilate. However, some microorganisms are capable of solubilizing inorganic phosphates through the production of low molecular weight organic acids that act on the inorganic phosphates making them available so that they can be used by plants [35]. Other microorganisms contain enzymes that can break down organic phosphates into a plant usable form [33].

Another important nutrient for plants is iron. The predominant form of iron in nature is Fe^{2+}, which is not assimilable by plants. Some microorganisms can synthesize complex peptide molecules with a high affinity for Fe^{3+}; these peptides are known as siderophores. The siderophores trap iron forming a complex; this complex may be taken up by membrane receptors of microorganisms and thus facilitates its acquisition. These iron–siderophore complexes can also be assimilated by plants and subsequently broken down inside of the plant, thus providing plants with the iron they need [37].

Nitrogen is one of the nutrients that plants require in larger concentrations, and it is found primarily in organic form in the soil. Nonetheless, plants take up inorganic nitrogen as ammonium and nitrates, rather than the organic form; thus, nitrogen mineralization from organic to inorganic form is crucial for plant growth and crop production [38,39]. Nitrogen fixing bacteria have gained attention in this regard, due to their capability to convert atmospheric nitrogen (N_2) into ammonia (NH_3), which plants can use, in a process called biological nitrogen fixation. Bacteria capable of such conversion encode the enzyme nitrogenase (a highly conserved enzyme complex), which catalyzes the conversion of N_2 to NH_3 [40–42].

Biological mechanisms, such as nitrogen, sulfur or phosphorous fixation [43], production of siderophores to increase iron bioavailability [44,45], phosphate and sulfates solubilization [46,47] and iron sequestration [33] help to incorporate or increase nutrients in the soil, along with their bioavailability to the plants. This increased provision of nutrients is provided by the following organisms: *Rhizobium* spp., *Sinorhizobium* spp., *Mesorhizobium* spp., *Azotobacter* spp., *Azospirillum* spp., *Pseudomonas* spp., *Bacillus* spp., *Aureobasidium pullulans*, *Epicoccum nigrum*, *Scolecobasidium constrictum*, *Myrothecium cinctum* and *Acidianus* spp., among others [43,44,46,47].

Some rhizospheric and endophytic microorganisms can produce plant hormones or induce their synthesis in plants. Many soil bacteria can produce hormones, such as cytokinins, gibberellins or auxins. In addition, some rhizosphere microorganisms produce the enzyme 1-aminocyclopropane-1-carboxylate (ACC) deaminase, where ACC is the immediate precursor of ethylene, a hormone related to the senescence of plants and the ripening of fruits and that very high levels of this hormone inhibit plant growth. The enzyme ACC deaminase converts ACC into α-ketobutyrate and ammonia, thus ethylene is no longer produced, and ammonia and α-ketobutyrate are compounds that can be assimilated by plants [42,48,49]. ACC deaminase production has been observed in plant beneficial organisms, such as *B. subtilis*, *P. fluorescens*, *B. amyloliquefaciens*, *Enterobacter cloacae* and *Trichoderma* sp., among others [50–52].

3.2. Indirect Mechanisms of Action

Other microorganisms can promote plant growth indirectly. A common indirect mechanism is competition for space and nutrients, where the beneficial microorganism competes with a pathogen and the one with the greatest capacity to take up nutrients and the fastest growth rate proliferating in the soil displaces the pathogen and prevents it from colonizing and infecting plants [34,53,54]. The successful competition of PGPB with pathogens provides plants with a greater opportunity to grow and develop. It is worth mentioning that the production of siderophores can also be classified as an indirect mechanism, since the microorganisms with the capacity to produce these molecules and take up siderophore–iron complexes will limit the growth of the pathogen competing for this nutrient [45,55,56].

The production of antibiotics or antimicrobial compounds is a common and often studied indirect mechanism of plant growth promotion. For example, pyrrolnitrin produced by *Pseudomanas* spp. [57], iturine produced by *Bacillus* spp. [58] and syringomycin, produced by *P. syringae* [59] are some of the most common antipathogen antibiotics.

In addition to antibiotics, various microorganisms produce volatile organic compounds (VOCs) that can also be toxic to pathogens, preventing their growth or leading to their death, such as dimethyl disulfide produced by *Bacillus* sp. E25 [60] and *B. thuringiensis* CR71 [61]. Some of these VOCs, in addition to having antimicrobial action, can also promote plant growth directly, as is the case of dimethyl-hexa-decylamine [62]. Therefore, the production of VOCs can be classified as both a direct and an indirect mechanism.

Another indirect mechanism of plant growth promotion is the production of lytic enzymes related to the ability of the microorganism to parasitize and destroy the pathogen, as is the case of mycoparasitism carried out by fungi of the genus *Trichoderma*, adhering to the hyphae of the pathogen, where it secretes enzymes that degrade the cell wall, resulting in the death of the pathogen, or the destruction of its structure, preventing its development [50,63–65].

Finally, induced systemic resistance, initiated by the PGPB, is an indirect mechanism that involves activating the biochemical and molecular defense responses of the host plant, these may include the production of reactive oxygen species (ROS), phytoalexins, synthesis of proteins related to pathogenesis (PR), lignin accumulation at the site of infection, among others [66]. Some microorganisms that promote plant growth induce this defense response, preventing the colonization or the development of infection by phytopathogens [45,67,68].

4. Formulation of Microbial Inoculants

To enable plant growth-promoting microorganisms to be used and applied in agricultural practices, it is necessary to develop formulations based on these bioinoculants. Formulating a bioinoculant includes the entire series of procedures and technologies after the growth in culture of the microorganisms that promote plant growth. Bioinoculant formulation includes the mixture of a selected beneficial strain with a suitable vehicle that preserves the viability of the microorganism in either a dormant or metabolically active state during transport, storage and application [69]. To obtain a successful formulation, the microorganism must overcome the conditions of temperature, humidity, salinity, UV radiation and water stress present in the soil and during its formulation, in addition to being effective and competitive against the native microbial populations of the soil [1,70].

The compatibility of the physical form of the bioinoculant (solids in the form of powder, granules or capsules and liquids) and its incorporation through agricultural practices is a key factor determining the durability of the product and its ability to colonize plant roots [71]. According to the physical form of the inoculant, it is classified as a liquid formulation, a solid formulation or bioencapsulated [21,71].

4.1. Liquid Formulation

Liquid formulations use culture broths or formulations based mainly on water, mineral or organic oils. Seeds and seedlings can be immersed in the inoculant before sowing or transplanting [70,72], and for the biocontrol of pathogens or physiological stimulation, they can be sprinkled on the foliage of already established plants or applied directly to the soil [73]. This formulation method is the most commonly used. It is directly applied to crops without going through other processes following fermentation; most microorganisms can survive for more than one year if the containers are kept at ~4 °C, they are easy to inoculate and their application is very practical when implemented in irrigation or sprinkler systems. Liquid formulations are relatively low cost. However, even when its efficacy has been proven, its stability during storage is often limited due to its susceptibility to contamination with other microorganisms [69,74].

Liquid formulations of different microorganisms, or even in microbial consortia, and with the use of various additives, have managed to increase yields in agricultural fields. For example, when a liquid bioinoculant based on sugar and coconut water, including *Pseudomonas* spp., *Bacillus* spp., *Klebsiella* spp., *Aspergillus* spp., and *Azotobacter* spp., was used to inoculate soybean plants, the result was improved nutrient solubility and increased crop yield [75].

It has also been shown that the phosphate solubilization capacity and the survival rate of *Pseudomonas* and *Pantonea* strains increases and are preserved when they are used in liquid formulations up to three months after their formulation containing diluted concentrations of phosphate buffer and nutrient broth with glycerol [76]. In addition, Camelo-Rusinque et al. [77] assessed the population dynamics of the *Azotobacter chrocoocum* strain AC1 in MBR culture medium under bioreactor conditions after 105 days and found that both the cell viability and the biological activity of the strain was maintained, regardless of the storage temperature. This indicates that some liquid formulations can be used for a specific time, with the organisms retaining their activity and continuing to be viable for use as bioinoculants.

4.2. Solid Formulation

Solid formulations are used widely in the agricultural industry because of the advantages they offer during storage and transportation. A simple technique used for the preparation of solid formulations is adsorption, which consists of mixing the microorganisms with a solid support, such as vermiculite, perlite, sepiolite, kaolin, diatomaceous earth, natural zeolite, peat or clay, the latter being of great interest in agriculture thanks to its ability to act as a desiccant and provide excellent storage conditions for various inocula, as it has a good ability to adsorb agents dispersed or suspended in it [78,79].

Peat is one of the supports most used worldwide in commercial crops due to its low cost. However, being a complex organic matter, different batches present great chemical variability and, consequently, it is difficult to maintain the same quality in all batches. In addition, its storage is very susceptible to humidity, which decreases the inoculum cell survival [78]. According to Rose et al. [80], it is essential to be able to quantify the number of viable cells of each microorganism per unit weight of inoculant, to determine the inoculum potential at different application doses and for field results to be properly interpreted.

In a study by Quiroz Sarmiento et al. [81], the effectiveness of peat was evaluated with the following bacteria: *Serratia liquefaciens* CPAC53, *S. plymuthica* CPPC55, *P. tolaasii* P61 and *P. yamanorum* OLsSf5, in comparison with the encapsulation of the strains using alginate beads. Following a storage period of 150 days, the results showed that the encapsulated strains maintained the highest population. The effect of both types of bioinoculants on poblano chili seedlings (*Capsicum annum* L.) was also evaluated. In this case, the best results were observed with the encapsulated strains [81]. This suggests that the success in using peat as a support material for solid formulations depends on the conditions in which the bioinoculum will be used and the availability of other strategies.

A common technique of solid formulation is spraying or lyophilization. This technique allows for the realization of high microbial survival rates without the need to use any support, allowing for easy inoculum storage for long periods, at room temperature, without the need for refrigeration. One of the disadvantages of lyophilization is that it is necessary to protect the cell membrane and cytoplasm against dehydration during the storage period, using a cryoprotectant, such as mannitol and microcrystalline cellulose [82]. In this way, the cells remain viable and can be used long after lyophilization, for at least a year [83].

Lyophilized microorganisms may be mixed with a solid support or used directly. For example, in the laboratory, Grzegorczyk et al. [84] studied the survival and storage stability of a strain of *Trichoderma hariazum*, four strains of *Trichoderma atroviride* and two strains of *Trichoderma virens*, after culture lyophilization in solid wheat straw medium with and without the addition of maltodextrin. It was observed that the strains had a higher survival capacity (except for strain *T. atroviride* TRS40), compared to the addition of distilled water only, and in comparison with the bioformulation containing just maltodextrin. Three months after lyophilization, the strains remained stable and most still showed cellulolytic and xylanolytic activity.

Wessman et al. [85] studied the survival of the bacterial strains *P. putida* KT2440 and *A. chlorophenolicus* A6 after lyophilization in four different formulations, including (i) sucrose, (ii) Ficoll PM400 a sucrose polymer, (iii) hydroxyethylcellulose (HEC), and (iv) hydroxypropylmethylcellulose (HPMC). The polymers were chosen to obtain a monomeric structure, such as sucrose. The results of this study indicated that a key factor to help cell survival is the ability of the added ingredients to replace water during dehydration, thereby maintaining the structure of proteins and cell membranes in a dry state. Disaccharides, such as sucrose, show this property, while polymers, such as starch-based polysaccharide, do not. Thus, some polymers can facilitate cell survival to the same extent as disaccharides provided that certain physical properties of the formulation are controlled [85].

In fact, one of the techniques that have gained great importance in recent years thanks to its advantages, is the solid formulations developed based on polymers. These polymers, in the presence of ions or changes in chemical conditions, form complex matrices so that microorganisms become immobilized and encapsulated in the matrix and are gradually released as the polymer degrades. The technique of microorganism immobilization ultimately creates barriers between the microbes and the environment, improving their bioavailability and preserving their biological stability [86].

5. Bioencapsulation of Plant Growth-Promoting Microorganisms

The growing demand for the use of microorganisms as bioinoculants for use in agriculture has facilitated the use of technological tools that allow compliance with the need to develop products that can promulgate agricultural sustainability. When plants are commercially inoculated with plant growth-promoting microbes, the formulation process generates certain problems when applied in the field, given that the main physical form in which the formulation is presented as a liquid or as powders that fails to protect the survival of the microbial strains in the face of abiotic conditions (temperature, humidity, salinity, UV radiation, pH) present in the application process [87].

The use of polymers through the encapsulation technique has been proven as a highly effective alternative to increase the viability of microorganisms and, in turn, provide protection against the environmental conditions present [88,89]. Figure 1 represents the various biotic and abiotic factors that bioencapsulated microbes face when used in the field.

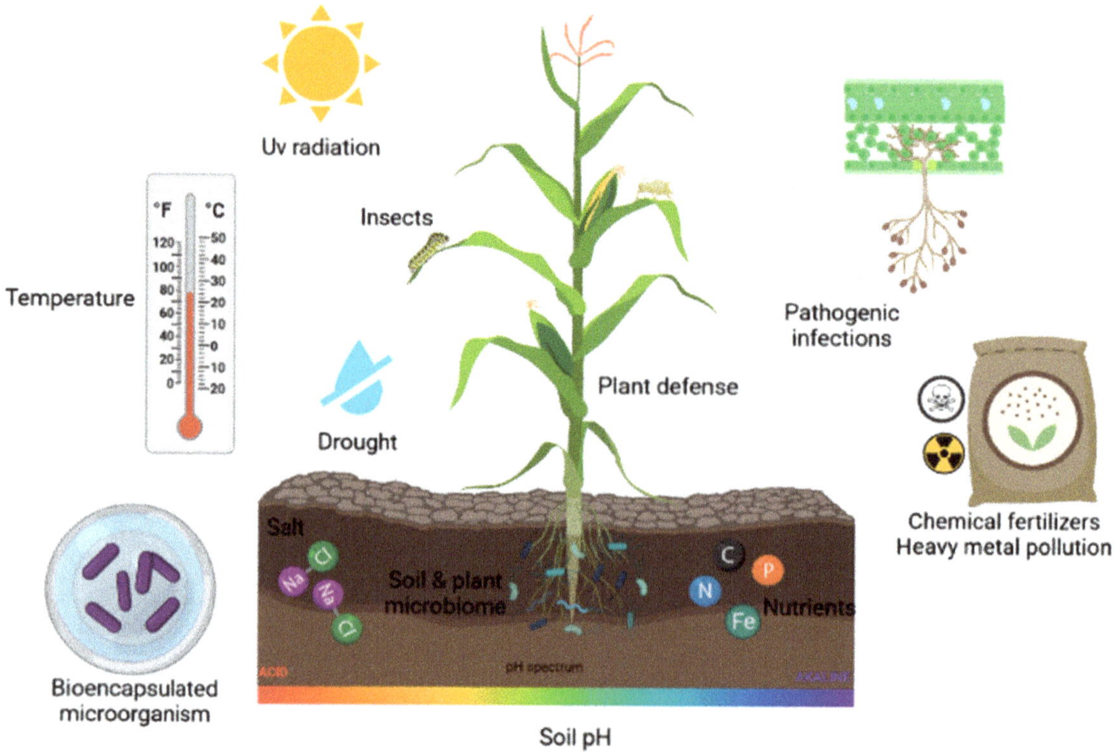

Figure 1. Biotic and abiotic factors that cause stress in plants and affect bioencapsulated microorganisms.

6. Classification of the Well-Encapsulated

Bioencapsulated microbes can be classified according to the type of mechanism used by the microorganism to promote plant growth or according to the type of microorganisms used in their formulation, whether they are bacteria, fungi or a combination of both [69,86].

6.1. Bioencapsulated Bacteria

Bacteria, particularly PGPB, have been widely used in agricultural research and have proven to be a tool for improving plant health and growth without causing environmental pollution [89,90]. PGPB manage to mitigate abiotic stress in the soil through the production of phytohormones and associated metabolites, as well as through significant morphological changes in the roots [91–93]. These changes result in a better nutritional status for plants and, in turn, stimulate plant defense mechanisms to overcome unfavorable environmental conditions [10,94].

Some bacteria can migrate from the rhizosphere to the internal tissues of the plant. They do this through cracks that form in the roots as a consequence of growth, they can also enter through lenticels, due to the emergence of lateral roots or by the cells of the root hairs, among other ways. Once inside the plant, they can exert their action of promoting plant growth either by direct or indirect mechanisms and, in turn, the bacteria are protected from abiotic stress and are exempt from competing for resources and space with other soil microorganisms. The bacteria capable of colonizing the internal tissues of a plant without causing damage are known as endophytes [30,95,96].

Some bacteria can form highly stable dormant spores that can germinate, forming active bacteria when conditions are favorable [97]. Fully formed spores are recognized as one of the most resistant life forms on the planet, they protect the bacterial genome against heat, desiccation, radiation and oxidation, as well as being an efficient way to escape predation by higher organisms. Spore germination is triggered by the presence of nutrients in the environment, these are detected by membrane receptors and in a matter of minutes, the nucleus of the spores is activated, the spore rehydrates, the cortex hydrolyzes and its surface cover changes [98–100]. This naturally occurring process is considered to be an effective technique for the development of inoculants as it allows greater survival of the strains during the storage process, making it possible to develop encapsulation on a large scale and, in turn, is more profitable [101].

Among the bacteria that form spores are *Clostridium*, *Sporosarcina*, *Thermoactinomyces* and *Bacillus*; there are numerous reports of the beneficial effects of *Bacillus* strains on crops of agricultural interest. *Bacillus* strains promote plant growth, increase phosphorus solubilization and increase the production of growth regulators, and they are highly efficient for pest control. Within this group is the *B. thuringensis* species, which is widely applied worldwide as a consequence of its activity as a biological insecticide [102,103].

Many strains of bacteria have been used to make bioencapsulated bacteria based on various polymers, and their effect as biostimulants and/or biopesticides has been studied in several field or greenhouse experiments (Table 1). The bioencapsulation of spore-forming bacteria provides them with additional resistance to environmental factors; however, this type of formulation represents a greater advantage for those organisms that do not form spores, also preserving their viability, which opens the possibility of using a greater diversity of organisms that promote plant growth or biocontrol in the field.

Table 1. Encapsulated microorganisms of agricultural importance, their beneficial effect on plant crop and application technique.

Encapsulated Microorganism	Encapsulation Polymer	Crop	Effect	Type of Assay	Reference
Streptomyces fulvissimus **Uts22**	Chitosan + gellan gum	*Triticum* sp.	Biocontrol against *Gaeumannomyces graminis*; Growth promotion of root and branches systems in wheat plants.	Greenhouse assay	[104]
Pseudomonas fluorescens **Ms-01 (Pf)** ***Azospirillum brasilense*** **DSM1690 (Ab)**	Montmorillonite+alginate [Mt-Ag] y hallosite+alginate [Ha-Ag]	*Triticum* sp.	Increase in root and branches biomass; Increase in root nitrogen adsorption	Greenhouse assay	[105]
Azotobacter chroococcum **C26**	Carrageenan	-	Cell preservation	-	[106]
Bacillus subtilis **SL-13**	Alginate-bentonite-polyvinyl alcohol-SDS	*Gossypium hirsutum*	Increase in seed germination and plant biomass	Greenhouse assay	[107]
Pseudomonas putida **Rs-198**	Alginate/bentonite/starch	*Gossypium hirsutum*	Plant growth promotion under saline stress; increase in IAA and gibberellin production	Greenhouse assay	[108]
Ensifer fredii **LP2/20**	Agar and alginate	-	Bacterial community structure modifications	Field assay	[109]
Pseudomonas putida	Alginate	*Capsicum annuum* L.	Plant growth promotion	Greenhouse assay	[110]
Pantonea agglomerans **KL**	Alginate	*Oryza sativa*	Reduction in saline stress	Greenhouse assay	[111]
Paenibacillus polymyxa **MSRH5,** ***Bacillus nakamurai*** **MSRH1 and** ***Bacillus pacificus*** **MSR H3**	Alginate	*Triticum aestivum* L.	Reduction in saline stress and increase in plant growth	Field assay	[112]
Pseudomonas fluorescens **strains VUPF5 y T17-4**	Alginate-gelatin	*Solanum tuberosum*	Biocontrol against *Fusarium solani*; Plant growth promotion	Greenhouse assay	[113]
Bacillus subtilis **cbrf24**	Carboxymethylcellulose Xanthan	*Solanum lycopersicum*	Nematicide against *Meloidogyne incognita*	Greenhouse assay	[114]
Pseudomonas fluorescens **(KY823007),** ***P. taiwanensis*** **(KY823006),** ***P. monteilii*** **(KY823008),** ***P. rhodesiae*** **(KY823010),** ***P. putida*** **(KY823009).**	Laponite	*Vigna unguiculata*	Overall plant growth promotion	Greenhouse assay	[115]
Pseudomonas libenensis **TR1**	Alginate	*Vigna unguiculata*	Plant growth promotion and reduction in drought stress	Greenhouse assay	[116]
Pseudomonas **sp. DN18**	Alginate supplemented with salicylic acid and zinc oxide	*Oryza sativa*	Plant growth promotion and biocontrol against *S. rolfsii*	Greenhouse assay	[117]
Bacillus licheniformis	Alginate supplemented with chitosan	*Capsicum annuum* L.	Plant growth promotion and biocontrol against *S. rolfsii*	Greenhouse assay	[118]

Table 1. *Cont.*

Encapsulated Microorganism	Encapsulation Polymer	Crop	Effect	Type of Assay	Reference
***Azospirillum brasilense*, *Burkholderia cepacia*, *Bacillus thuringiensis*, *B. megaterium*, *B. cereus*, *B. subtilis*, *B. subtilis* 1411 and *Trichoderma* sp.**	Alginate Clay	*Eugenia stipitata*	Increase in nitrate and phosphorous concentration *in planta*	Field assay	[119]
***Bacillus megaterium* MTCC 2412, *Azotobacter chroococcum* MTCC 3853 *Pseudomonas flourescens* MTCC*Trichoderma viride* MTCC 793**	Alginate	*Cajanus cajan*	Plant growth promotion	Greenhouse assay	[120]
***Glomus* sp. y *Acaulospora* sp.**	Alginate	*Zea mays*	Increase in root colonization of maize plants and increase resistance to drought stress	Greenhouse assay	[121]
***Mesorhizobium ciceri* ST-282 *Bradyrhizobium japonicum* M8**	Alginate + gelatin/pectin/kaolin/bentonite	*Cicer arietinum* and *Glycine max*	Increase in plant nodules	Field assay	[122]
***Metarhizium brunneum* CB15**	Pectin/starch, cellulose, and yeast	*Solanum tuberosum*	Plant growth promotion; increase in nitrogen and phosphorous content	Greenhouse assay	[123]
***Metarhizium brunneum* BIPESCO5**	Alginate	*Solanum lycopersicum* L. cv. Ruthje	Increase in endophytic behavior	Greenhouse effect	[124]
Trichoderma viride	Alginate in combination with calcium or copper	*Lactuca sativa* L.	Increase in yield and secondary metabolite production	Field assay	[125]
***Candida tropicals* CY4, *Cryptococcus tephrensis* TY17 and *Saccharomyces cerevisiae* CuY10**	Alginate	Cucumber cv. *Beta alpha*	Biocontrol againts *F. oxysporum*; Increase in plant growth	Greenhouse assay	[126]

6.2. Bioencapsulated Fungi

Among the fungi that have been bioencapsulated, the entomopathogenic fungi, mycorrhizal fungi and fungi that promote plant growth and biocontrol stand out.

Entomopathogenic fungi (HEP) are part of the most important biological formulations in the microbial control of pest insects; they have also gained importance in their use as plant growth promoters and even as good plant tissue colonizers. For example, the endophyte, *Metarhizum brunneum*, encapsulated in alginate, was able to preserve its viability, in addition to being able to efficiently colonize tomato plants [124].

Arbuscular mycorrhizal fungi (AMF) are plant growth-promoting organisms that, following the process of absorbing mineral nutrients from the soil, transfer these nutrients to the plant. Arbuscular mycorrhizal hyphae spread widely in the soil and function as an extension of the roots, increasing the ability of plants to absorb water and nutrients from the soil [127]. Among the beneficial effects they offer are stability to the soil structure, greater host plant tolerance to water stress [128] and, as a consequence of its mycelium increasing the absorption range of the roots, it mitigates aluminum toxicity stress generated in the soil [129].

The use of encapsulation for AMF spores improves the efficiency and stability of fungal bioinocula and, therefore, it is possible to meet the optimal mechanical properties for their handling, transport and stability, following their formulation [121].

Trichoderma and *Metarhizum* are among the genera of fungi that promote plant growth of agronomic importance, which, through their encapsulation based on polymers, have managed to promote the plant growth of crops, such as *Cajanus cajan*, *Lactuca sativa* and *Solanum tuberosum* [120,123,125].

7. Bioencapsulation Process

Bioencapsulation is a cell immobilization process that consists of trapping microorganisms in a polymeric material, capable of allowing the passage of metabolites and gases to preserve cell viability and forming small capsules [86,130]. The bioencapsulation process uses different techniques, depending on the purpose and the type of microorganism being used. However, for each bioencapsulation, it is necessary to take into account: the selection of the material or polymer used, the desired size of the capsule, and the most appropriate technique to use [86]. Figure 2 illustrates the general steps of the bioencapsulation process of plant growth-promoting microorganisms (PGPM), and the most relevant points to consider for bioencapsulation are described below.

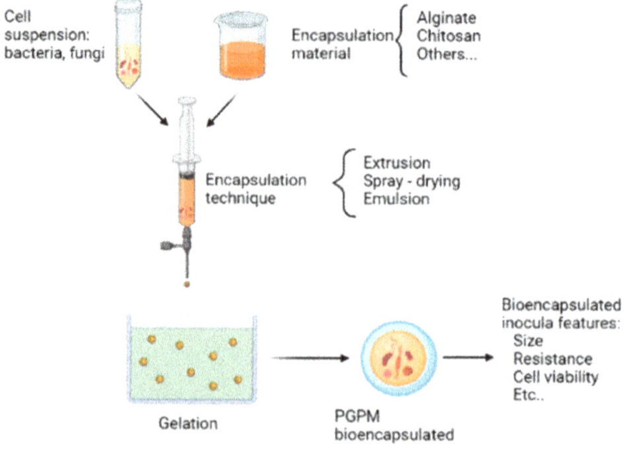

Figure 2. General bioencapsulation process of plant growth-promoting microorganisms (PGPM).

7.1. Coating Materials in Encapsulation

The materials used for the encapsulation of the microbial cells are a key part of the formulation process. It is necessary that these materials have certain properties or characteristics, such as the ability to (i) protect cells against environmental conditions, (ii) disperse with the material to be encapsulated, (iii) release their contents under specific conditions, and (iv) cover and maintain the encapsulated organisms within its structure [78,86,131,132]. These coating materials include hydrogenated oils, waxes, maltodextrins, celluloses, starches, gums and various polymers, the latter of which play a dominant role in determining and forming capsules [69,86].

Polymers in general are chemical substances made up of many repeating units called monomers, with multiple bonds chemically linked or polarized together to build the polymer chain. Natural polymers (biopolymers) are compounds typically formed by polysaccharides, such as cellulose, chitosan or starch, and proteins, such as keratin or collagen [133]. These biopolymers have gained importance in the manufacture of bioencapsulated microorganisms because, compared to synthetic polymers, they have greater benefits for the microorganisms that are encapsulated within them. For example, greater resistance to environmental factors and increased cell viability was observed with biopolymers, compared to synthetic polymers [78,86]. Among the main polymers used in agricultural bacterial encapsulation are alginate and chitosan. In addition, other biopolymers, such as carrageenan, gelatin and laponite, have been used [106,115,118], although their use is less frequent, or they are used in combination with other polymers.

7.1.1. Alginate

Alginate is a linear polysaccharide of D-mannuronic and L-guluronic acids that is naturally present in various species of algae and some bacteria [134,135]. Alginate is the most widely used polysaccharide in the encapsulation process because it is a non-toxic, biocompatible, inexpensive and readily available material that allows the encapsulation of microorganisms in a simple way [136]. Among its main advantages is the fact that it can gel easily, has good solubility and low viscosity and can be used in mild conditions, allowing cells to be trapped with minimal losses of viability [134,137].

Calcium alginate capsules (pearls) are structured as a flexible network, and if they have been used for encapsulation, they are filled with a large amount of water (97–98% by weight). Using only alginate for cell bioencapsulation does not adequately protect the cells during the drying or solidification process and, consequently, the pearls are slightly deformed, so it is generally used in combination with other organic compounds. These compounds include starch, glycerol, chitin, skim milk or humic acids [69,131,138]. Alternatively, alginate may be mixed with clays, such as montmorillonite or halloysite, thereby increasing the mechanical resistance and improving the quality of the capsule, in addition to the survival of microorganisms [105].

Biocapsules made with alginate have many advantages for microorganisms. *Glomus* sp. and *Acaulospora* sp. alginate biocapsules increase water adsorption, thereby favoring the germination of these mycorrhizal fungi; increase colonization of plant roots; and plant resistance to water stress [121]. The concentration of nitrate and phosphorus in *Eugenia stipitata* plants, as well as plant biomass, increases when plants are treated with encapsulated inocula (with alginate or clay) of *Azospirillum brasilense, Burkholderia cepacia, B. thuringiensis, B. megaterium, B. cereus, B. subtilis, B. subtilis* strain 1411 and *Trichoderma* sp. [139]. Using alginate with gelatin, pectin, kaolin or bentonite to encapsulate *Mesorhizobium ciceri* and *Bradyrhizobium japonicum* increased the number of nodules formed in plants of *Cicer arietinum* (chickpea) and *Glycine max* (soybean), in comparison with non-inoculated plants [122].

The use of alginate to encapsulate microorganisms of agricultural interest improves their characteristics as biocontrol agents and as growth promoters, promoting endophytism and the content of nitrogen and phosphorus in plants [112,113,116,123,140].

7.1.2. Chitosan

Chitosan is made from glucosamine and N-acetyl-glucosamine units, is biodegradable, abundant and easy to obtain from the deacetylation of chitin, a component of the exoskeleton of crustaceans, mollusks, insects and fungi [141].

In the presence of anions and polyanions, such as alginate, chitosan can polymerize through cross-links, reducing the porosity of the alginate, and, therefore, improving its protective effect of the encapsulated microbe [141]. By combining both biopolymers to encapsulate the bacterium *Bacillus licheniformis*, the promotion of plant growth of *Capsicum annum* L. plants was improved, as well as its capacity to act as a biocontrol agent of the pathogen *S. rolfsii* [118]. *Streptomyces fulvissimus* Uts22 encapsulated in a mixture of chitosan with gellan gum promotes the development of roots and lateral shoots in wheat plants, in addition to controlling the growth of the pathogen *Gaeumannomyces graminis* [104] (Table 1).

In general, the use of chitosan, whether applied alone or for the introduction of other particles or microorganisms in agriculture, is highly efficient in controlling both biotic and abiotic stress, in addition to promoting growth of various plant species, thus, when used to encapsulate microbes, chitosan can have an additive effect to the benefits of the microbe [142].

7.1.3. Other Biopolymers

In addition to alginate and chitosan, there are other polymers that can be used to improve the stability and/or quality of capsules. Among these polymers is gum Arabic, which is obtained from *Acacia senegal* and *A. seyal* trees and is made up of a mixture of complex polysaccharides. In addition to oligosaccharides and glycoproteins, gum Arabic is rich in essential elements and trace elements, such as aluminum, phosphorus, magnesium, copper, zinc and iron, and acts as an emulsifier, stabilizer and protector from chemical decomposition [143–145].

Starch is one of the most widely used accompanying polymers with alginate, as it provides good protection to bacterial cells and allows optimal diffusion of micronutrients and metabolites in various formulations [146]. The starch is mainly obtained from corn, potato, barley and oats, and its amylose and amylopectin units are linked by glycosidic bonds [147]. In combination with maltodextrin and sodium alginate, the encapsulation of *B. subtilis* with these materials maintain cell viability and is an efficient way to control the pathogen *Fusarium oxysporum* f. sp. *lycopersici* [148].

Maltodextrin, which is obtained from starch, is a linear polysaccharide of glucosamine and N-acetyl-D-glucosamine units. It is used at low concentrations as a coating material for the production of microparticles [148].

Gelatin, which is derived from collagen and consists of glycine, proline and 4-hydroxyproline residues, is very useful as a thermoreversible gelling agent for encapsulation, either alone or in combination with other polymers. Due to its amphoteric nature, it can also form a strong interaction with anionic polymers, providing greater stability to the capsules. Gelatin, in combination with gum Arabic, has been used to encapsulate *Metarhizium anisopliae*, which is used to biocontrol fire ants (*Solenopsis invicta*) [145] or in combination with alginate to encapsulate *B. subtilis* SL-13, greatly increasing cell viability [74].

Carrageenan, which is extracted from red seaweed (Irish moss) and some bacteria, are polymers that have a linear structure made up of D-galactose units alternately linked by α-(1,3) and β-(1,4) bonds. These polymers can form a gel that traps microbial cells and can be used in combination with alginate. However, the gelation of carrageenan is induced by changes in temperature, which should be considered when using it to make bioencapsulated microbes, especially when it comes to organisms that may be sensitive to temperature [149,150].

Another polymer that has been used is agar/agarose. It is mainly extracted from marine red algae and is composed of alternating β-D-galactopyranosyl and 3,6-anhydro-α-L-galactopyranosyl units. One of its important characteristics is that it is resistant to degradation by most known microorganisms and is a thermosetting hydrogel, gelling in response to a reduction in temperature. However, it has the disadvantages of low mechanical strength and high cost [151,152], which should be considered when using it as a material to produce bioencapsulated microbes.

An important factor for the success of the bioencapsulated microbe formulation is the correct selection of the appropriate carrier for the microorganism of interest, providing stability and protection against environmental factors, such as UV radiation, dryness and high temperature [152]. Therefore, when choosing the appropriate polymer, factors, such as availability of the polymer, resistance to environmental factors, ability to allow cell viability, whether it will be used alone or in combination with other polymers and costs, should be considered. Table 1 shows examples of studies where various biopolymers have been used for microbial encapsulations.

7.2. Capsule Size Selection

According to the size of the capsule that is formed, capsules can be classified as macrocapsules or microcapsules. Macrocapsules range in size from millimeters to centimeters, while microcapsules range in size from 1 to 1000 μm [86]. For the preparation of bioencapsulated microorganisms of agricultural interest, the use of microcapsules is preferred, because the smaller size increases the cell concentration, is more resistant and can be better dispersed in the soil or in pots [86,131].

The size of the capsule is important at the time of application in an agricultural field; normally the most common formulation is 1–4 mm in size. However, when freely mixed with seeds and sown together, the spheres can fall far from the seed and these distances can be restrictive for many beneficial bacteria, even though their mobility in the soil has been proven. To produce smaller spheres with sizes ranging from 50 to 200 μm, it is necessary to use the appropriate technology that ensures the concentration of the biomass of the microorganisms to be encapsulated [88,131], Therefore, a small capsule size is preferred to favor its dispersion close to the plant and to ensure the interaction of beneficial microorganisms with the plants (Figure 2).

7.3. Encapsulation Techniques

There are various encapsulation techniques that allow microcapsules to be produced. However, their choice depends on various factors, such as the microorganism, to be encapsulated, the temperature, humidity, and agitation (all of which affect the microorganism's survival), the polymer to be used and the purpose of the bioencapsulation. The most commonly used techniques for the formulation of bioencapsulated microbes for use in agriculture is extrusion, although there are other techniques that could also be used, such as the spray or emulsification technique.

7.3.1. Extrusion Encapsulation

Extrusion encapsulation is the most studied and oldest technique for producing capsules with polysaccharides, such as alginate, as it has the advantages of low cost and ease of implementation. This technique consists of mixing microbial cells in an aqueous solution of biopolymer that has gelling capabilities, and then this mixture is extruded in a gelling environment through a small nozzle or syringe to create small droplets of biopolymer containing microbial cells [86]. This technique also incorporates some methods to coat or gel the capsules, thus stabilizing the biopolymer droplets to prevent their dissociation or aggregation and provides them with better quality [133].

There are two mechanisms to carry out the gelation of capsules. The first is called external gelation and is obtained when the solution of the compound to be encapsulated is added together with the selected coating material, the mixture is forced through nozzles generating drops, and these fall into a bath of calcium ions, thereby forming a gel capsule. This mechanism is quite common and is simple to perform; however, it produces heterogeneous gels because surface gelation often occurs before core gelation, resulting in a rigid surface and a soft core [153,154]. The second mechanism, internal gelation, consists of the preparation of a solution of calcium ions and the compound of interest to be encapsulated, this mixture is forced through a nozzle and poured into a bath of sodium alginate. To carry out the release of calcium ions, the medium is acidified by adding an organic acid, such as acetic, adipic or glucono δ-lactone. In addition, a sequestering agent is added, which binds with free calcium, thus slowing down the gelling process [155–157].

External gelation provides larger capsule sizes (>2000 μm) and better encapsulation efficiency, therefore, it is used for the encapsulation of essential oils, bioactive compounds and plant-derived extracts [158,159]. On the other hand, internal gelation is used more frequently for the encapsulation of microorganisms due to the fact that a uniform capsule size and a smooth surface are obtained, through which agglutinations, cracks and pores are less likely to occur [154,160], resulting in better quality and more resistant capsules, without the need to use organic solvents or high temperatures to harden or coat the capsules.

The extrusion technique is gentle on microorganisms and does not require toxic solvents, and, therefore, does not cause cell damage, ensuring greater cell viability. However, this technique has the disadvantages that the production capacity is low if it is required on an industrial scale, and the size of the particles can be too large for some uses, such as agriculture [161]. However, other techniques can be adapted, along with extrusion, to obtain the desired particle size, such as the precision particle manufacturing (PPF) technique, and rotary atomization discs, among others [86,161].

Other bioencapsulation techniques have been used, such as spray-drying or emulsification, to obtain capsules of beneficial microorganisms for the food industry; however, the use and study of these techniques for the encapsulation of microorganisms of agricultural interest is not well studied.

7.3.2. Spray Drying

Spray drying is a technique that has been employed mainly in the encapsulation of microorganisms for the food industry [86,161]. It is a technique that consists of atomizing a suspension that contains the microorganism to be encapsulated and a polymeric material inside a chamber with hot gas (>100 °C up to 170 °C), generally air, which promotes the evaporation of water, causing the microorganisms to remain trapped inside the encapsulating material, giving rise to the formation of microparticles [161]. However, because this technique simultaneously dehydrates and raises the temperature of microorganisms, various microorganisms, such as non-spore-forming bacteria can suffer high mortality [69]. In addition to temperature, other variables that need to be controlled are feed flow and air flow, being key to determining the viability of microorganisms and the success of this technique for the production of bioencapsulated microbes [162,163].

7.3.3. Emulsification

Emulsification is a technique used to encapsulate different microorganisms, and it consists of mixing a disperse phase, formed by the cells and the encapsulating polymer, in a continuous phase, which is generally oil or an organic solvent. In this way, an emulsion of water in oil is obtained that is homogenized using a surfactant, such as Tween, and with constant agitation, which promotes the stability of the capsules. Alginate, carrageenan and pectin are ideal for use in this technique as encapsulating materials. Then a solidifying agent, such as calcium chloride, is added to the emulsion to form the capsules that will later be filtered. The capsules obtained can vary in size between 25 μm and 2 mm, depending on the speed of agitation, so this technique can be used to make microcapsules. However, one of the main disadvantages of this technique is the use of organic solvents that could be toxic in the subsequent use of the encapsulated microbes and removing the oil from the capsules can be difficult [86,164]. On the other hand, with this technique other types of low molecular weight materials can also be used to coat the capsules, however, the microorganisms tend to be rapidly released through the gel [133].

8. Benefits of Bioencapsulation in the Field

One of the greatest challenges in the use of plant growth promoting microorganisms in the field is their distribution in the soil and how to apply them. By directly applying microorganisms to the soil, they are exposed to environmental conditions that can be harmful to them, such as lack of water, changes in pH and temperature, in addition to the fact that their distribution area may be limited, all of which can result in the mortality of organisms and a decrease in their ability to benefit crops [69]. The use of techniques to immobilize and distribute microorganisms, such as the production of bioencapsulated microbes, has become essential to overcome the limitations encountered with the direct use of these in the field. Thus, bioencapsulation has many advantages for and the deliberate release of microorganisms to the soil. These advantages include being able to provide protection against environmental factors, increasing cell viability in the soil, favoring cell dispersion and facilitating microbial cell contact with plants, thereby increasing their effectiveness [69,86].

Among the microorganisms evaluated in the field or in greenhouse experiments, there are fungi, such as *Trichoderma* and *Metarhizum*, and a variety of bacteria, including *Bacillus*, *Pseudomonas*, *Azotobacter*, among others, which have been widely used as biofertilizers, biostimulators or as biopesticides and insecticides in crops of interest, such as *Zea mays*, *Triticum* sp., *Solanum lycopersicum*, *Oryza sativa*, etc., [90,105,114,121].

In a greenhouse experiment, the inoculation of wheat seeds with *Streptomyces fulvissimus* Uts22 microcapsules, prepared with chitosan and gellan gum, promoted the growth of plants and increased their resistance to the pathogen *Gaeumannomyces graminis* var. *tritici*, to a greater extent than inoculation with the free bacteria [104]. In a separate greenhouse experiment, with bell pepper plants, the inoculation of *P. putida* microcapsules significantly promoted plant growth, compared to uninoculated plants or plants inoculated with unencapsulated bacterial cultures [110]. These examples are consistent with the benefits of encapsulated microbes, compared to the application of liquid or unencapsulated cultures.

In the field, the inoculation of capsules of different bacteria of the genus *Bacillus*, *Azospirillum* and *Burkholderia* increased the concentration of nutrients, such as nitrogen and phosphorus, in *Eugenia stipitata* plants [119]. Microcapsules containing *Mesorhizobium ciceri* ST-282 and *Bradyrhizobium japonicum* M8 increased the number of nodules in roots of chickpea and soybean plants in a field experiment [122]. The encapsulation of *T. viride* increased the content of secondary metabolites in lettuce plants, when grown in the field and also in hydroponic culture [125], highlighting the importance of encapsulating beneficial microorganisms, including both fungi bacteria in different cultivation techniques.

The encapsulation of *Ensifer fredii* LP2/20 applied to soil cultivated with kale significantly modified the composition of the microbial community, also increasing the biomass of plants [109], which suggests an interaction of the encapsulated microorganisms with the soil microbiota, resulting in a positive effect for crops in general.

Beneficial microorganisms can also protect plants against different types of stress. The encapsulation of *Paenibacillus polymyxa* MSRH5, *B. nakamurai* MSRH1 and *B. pacificus* MSR H3 reduce the effects caused by salt stress in wheat plants, in addition to increasing plant biomass [112]. The inoculation of microcapsules of *P. putida* Rs-198 promotes plant growth of cotton plants when they are subjected to salt stress [108], providing resistance to this type of abiotic stress, and suggesting that encapsulated microorganisms can resist environmental factors, such as salt stress.

It is worth mentioning that most of the bioencapsulation experiments have been carried out under greenhouse conditions (Table 1) with favorable results. However, the application of bioencapsulated microbes in field experiments poses other challenges, such as greater exposure to environmental factors, that cannot be controlled. The studies mentioned above and others referring to organisms of agricultural interest that have been encapsulated and used in greenhouse or field experiments are summarized in Table 1.

9. Conclusions and Perspectives

Today the growing world population is accompanied by an increased demand for agricultural products. Thus, it is necessary to maintain food security for humans and animals without neglecting the conservation and improvement of ecosystems. The excessive use of agrochemicals to increase agrarian production has caused great damage to human health and ecosystems, deteriorating soil and water quality and in general, altering the environment. However, one of the alternatives to agrochemicals in the field is the use of beneficial microorganisms. Its use favorably contributes to increased crop yields, increased tolerance to water and saline stress conditions, and increased resistance to phytopathogens, thus potentially replacing the excessive use of chemical fertilizers and pesticides. However, the employment of microorganisms in the soil poses various challenges, such as maintaining cell viability and microbial resistance to different environmental conditions. Therefore, it is important to study tools and techniques, such as bioencapsulation, which allow these difficulties to be overcome.

Microbial encapsulation is one of the bioinoculum formulation methods that has attracted great interest in the agricultural area, thanks to the benefits it offers by safeguarding microbial cell viability during its formulation and later in its application and release in the field. To hasten the adoption of this technology, it is important to carry out in-depth studies on the appropriate bioencapsulation technique to be used, which is adjusted to the needs of the crop, the microorganism used and the specific environmental conditions. This approach should help to achieve the preservation of beneficial microorganisms and their efficient distribution in the soil, thus guaranteeing their efficacy as biostimulants, biofertilizers and biopesticides.

Author Contributions: Writing—original draft preparation, B.R.-S., L.R.M.-C. and P.G.-G.; writing—review and editing, M.d.C.O.-M., B.C.S.-M., J.M.S.-Y., A.E.F., O.O.B., B.R.G. and G.S.; Conceptualization, B.R.-S., M.d.C.O.-M. and G.S.; Supervision, Funding acquisition, M.d.C.O.-M. and G.S. All authors have read and agreed to the published version of the manuscript.

Funding: G.S. thanks CONACYT-México (Proposal: A1-S-15956) and CIC-UMSNH (2021-2022) for financial support to our research projects.

Institutional Review Board Statement: Not applicable.

Informed Consent Statement: Not applicable.

Data Availability Statement: Not applicable.

Acknowledgments: Blanca Rojas-Sánchez and Luzmaría R. Morales-Cedeño were recipients of doctoral fellowships from CONACYT, México. Paulina Guzmán-Guzmán received a postdoctoral fellowship under the research supported by CONACYT-México.

Conflicts of Interest: The authors declare no conflict of interest.

References

1. Glare, T.R.; Moran-Diez, M.E. *Microbial-Based Biopesticides: Methods and Protocols*; Springer: Cham, Switzerland, 2016; ISBN 9781493963652.
2. Vejan, P.; Abdullah, R.; Khadiran, T.; Ismail, S.; Nasrulhaq Boyce, A. Role of Plant Growth Promoting Rhizobacteria in Agricultural Sustainability—A Review. *Molecules* **2016**, *21*, 573. [CrossRef]
3. Bhardwaj, D.; Ansari, M.W.; Sahoo, R.K.; Tuteja, N. Biofertilizers Function as Key Player in Sustainable Agriculture by Improving Soil Fertility, Plant Tolerance and Crop Productivity. *Microb. Cell Fact.* **2014**, *13*, 66. [CrossRef]
4. Moreno Reséndez, A.; García Mendoza, V.; Reyes Carrillo, J.L.; Vásquez Arroyo, J.; Cano Ríos, P. Rizobacterias Promotoras del Crecimiento Vegetal: Una Alternativa de Biofertilización Para la Agricultura Sustentable. *Rev. Colomb. Biotecnol.* **2018**, *20*, 68–83. [CrossRef]
5. Massa, F.; Defez, R.; Bianco, C. Exploitation of Plant Growth Promoting Bacteria for Sustainable Agriculture: Hierarchical Approach to Link Laboratory and Field Experiments. *Microorganisms* **2022**, *10*, 865. [CrossRef]
6. Maisarah Nur Sarbani, M.; Yahaya, N. Advanced Development of Bio-Fertilizer Formulations Using Microorganisms as Inoculant for Sustainable Agriculture and Environment—A Review. *Malays. J. Sci. Health Technol.* **2022**, *8*, 92–101.

7. Hakim, S.; Naqqash, T.; Nawaz, M.S.; Laraib, I.; Siddique, M.J.; Zia, R.; Mirza, M.S.; Imran, A. Rhizosphere engineering with plant growth-promoting microorganisms for agriculture and ecological sustainability. *Front. Sustain. Food Syst.* **2021**, *5*, 617157. [CrossRef]
8. Hossain, M.M.; Sultana, F.; Islam, S. Plant Growth-Promoting Fungi (PGPF): Phytostimulation and Induced Systemic Resistance. In *Plant-Microbe Interactions in Agro-Ecological Perspectives*; Singh, D.P., Singh, H.B., Prabha, R., Eds.; Springer: Singapore, 2017; Volume 2, pp. 135–191, ISBN 9789811065934.
9. Dos Santos Lopes, M.J.; Dias-Filho, M.B.; Gurgel, E.S.C. Successful Plant Growth-Promoting Microbes: Inoculation Methods and Abiotic Factors. *Front. Sustain. Food Syst.* **2021**, *5*, 606454. [CrossRef]
10. Eichmann, R.; Richards, L.; Schäfer, P. Hormones as Go-Betweens in Plant Microbiome Assembly. *Plant J.* **2021**, *105*, 518–541. [CrossRef]
11. Lynch, J.M. Resilience of the Rhizosphere to Anthropogenic Disturbance. *Biodegradation* **2002**, *13*, 21–27. [CrossRef]
12. Larsen, J.; Jaramillo-López, P.; Nájera-Rincon, M.; Gonzaléz-Esquivel, C.E. Biotic Interactions in the Rhizosphere in Relation to Plant and Soil Nutrient Dynamics. *J. Soil Sci. Plant Nutr.* **2015**, *15*, 449–463. [CrossRef]
13. Olanrewaju, O.S.; Ayangbenro, A.S.; Glick, B.R.; Babalola, O.O. Plant Health: Feedback Effect of Root Exudates-Rhizobiome Interactions. *Appl. Microbiol. Biotechnol.* **2019**, *103*, 1155–1166. [CrossRef]
14. Schirawski, J.; Perlin, M. Plant–Microbe Interaction 2017—The Good, the Bad and the Diverse. *Int. J. Mol. Sci.* **2018**, *19*, 1374. [CrossRef]
15. Nadeem, S.M.; Ahmad, M.; Zahir, Z.A.; Javaid, A.; Ashraf, M. The Role of Mycorrhizae and Plant Growth Promoting Rhizobacteria (PGPR) in Improving Crop Productivity under Stressful Environments. *Biotechnol. Adv.* **2014**, *32*, 429–448. [CrossRef]
16. Berruti, A.; Lumini, E.; Balestrini, R.; Bianciotto, V. Arbuscular Mycorrhizal Fungi as Natural Biofertilizers: Let's Benefit from Past Successes. *Front. Microbiol.* **2016**, *6*, 1559. [CrossRef]
17. Albuquerque da Silva Campos, M. Bioprotection by Arbuscular Mycorrhizal Fungi in Plants Infected with Meloidogyne Nematodes: A Sustainable Alternative. *Crop Prot.* **2020**, *135*, 105203. [CrossRef]
18. Naziya, B.; Murali, M.; Amruthesh, K.N. Plant Growth-Promoting Fungi (Pgpf) Instigate Plant Growth and Induce Disease Resistance in *Capsicum annuum* L. upon Infection with *Colletotrichum capsici* (Syd.) Butler & Bisby. *Biomolecules* **2020**, *10*, 41. [CrossRef]
19. Ghorbanpour, M.; Omidvari, M.; Abbaszadeh-Dahaji, P.; Omidvar, R.; Kariman, K. Mechanisms Underlying the Protective Effects of Beneficial Fungi against Plant Diseases. *Biol. Control* **2018**, *117*, 147–157. [CrossRef]
20. Rai, M.; Zimowska, B.; Shinde, S.; Tres, M.V. Bioherbicidal Potential of Different Species of *Phoma*: Opportunities and Challenges. *Appl. Microbiol. Biotechnol.* **2021**, *105*, 3009–3018. [CrossRef]
21. Backer, R.; Rokem, J.S.; Ilangumaran, G.; Lamont, J.; Praslickova, D.; Ricci, E.; Subramanian, S.; Smith, D.L. Plant Growth-Promoting Rhizobacteria: Context, Mechanisms of Action, and Roadmap to Commercialization of Biostimulants for Sustainable Agriculture. *Front. Plant Sci.* **2018**, *9*, 1473. [CrossRef]
22. Ke, X.; Feng, S.; Wang, J.; Lu, W.; Zhang, W.; Chen, M.; Lin, M. Effect of Inoculation with Nitrogen-Fixing Bacterium Pseudomonas Stutzeri A1501 on Maize Plant Growth and the Microbiome Indigenous to the Rhizosphere. *Syst. Appl. Microbiol.* **2019**, *42*, 248–260. [CrossRef]
23. Pace, L.; Pellegrini, M.; Palmieri, S.; Rocchi, R.; Lippa, L.; del Gallo, M. Plant Growth-Promoting Rhizobacteria for in vitro and ex vitro Performance Enhancement of Apennines' Genepì (*Artemisia umbelliformis* subsp. *Eriantha*), an Endangered Phytotherapeutic Plant. *Vitr. Cell. Dev. Biol.-Plant* **2020**, *56*, 134–142. [CrossRef]
24. Ali, M.; Ahmad, Z.; Ashraf, M.F.; Dong, W. Maize Endophytic Microbial-Communities Revealed by Removing PCR and 16S RRNA Sequencing and Their Synthetic Applications to Suppress Maize Banded Leaf and Sheath Blight. *Microbiol. Res.* **2021**, *242*, 126639. [CrossRef] [PubMed]
25. Vassilev, N.; Vassileva, M.; Martos, V.; Garcia del Moral, L.F.; Kowalska, J.; Tylkowski, B.; Malusá, E. Formulation of Microbial Inoculants by Encapsulation in Natural Polysaccharides: Focus on Beneficial Properties of Carrier Additives and Derivatives. *Front. Plant Sci.* **2020**, *11*, 270. [CrossRef]
26. Nayana, A.R.; Joseph, B.J.; Jose, A.; Radhakrishnan, E.K. Nanotechnological Advances with PGPR Applications. In *Sustainable Agriculture Review*; Hayat, S., Pichtel, J., Faizan, M., Fariduddin, Q., Eds.; Springer: Cham, Switzerland, 2020; pp. 163–180, ISBN 9783030339968.
27. Muñoz Rojas, J.; Molina-Romero, D.; del Rocío Bustillos, M.; Rodríguez-Andrade, O.; Morales-García, Y.E.; Santiago-Saenz, Y.; Castañeda-Lucio, M.; Muñoz-Rojas, J. Mecanismos de Fitoestimulación por Rizobacterias, Aislamientos en América y Potencial Biotecnológico. *Biológicas* **2015**, *17*, 24–34.
28. Agbodjato, N.A.; Assogba, S.A.; Babalola, O.O.; Koda, A.D.; Aguegue, R.M.; Sina, H.; Dagbenonbakin, G.D.; Adjanohoun, A.; Baba-Moussa, L. Formulation of Biostimulants Based on Arbuscular Mycorrhizal Fungi for Maize Growth and Yield. *Front. Agron.* **2022**, *4*, 894489. [CrossRef]
29. Glick, B.R. Plant Growth-Promoting Bacteria: Mechanisms and Applications. *Scientifica* **2012**, *2012*, 963401. [CrossRef]
30. Adeleke, B.S.; Babalola, O.O.; Glick, B.R. Plant Growth-Promoting Root-Colonizing Bacterial Endophytes. *Rhizosphere* **2021**, *20*, 100433. [CrossRef]
31. Olanrewaju, O.S.; Glick, B.R.; Babalola, O.O. Mechanisms of Action of Plant Growth Promoting Bacteria. *World J. Microbiol. Biotechnol.* **2017**, *33*, 197. [CrossRef]

32. Defez, R.; Valenti, A.; Andreozzi, A.; Romano, S.; Ciaramella, M.; Pesaresi, P.; Forlani, S.; Bianco, C. New Insights into Structural and Functional Roles of Indole-3-Acetic Acid (IAA): Changes in DNA Topology and Gene Expression in Bacteria. *Biomolecules* **2019**, *9*, 522. [CrossRef]
33. Glick, B.R. *Beneficial Plant-Bacterial Interactions*, 2nd ed.; Springer: Berlin/Heidelberg, Germany, 2020; ISBN 3030443671.
34. Orozco-Mosqueda, M.d.C.; Flores, A.; Rojas-Sánchez, B.; Urtis-Flores, C.A.; Morales-Cedeño, L.R.; Valencia-Marin, M.F.; Chávez-Avila, S.; Rojas-Solis, D.; Santoyo, G. Plant Growth-Promoting Bacteria as Bioinoculants: Attributes and Challenges for Sustainable Crop Improvement. *Agronomy* **2021**, *11*, 1167. [CrossRef]
35. Billah, M.; Khan, M.; Bano, A.; Hassan, T.U.; Munir, A.; Gurmani, A.R. Phosphorus and Phosphate Solubilizing Bacteria: Keys for Sustainable Agriculture. *Geomicrobiol. J.* **2019**, *36*, 904–916. [CrossRef]
36. Mahdi, S.S.; Talat, M.A.; Dar, M.H.; Hamid, A.; Ahmad, L. Soil Phosphorus Fixation Chemistry and Role of Phosphate Solubilizing Bacteria in Enhancing Its Efficiency for Sustainable Cropping—A Review. *J. Pure Appl. Microbiol.* **2012**, *66*, 1905–1911.
37. Kramer, J.; Özkaya, Ö.; Kümmerli, R. Bacterial Siderophores in Community and Host Interactions. *Nat. Rev. Microbiol.* **2020**, *18*, 152–163. [CrossRef]
38. Schulten, H.R.; Schnitzer, M. The Chemistry of Soil Organic Nitrogen: A Review. *Biol. Fertil. Soils* **1997**, *26*, 1–15. [CrossRef]
39. Keuper, F.; Dorrepaal, E.; van Bodegom, P.M.; van Logtestijn, R.; Venhuizen, G.; van Hal, J.; Aerts, R. Experimentally Increased Nutrient Availability at the Permafrost Thaw Front Selectively Enhances Biomass Production of Deep-Rooting Subarctic Peatland Species. *Glob. Chang. Biol.* **2017**, *23*, 4257–4266. [CrossRef] [PubMed]
40. Igiehon, N.O.; Babalola, O.O. Rhizosphere Microbiome Modulators: Contributions of Nitrogen Fixing Bacteria towards Sustainable Agriculture. *Int. J. Environ. Res. Public Health* **2018**, *15*, 574. [CrossRef]
41. Li, Z.; Tian, D.; Wang, B.; Wang, J.; Wang, S.; Chen, H.Y.H.; Xu, X.; Wang, C.; He, N.; Niu, S. Microbes Drive Global Soil Nitrogen Mineralization and Availability. *Glob. Chang. Biol.* **2019**, *25*, 1078–1088. [CrossRef] [PubMed]
42. Pandey, S.; Gupta, S. Evaluation of *Pseudomonas* sp. for Its Multifarious Plant Growth Promoting Potential and Its Ability to Alleviate Biotic and Abiotic Stress in Tomato (*Solanum lycopersicum*) Plants. *Sci. Rep.* **2020**, *10*, 20951. [CrossRef]
43. Barney, B.M. Aerobic Nitrogen-Fixing Bacteria for Hydrogen and Ammonium Production: Current State and Perspectives. *Appl. Microbiol. Biotechnol.* **2020**, *104*, 1383–1399. [CrossRef]
44. Fedele, G.; Brischetto, C.; Rossi, V. Biocontrol of *Botrytis cinerea* on Grape Berries as Influenced by Temperature and Humidity. *Front. Plant Sci.* **2020**, *11*, 1232. [CrossRef]
45. Morales-Cedeño, L.R.; Flores, A.; de los Santos-Villalobos, S.; Santoyo, G. Bacterias Endófitas Promotoras del Crecimiento Vegetal Como Agentes Biocontrol de Patógenos Postcosecha. In *Bacterias Promotoras del Crecimiento Vegetal: Aspectos Básicos y Aplicaciones para una Agricultura Sustentable;* Orzoco-Mosqueda, M.d.C., Santoyo, G., Eds.; Fontamara: Mexico City, Mexico, 2020; pp. 111–130, ISBN 9786077366591.
46. Bargaz, A.; Elhaissoufi, W.; Khourchi, S.; Benmrid, B.; Borden, K.A.; Rchiad, Z. Benefits of Phosphate Solubilizing Bacteria on Belowground Crop Performance for Improved Crop Acquisition of Phosphorus. *Microbiol. Res.* **2021**, *252*, 126842. [CrossRef]
47. Monachon, M.; Albelda-Berenguer, M.; Joseph, E. Biological Oxidation of Iron Sulfides. *Adv. Appl. Microbiol.* **2019**, *107*, 1–27. [CrossRef]
48. Nascimento, F.X.; Rossi, M.J.; Soares, C.R.F.S.; McConkey, B.J.; Glick, B.R. New Insights into 1-Aminocyclopropane-1-Carboxylate (ACC) Deaminase Phylogeny, Evolution and Ecological Significance. *PLoS ONE* **2014**, *9*, e99168. [CrossRef]
49. Orozco-Mosqueda, M.d.C.; Glick, B.R.; Santoyo, G. ACC Deaminase in Plant Growth-Promoting Bacteria (PGPB): An Efficient Mechanism to Counter Salt Stress in Crops. *Microbiol. Res.* **2020**, *235*, 126439. [CrossRef]
50. Guzmán-Guzmán, P.; Porras-Troncoso, M.D.; Olmedo-Monfil, V.; Herrera-Estrella, A. Trichoderma Species: Versatile Plant Symbionts. *Phytopathology* **2019**, *109*, 6–16. [CrossRef]
51. Khatoon, Z.; Huang, S.; Rafique, M.; Fakhar, A.; Kamran, M.A.; Santoyo, G. Unlocking the Potential of Plant Growth-Promoting Rhizobacteria on Soil Health and the Sustainability of Agricultural Systems. *J. Environ. Manag.* **2020**, *273*, 111118. [CrossRef]
52. Rieusset, L.; Rey, M.; Muller, D.; Vacheron, J.; Gerin, F.; Dubost, A.; Comte, G.; Prigent-Combaret, C. Secondary Metabolites from Plant-Associated Pseudomonas Are Overproduced in Biofilm. *Microb. Biotechnol.* **2020**, *13*, 1562–1580. [CrossRef]
53. Rillig, M.C.; Lehmann, A.; Lehmann, J.; Camenzind, T.; Rauh, C. Soil Biodiversity Effects from Field to Fork. *Trends Plant Sci.* **2018**, *23*, 17–24. [CrossRef]
54. Vacheron, J.; Desbrosses, G.; Bouffaud, M.L.; Touraine, B.; Moënne-Loccoz, Y.; Muller, D.; Legendre, L.; Wisniewski-Dyé, F.; Prigent-Combaret, C. Plant Growth-Promoting Rhizobacteria and Root System Functioning. *Front. Plant Sci.* **2013**, *4*, 356. [CrossRef]
55. Fröhlich, A.; Buddrus-Schiemann, K.; Durner, J.; Hartmann, A.; von Rad, U. Response of Barley to Root Colonization by *Pseudomonas* sp. DSMZ 13134 under Laboratory, Greenhouse, and Field Conditions. *J. Plant Interact.* **2012**, *7*, 1–9. [CrossRef]
56. Zhao, S.; Wei, H.; Lin, C.Y.; Zeng, Y.; Tucker, M.P.; Himmel, M.E.; Ding, S.Y. *Burkholderia phytofirmans* Inoculation-Induced Changes on the Shoot Cell Anatomy and Iron Accumulation Reveal Novel Components of *Arabidopsis*-Endophyte Interaction That Can Benefit Downstream Biomass Deconstruction. *Front. Plant Sci.* **2016**, *7*, 24. [CrossRef] [PubMed]
57. Pawar, S.; Chaudhari, A.; Prabha, R.; Shukla, R.; Singh, D.P. Microbial Pyrrolnitrin: Natural Metabolite with Immense Practical Utility. *Biomolecules* **2019**, *9*, 443. [CrossRef] [PubMed]
58. Wan, C.; Fan, X.; Lou, Z.; Wang, H.; Olatunde, A.; Rengasamy, K.R.R. Iturin: Cyclic Lipopeptide with Multifunction Biological Potential. *Crit. Rev. Food Sci. Nutr.* **2021**, *1*, 13. [CrossRef]

59. Bender, C.L.; Alarcón-Chaidez, F.; Gross, D.C. *Pseudomonas syringae* Phytotoxins: Mode of Action, Regulation, and Biosynthesis by Peptide and Polyketide Synthetases. *Microbiol. Mol. Biol. Rev.* **1999**, *63*, 266–292. [CrossRef]
60. Pérez-Equihua, A.; Santoyo, G. Draft Genome Sequence of *Bacillus* sp. Strain E25, a Biocontrol and Plant Growth-Promoting Bacterial Endophyte Isolated from Mexican Husk Tomato Roots (*Physalis ixocarpa* Brot. Ex Horm.). *Microbiol. Resour. Announc.* **2021**, *10*, e01112-20. [CrossRef]
61. Flores, A.; Diaz-Zamora, J.T.; Orozco-Mosqueda, M.D.C.; Chávez, A.; de los Santos-Villalobos, S.; Valencia-Cantero, E.; Santoyo, G. Bridging Genomics and Field Research: Draft Genome Sequence of *Bacillus thuringiensis* CR71, an Endophytic Bacterium That Promotes Plant Growth and Fruit Yield in *Cucumis sativus* L. *3 Biotech* **2020**, *10*, 220. [CrossRef] [PubMed]
62. Vázquez-Chimalhua, E.; Valencia-Cantero, E.; López-Bucio, J.; Ruiz-Herrera, L.F. *N,N*-Dimethyl-Hexadecylamine Modulates Arabidopsis Root Growth through Modifying the Balance between Stem Cell Niche and Jasmonic Acid-Dependent Gene Expression. *Gene Expr. Patterns* **2021**, *41*, 119201. [CrossRef]
63. Mukherjee, P.K.; Mendoza-Mendoza, A.; Zeilinger, S.; Horwitz, B.A. Mycoparasitism as a Mechanism of Trichoderma-Mediated Suppression of Plant Diseases. *Fungal Biol. Rev.* **2022**, *39*, 15–33. [CrossRef]
64. Adeleke, B.S.; Ayilara, S.; Akinola, S.A.; Babalola, O.O. Biocontrol Mechanisms of Endophytic Fungi. *Egypt. J. Biol. Pest Control* **2022**, *32*, 46. [CrossRef]
65. Olowe, O.M.; Nicola, L.; Asemoloye, M.D.; Akanmu, A.O.; Babalola, O.O. Trichoderma: Potential Bio-Resource for the Management of Tomato Root Rot Diseases in Africa. *Microbiol. Res.* **2022**, *257*, 126978. [CrossRef]
66. Stael, S.; Kmiecik, P.; Willems, P.; van der Kelen, K.; Coll, N.S.; Teige, M.; van Breusegem, F. Plant Innate Immunity–Sunny Side up? *Trends Plant Sci.* **2015**, *20*, 3–11. [CrossRef]
67. Oleńska, E.; Małek, W.; Wójcik, M.; Swiecicka, I.; Thijs, S.; Vangronsveld, J. Beneficial Features of Plant Growth-Promoting Rhizobacteria for Improving Plant Growth and Health in Challenging Conditions: A Methodical Review. *Sci. Total Environ.* **2020**, *743*, 140682. [CrossRef]
68. Fadiji, A.E.; Babalola, O.O. Elucidating Mechanisms of Endophytes Used in Plant Protection and Other Bioactivities with Multifunctional Prospects. *Front. Bioeng. Biotechnol.* **2020**, *8*, 467. [CrossRef]
69. Schoebitz, M.; López, M.D.; Roldán, A. Bioencapsulation of Microbial Inoculants for Better Soil-Plant Fertilization: A Review. *Agron. Sustain. Dev.* **2013**, *33*, 751–765. [CrossRef]
70. Malusá, E.; Sas-Paszt, L.; Ciesielska, J. Technologies for Beneficial Microorganisms Inocula Used as Biofertilizers. *Sci. World J.* **2012**, *2012*, 491206. [CrossRef]
71. Bashan, Y.; de-Bashan, L.E.; Prabhu, S.R.; Hernandez, J.P. Advances in Plant Growth-Promoting Bacterial Inoculant Technology: Formulations and Practical Perspectives (1998–2013). *Plant Soil* **2014**, *378*, 1–33. [CrossRef]
72. Mendis, H.C.; Thomas, V.P.; Schwientek, P.; Salamzade, R.; Chien, J.T.; Waidyarathne, P.; Kloepper, J.; de La Fuente, L. Strain-Specific Quantification of Root Colonization by Plant Growth Promoting Rhizobacteria *Bacillus firmus* I-1582 and *Bacillus amyloliquefaciens* QST713 in Non-Sterile Soil and Field Conditions. *PLoS ONE* **2018**, *13*, e0193119. [CrossRef]
73. Wong, C.K.F.; Saidi, N.B.; Vadamalai, G.; Teh, C.Y.; Zulperi, D. Effect of Bioformulations on the Biocontrol Efficacy, Microbial Viability and Storage Stability of a Consortium of Biocontrol Agents against *Fusarium* Wilt of Banana. *J. Appl. Microbiol.* **2019**, *127*, 544–555. [CrossRef]
74. Tu, L.; He, Y.; Yang, H.; Wu, Z.; Yi, L. Preparation and Characterization of Alginate-Gelatin Microencapsulated *Bacillus subtilis* SL-13 by Emulsification/Internal Gelation. *J. Biomater. Sci. Polym. Ed.* **2015**, *26*, 735–749. [CrossRef]
75. Neneng, L. Formulation of Liquid Biofertilizer for Enhance of Soil Nutrients in Peatland. *Birex J.* **2020**, *2*, 314–322.
76. Goljanian-Tabrizi, S.; Amiri, S.; Nikaein, D.; Motesharrei, Z. The Comparison of Five Low Cost Liquid Formulations to Preserve Two Phosphate Solubilizing Bacteria from the Genera Pseudomonas and Pantoea. *Iran. J. Microbiol.* **2016**, *8*, 377–382. [PubMed]
77. Camelo-Rusinque, M.; Moreno-Galván, A.; Romero-Perdomo, F.; Bonilla-Buitrago, R. Development of a Liquid Fermentation System and Encystment for a Nitrogen-Fixing Bacterium Strain Having Biofertilizer Potential. *Rev. Argent. Microbiol.* **2017**, *49*, 289–296. [CrossRef] [PubMed]
78. Chaudhary, T.; Dixit, M.; Gera, R.; Shukla, A.K.; Prakash, A.; Gupta, G.; Shukla, P. Techniques for Improving Formulations of Bioinoculants. *3 Biotech* **2020**, *10*, 199. [CrossRef] [PubMed]
79. John, R.P.; Tyagi, R.D.; Brar, S.K.; Surampalli, R.Y.; Prévost, D. Bio-Encapsulation of Microbial Cells for Targeted Agricultural Delivery. *Crit. Rev. Biotechnol.* **2011**, *31*, 211–226. [CrossRef]
80. Rose, M.T.; Deaker, R.; Potard, S.; Tran, C.K.T.; Vu, N.T.; Kennedy, I.R. The Survival of Plant Growth Promoting Microorganisms in Peat Inoculant as Measured by Selective Plate Counting and Enzyme-Linked Immunoassay. *World J. Microbiol. Biotechnol.* **2011**, *27*, 1649–1659. [CrossRef]
81. Quiroz Sarmiento, V.F.; Almaraz Suarez, J.J.; Sánchez Viveros, G.; Argumedo Delira, R.; González Mancilla, A. Biofertilizantes de Rizobacterias en el Crecimiento de Plántulas de Chile Poblano. *Rev. Mex. Cienc. Agríc.* **2019**, *10*, 1733–1745. [CrossRef]
82. Berger, B.; Patz, S.; Ruppel, S.; Dietel, K.; Faetke, S.; Junge, H.; Becker, M. Successful formulation and application of plant growth-promoting *Kosakonia radicincitans* in maize cultivation. *BioMed Res. Int.* **2018**, *2018*, 6439481. [CrossRef]
83. King, V.A.E.; Lin, H.J.; Liu, C.F. Accelerated Storage Testing of Freeze-Dried and Controlled Low-Temperature Vacuum Dehydrated *Lactobacillu acidophilus*. *J. Gen. Appl. Microbiol.* **1998**, *44*, 161–165. [CrossRef]
84. Grzegorczyk, M.; Kancelista, A.; Łaba, W.; Piegza, M.; Witkowska, D. The Effect of Lyophilization and Storage Time on the Survival Rate and Hydrolytic Activity of *Trichoderma* Strains. *Folia Microbiol.* **2018**, *63*, 433–441. [CrossRef]

85. Wessman, P.; Håkansson, S.; Leifer, K.; Rubino, S. Formulations for Freeze-Drying of Bacteria and Their Influence on Cell Survival. *J. Vis. Exp.* **2013**, *78*, 4058. [CrossRef]
86. Rathore, S.; Desai, P.M.; Liew, C.V.; Chan, L.W.; Heng, P.W.S. Microencapsulation of Microbial Cells. *J. Food Eng.* **2013**, *116*, 369–381. [CrossRef]
87. Msimbira, L.A.; Smith, D.L. The Roles of Plant Growth Promoting Microbes in Enhancing Plant Tolerance to Acidity and Alkalinity Stresses. *Front. Sustain. Food Syst.* **2020**, *4*, 106. [CrossRef]
88. Bashan, Y. Alginate Beads as Synthetic Inoculant Carriers for Slow Release of Bacteria That Affect Plant Growth. *Appl. Environ. Microbiol.* **1986**, *51*, 1089–1098. [CrossRef] [PubMed]
89. Vejan, P.; Khadiran, T.; Abdullah, R.; Ismail, S.; Dadrasnia, A. Encapsulation of Plant Growth Promoting Rhizobacteria—Prospects and Potential in Agricultural Sector: A Review. *J. Plant Nutr.* **2019**, *42*, 2600–2623. [CrossRef]
90. Reed, M.L.E.; Glick, B.R. Applications of Plant Growth-Promoting Bacteria for Plant and Soil Systems. In *Applications of Microbial Engineering*; Gupta, V.K., Schmoll, M., Maki, M., Tuohy, M., Mazutti, M.A., Eds.; Taylor & Francis: Enfield, CT, USA, 2013; pp. 181–229, ISBN 9781466585782.
91. Gamalero, E.; Glick, B.R. Ethylene and Abiotic Stress Tolerance in Plants. In *Environmental Adaptations and Stress Tolerance of Plants in the Era of Climate Change*; Springer: New York, NY, USA, 2012; pp. 395–412, ISBN 9781461408154.
92. Gepstein, S.; Glick, B.R. Strategies to Ameliorate Abiotic Stress-Induced Plant Senescence. *Plant Mol. Biol.* **2013**, *82*, 623–633. [CrossRef]
93. Ali, S.; Glick, B.R. Plant-Bacterial Interactions in Management of Plant Growth under Abiotic Stresses. In *New and Future Developments in Microbial Biotechnology and Bioengineering: Microbes in Soil, Crop and Environmental Sustainability*; Singh, J.S., Ed.; Elsevier: Amsterdam, The Netherlands, 2019; pp. 21–45, ISBN 9780128182581.
94. Goswami, M.; Deka, S. Plant Growth-Promoting Rhizobacteria—Alleviators of Abiotic Stresses in Soil: A Review. *Pedosphere* **2020**, *30*, 40–61. [CrossRef]
95. Santoyo, G.; Moreno-Hagelsieb, G.; del Carmen Orozco-Mosqueda, M.; Glick, B.R. Plant Growth-Promoting Bacterial Endophytes. *Microbiol. Res.* **2016**, *183*, 92–99. [CrossRef]
96. Morales-Cedeño, L.R.; Orozco-Mosqueda, M.d.C.; Loeza-Lara, P.D.; Parra-Cota, F.I.; de los Santos-Villalobos, S.; Santoyo, G. Plant Growth-Promoting Bacterial Endophytes as Biocontrol Agents of Pre- and Post-Harvest Diseases: Fundamentals, Methods of Application and Future Perspectives. *Microbiol. Res.* **2021**, *242*, 126612. [CrossRef]
97. Errington, J. Regulation of Endospore Formation in Bacillus Subtilis. *Nat. Rev. Microbiol.* **2003**, *1*, 117–126. [CrossRef]
98. Klobutcher, L.A.; Ragkousi, K.; Setlow, P. The *Bacillus subtilis* Spore Coat Provides "Eat Resistance" during Phagocytic Predation by the Protozoan *Tetrahymena thermophila*. *Proc. Natl. Acad. Sci. USA* **2006**, *103*, 165–170. [CrossRef]
99. Nicholson, W.L.; Munakata, N.; Horneck, G.; Melosh, H.J.; Setlow, P. Resistance of *Bacillus* Endospores to Extreme Terrestrial and Extraterrestrial Environments. *Microbiol. Mol. Biol. Rev.* **2000**, *64*, 548–572. [CrossRef] [PubMed]
100. Setlow, P. Spore Germination. *Curr. Opin. Microbiol.* **2003**, *6*, 550–556. [CrossRef] [PubMed]
101. Sosnik, A.; Menaker Raskin, M. Polymeric Micelles in Mucosal Drug Delivery: Challenges towards Clinical Translation. *Biotechnol. Adv.* **2015**, *33*, 1380–1392. [CrossRef]
102. Sauka, D.H.; Benintende, G.B. *Bacillus thuringiensis*: Generalidades. Un Acercamiento a Su Empleo En El Biocontrol de Insectos Lepidópteros Que Son Plagas Agrícolas. *Rev. Argent. Microbiol.* **2008**, *40*, 124–140. [PubMed]
103. Corrales Ramírez, L.C.; Caycedo Lozano, L.; Gómez Méndez, M.A.; Ramos Rojas, S.J.; Rodríguez Torres, J.N. *Bacillus* spp.: Una Alternativa para la Promoción Vegetal por dos Caminos Enzimáticos. *Nova* **2017**, *15*, 45. [CrossRef]
104. Saberi-Riseh, R.; Moradi-Pour, M. A Novel Encapsulation of *Streptomyces fulvissimus* Uts22 by Spray Drying and Its Biocontrol Efficiency against *Gaeumannomyces graminis*, the Causal Agent of Take-All Disease in Wheat. *Pest Manag. Sci.* **2021**, *77*, 4357–4364. [CrossRef]
105. Meftah Kadmiri, I.; El Mernissi, N.; Azaroual, S.E.; Mekhzoum, M.E.M.; Qaiss, A.E.K.; Bouhfid, R. Bioformulation of Microbial Fertilizer Based on Clay and Alginate Encapsulation. *Curr. Microbiol.* **2021**, *78*, 86–94. [CrossRef] [PubMed]
106. Rojas-Tapias, D.; Sierra, O.O.; Botía, D.R.; Bonilla, R. Preservation of *Azotobacter chroococcum* Vegetative Cells in Dry Polymers. *Univ. Sci.* **2015**, *20*, 201–207. [CrossRef]
107. Tu, L.; He, Y.; Shan, C.; Wu, Z. Preparation of Microencapsulated *Bacillus subtilis* SL-13 Seed Coating Agents and Their Effects on the Growth of Cotton Seedlings. *BioMed Res. Int.* **2016**, *2016*, 3251357. [CrossRef]
108. He, Y.; Wu, Z.; Ye, B.C.; Wang, J.; Guan, X.; Zhang, J. Viability Evaluation of Alginate-Encapsulated *Pseudomonas putida* Rs-198 under Simulated Salt-Stress Conditions and Its Effect on Cotton Growth. *Eur. J. Soil Biol.* **2016**, *75*, 135–141. [CrossRef]
109. Pongsilp, N.; Nimnoi, P. Inoculation of *Ensifer fredii* Strain LP2/20 Immobilized in Agar Results in Growth Promotion and Alteration of Bacterial Community Structure of Chinese Kale Planted Soil. *Sci. Rep.* **2020**, *10*, 15857. [CrossRef] [PubMed]
110. Hernández Montiel, L.G.; Chiquito Contreras, R.G.; Castillo Rocha, D.G.; Chiquito Contreras, C.J.; Vidal Hernández, L.; Beltrán Morales, F.A. Efecto de Microcápsulas de *Pseudomonas putida* Sobre Crecimiento y Rendimiento de Pimiento Morrón. *Rev. Mex. Cienc. Agríc.* **2018**, *20*, 4223–4233. [CrossRef]
111. Bhise, K.K.; Dandge, P.B. Alleviation of Salinity Stress in Rice Plant by Encapsulated Salt Tolerant Plant Growth Promoting Bacteria *Pantoea agglomerans* Strain KL and Its Root Colonization Ability. *Arch. Agron. Soil Sci.* **2019**, *65*, 1955–1968. [CrossRef]
112. Saad, M.M.; Abo-Koura, H.A.; Bishara, M.M.; Gomaa, I.M. Microencapsulation: Toward the Reduction of the Salinity Stress Effect on Wheat Plants Using NPK Rhizobacteria. *Biotechnol. J. Int.* **2020**, *23*, 1–18. [CrossRef]

113. Pour, M.M.; Saberi-Riseh, R.; Mohammadinejad, R.; Hosseini, A. Investigating the Formulation of Alginate-Gelatin Encapsulated *Pseudomonas fluorescens* (VUPF5 and T17-4 Strains) for Controlling *Fusarium solani* on Potato. *Int. J. Biol. Macromol.* **2019**, *133*, 603–613. [CrossRef]
114. Pacheco-Aguirre, J.; Ruiz-Sánchez, E.; Reyes-Ramírez, A.; Cristóbal-Alejo, J.; Tun-Suárez, J.; Borges-Gómez, L. Polymer-Based Encapsulation of *Bacillus subtilis* and Its Effect on *Meloidogyne incognita* in Tomato. *Phyton* **2016**, *85*, 1–6.
115. Snigdha, S.; Kalarikkal, N.; Thomas, S.; Radhakrishnan, E.K. Laponite® Clay/Poly(Ethylene Oxide) Gel Beads for Delivery of Plant Growth-Promoting Rhizobacteria. *Bull. Mater. Sci.* **2021**, *44*, 107. [CrossRef]
116. Souza-Alonso, P.; Rocha, M.; Rocha, I.; Ma, Y.; Freitas, H.; Oliveira, R.S. Encapsulation of *Pseudomonas libanensis* in Alginate Beads to Sustain Bacterial Viability and Inoculation of *Vigna unguiculata* under Drought Stress. *3 Biotech* **2021**, *11*, 293. [CrossRef]
117. Panichikkal, J.; Prathap, G.; Nair, R.A.; Krishnankutty, R.E. Evaluation of Plant Probiotic Performance of *Pseudomonas* sp. Encapsulated in Alginate Supplemented with Salicylic Acid and Zinc Oxide Nanoparticles. *Int. J. Biol. Macromol.* **2021**, *166*, 138–143. [CrossRef]
118. Panichikkal, J.; Puthiyattil, N.; Raveendran, A.; Nair, R.A.; Krishnankutty, R.E. Application of Encapsulated *Bacillus licheniformis* Supplemented with Chitosan Nanoparticles and Rice Starch for the Control of *Sclerotium rolfsii* in *Capsicum annuum* (L.) Seedlings. *Curr. Microbiol.* **2021**, *78*, 911–919. [CrossRef]
119. Nascimento, F.C.; Santos, C.H.B.; Kandasamy, S.; Rigobelo, E.C. Efficacy of Alginate- and Clay-Encapsulated Microorganisms on the Growth of Araçá-Boi Seedlings (*Eugenia stipitata*). *Acta Sci.-Biol. Sci.* **2019**, *41*, e43936. [CrossRef]
120. Venkata Raju, N.; Sukumar, K.; Reddy, G.B.; Busetty, M.; Paritala, V.; Praveena, K. Usage of Encapsulated Plant Growth Promoting-Microbial Consortia and Testing Its Efficacy on *Cajanus cajan*. *J. Microb. Biochem. Technol.* **2021**, *12*, 452. [CrossRef]
121. Pitaktamrong, P.; Kingkaew, J.; Yooyongwech, S.; Cha-Um, S.; Phisalaphong, M. Development of Arbuscular Mycorrhizal Fungi-Organic Fertilizer Pellets Encapsulated with Alginate Film. *Eng. J.* **2018**, *22*, 65–79. [CrossRef]
122. Shcherbakova, E.N.; Shcherbakov, A.V.; Rots, P.Y.; Gonchar, L.N.; Mulina, S.A.; Yahina, L.M.; Lactionov, Y.V.; Chebotar, V.K. Inoculation Technology for Legumes Based on Alginate Encapsulation. *Agron. Res.* **2018**, *16*, 2156–2168. [CrossRef]
123. Krell, V.; Unger, S.; Jakobs-Schoenwandt, D.; Patel, A.V. Endophytic *Metarhizium brunneum* Mitigates Nutrient Deficits in Potato and Improves Plant Productivity and Vitality. *Fungal Ecol.* **2018**, *34*, 43–49. [CrossRef]
124. Krell, V.; Jakobs-Schoenwandt, D.; Vidal, S.; Patel, A.V. Encapsulation of *Metarhizium brunneum* Enhances Endophytism in Tomato Plants. *Biol. Control* **2018**, *116*, 62–73. [CrossRef]
125. Jurić, S.; Sopko Stracenski, K.; Król-Kilińska, Ż.; Žutić, I.; Uher, S.F.; Đermić, E.; Topolovec-Pintarić, S.; Vinceković, M. The Enhancement of Plant Secondary Metabolites Content in *Lactuca sativa* L. by Encapsulated Bioactive Agents. *Sci. Rep.* **2020**, *10*, 3737. [CrossRef]
126. Kamel, S.M.; Ebtsam, M.M.; Massoud, O.N. Potentiality of Some Yeast Species as Biocontrol Agents against *Fusarium oxysporum* f. sp. *Cucumerinum the Causal Agent of Cucumber Wilt. Egypt. J. Biol. Pest Control* **2016**, *26*, 299–307.
127. Barrer, S.E. El Uso de Hongos *Micorrizicos arbusculares* como una Alternativa para la Agricultura. *Fac. Cienc. Agropecu.* **2009**, *7*, 124–132.
128. Brundrett, M. Diversity and Classification of Mycorrhizal Associations. *Biol. Rev. Camb. Philos. Soc.* **2004**, *79*, 473–495. [CrossRef]
129. Gomes, E.A.; Oliveira, C.A.; Lana, U.G.P.; Noda, R.W.; Marriel, I.E.; de Souza, F.A. Arbuscular Mycorrhizal Fungal Communities in the Roots of Maize Lines Contrasting for Al Tolerance Grown in Limed and Non-Limed Brazilian Oxisoil. *J. Microbiol. Biotechnol.* **2015**, *25*, 978–987. [CrossRef] [PubMed]
130. De Vos, P.; Bučko, M.; Gemeiner, P.; Navrátil, M.; Švitel, J.; Faas, M.; Strand, B.L.; Skjak-Braek, G.; Morch, Y.A.; Vikartovská, A.; et al. Multiscale Requirements for Bioencapsulation in Medicine and Biotechnology. *Biomaterials* **2009**, *30*, 2559–2570. [CrossRef] [PubMed]
131. Bashan, Y.; Hernandez, J.P.; Leyva, L.A.; Bacilio, M. Alginate Microbeads as Inoculant Carriers for Plant Growth-Promoting Bacteria. *Biol. Fertil. Soils* **2002**, *35*, 359–368. [CrossRef]
132. Nava Saucedo, J.E.; Barbotin, J.N. Bioencapsulation Revisited. *Biomater. Artif. Cells Immobil. Biotechnol.* **1993**, *21*, 383–389. [CrossRef]
133. Park, S.; Oh, K.K.; Lee, S.H. Biopolymer-Based Composite Materials Prepared Using Ionic Liquids. *Adv. Biochem. Eng. Biotechnol.* **2019**, *168*, 133–176. [CrossRef] [PubMed]
134. Szczech, M.; Maciorowski, R. Microencapsulation Technique with Organic Additives for Biocontrol Agents. *J. Hortic. Res.* **2016**, *24*, 111–122. [CrossRef]
135. Krasaekoopt, W.; Bhandari, B.; Deeth, H. Evaluation of Encapsulation Techniques of Probiotics for Yoghurt. *Int. Dairy J.* **2003**, *13*, 3–13. [CrossRef]
136. Uyen, N.T.T.; Hamid, Z.A.A.; Tram, N.X.T.; Ahmad, N. Fabrication of Alginate Microspheres for Drug Delivery: A Review. *Int. J. Biol. Macromol.* **2020**, *153*, 1035–1046. [CrossRef]
137. Dos Santos, G.F.; Locatelli, G.O.; Coêlho, D.A.; Botelho, P.S.; de Amorim, M.S.; de Vasconcelos, T.C.L.; Bueno, L.A. Factorial Design, Preparation and Characterization of New Beads Formed from Alginate, Polyphosphate and Glycerol Gelling Solution for Microorganism Microencapsulation. *J. Sol-Gel Sci. Technol.* **2015**, *75*, 345–352. [CrossRef]
138. Paques, J.P.; van der Linden, E.; van Rijn, C.J.; Sagis, L.M. Preparation methods of alginate nanoparticles. *Adv. Colloid Interface Sci.* **2014**, *209*, 163–171. [CrossRef]

139. Nascimento, F.X.; Tavares, M.J.; Franck, J.; Ali, S.; Glick, B.R.; Rossi, M.J. ACC Deaminase Plays a Major Role in *Pseudomonas fluorescens* YsS6 Ability to Promote the Nodulation of Alpha- and Betaproteobacteria Rhizobial Strains. *Arch. Microbiol.* **2019**, *201*, 817–822. [CrossRef] [PubMed]
140. Hernández-Montiel, L.G.; Chiquito-Contreras, C.J.; Murillo-Amador, B.; Vidal-Hernández, L.; Quiñones-Aguilar, E.E.; Chiquito-Contreras, R.G. Efficiency of Two Inoculation Methods of *Pseudomonas putida* on Growth and Yield of Tomato Plants. *J. Soil Sci. Plant Nutr.* **2017**, *17*, 1003–1012. [CrossRef]
141. Muxika, A.; Etxabide, A.; Uranga, J.; Guerrero, P.; de la Caba, K. Chitosan as a Bioactive Polymer: Processing, Properties and Applications. *Int. J. Biol. Macromol.* **2017**, *105*, 1358–1368. [CrossRef] [PubMed]
142. Malerba, M.; Cerana, R. Chitosan Effects on Plant Systems. *Int. J. Mol. Sci.* **2016**, *17*, 996. [CrossRef] [PubMed]
143. Ali, B.H.; Ziada, A.; Blunden, G. Biological Effects of Gum Arabic: A Review of Some Recent Research. *Food Chem. Toxicol.* **2009**, *47*, 1–8. [CrossRef]
144. Ma, X.; Wang, X.; Cheng, J.; Nie, X.; Yu, X.; Zhao, Y.; Wang, W. Microencapsulation of *Bacillus subtilis* B99-2 and Its Biocontrol Efficiency against *Rhizoctonia solani* in Tomato. *Biol. Control* **2015**, *90*, 34–41. [CrossRef]
145. Qiu, H.L.; Fox, E.G.P.; Qin, C.S.; Zhao, D.Y.; Yang, H.; Xu, J.Z. Microcapsuled Entomopathogenic Fungus against Fire Ants, *Solenopsis invicta*. *Biol. Control* **2019**, *134*, 141–149. [CrossRef]
146. Saberi Riseh, R.; Ebrahimi-Zarandi, M.; Gholizadeh Vazvani, M.; Skorik, Y.A. Reducing Drought Stress in Plants by Encapsulating Plant Growth-Promoting Bacteria with Polysaccharides. *Int. J. Mol. Sci.* **2021**, *22*, 12979. [CrossRef]
147. Cummings, J.H.; Stephen, A.M. Carbohydrate Terminology and Classification. *Eur. J. Clin. Nutr.* **2007**, *61*, S5–S18. [CrossRef]
148. Estefania, C.A.; Ligia, S.L. Immobilized Bacillus Subtilis by Ionic Gelation as Biocontrol Alternative of *Fusarium oxysporum* f. sp. *Lycopersici*. *Indian J. Anim. Res.* **2018**, *52*, 655–660. [CrossRef]
149. Gaaloul, S.; Turgeon, S.L.; Corredig, M. Influence of Shearing on the Physical Characteristics and Rheological Behaviour of an Aqueous Whey Protein Isolate–Kappa-Carrageenan Mixture. *Food Hydrocoll.* **2009**, *23*, 1243–1252. [CrossRef]
150. Choińska-Pulit, A.; Mituła, P.; Śliwka, P.; Łaba, W.; Skaradzińska, A. Bacteriophage Encapsulation: Trends and Potential Applications. *Trends Food Sci. Technol.* **2015**, *45*, 212–221. [CrossRef]
151. Mącik, M.; Gryta, A.; Frąc, M. Biofertilizers in Agriculture: An Overview on Concepts, Strategies and Effects on Soil Microorganisms. *Adv. Agron.* **2020**, *162*, 31–87. [CrossRef]
152. Vemmer, M.; Patel, A.V. Review of Encapsulation Methods Suitable for Microbial Biological Control Agents. *Biol. Control* **2013**, *67*, 380–389. [CrossRef]
153. Poncelet, D.; Lencki, R.; Beaulieu, C.; Halle, J.P.; Neufeld, R.J.; Fournier, A. Production of Alginate Beads by Emulsification/Internal Gelation. I. Methodology. *Appl. Microbiol. Biotechnol.* **1992**, *38*, 39–45. [CrossRef] [PubMed]
154. Liu, Q.; Rauth, A.M.; Wu, X.Y. Immobilization and Bioactivity of Glucose Oxidase in Hydrogel Microspheres Formulated by an Emulsification-Internal Gelation-Adsorption-Polyelectrolyte Coating Method. *Int. J. Pharm.* **2007**, *339*, 148–156. [CrossRef]
155. Secchi, E.; Munarin, F.; Alaimo, M.D.; Bosisio, S.; Buzzaccaro, S.; Ciccarella, G.; Vergaro, V.; Petrini, P.; Piazza, R. External and Internal Gelation of Pectin Solutions: Microscopic Dynamics versus Macroscopic Rheology. *J. Phys. Condens. Matter* **2014**, *26*, 464106. [CrossRef] [PubMed]
156. Dhamecha, D.; Movsas, R.; Sano, U.; Menon, J.U. Applications of Alginate Microspheres in Therapeutics Delivery and Cell Culture: Past, Present and Future. *Int. J. Pharm.* **2019**, *569*, 118627. [CrossRef]
157. Lin, Y.H.; Liang, H.F.; Chung, C.K.; Chen, M.C.; Sung, H.W. Physically Crosslinked Alginate/N,O-Carboxymethyl Chitosan Hydrogels with Calcium for Oral Delivery of Protein Drugs. *Biomaterials* **2005**, *26*, 2105–2113. [CrossRef]
158. De Moura, S.C.S.R.; Berling, C.L.; Germer, S.P.M.; Alvim, I.D.; Hubinger, M.D. Encapsulating Anthocyanins from *Hibiscus sabdariffa* L. Calyces by Ionic Gelation: Pigment Stability during Storage of Microparticles. *Food Chem.* **2018**, *241*, 317–327. [CrossRef]
159. Colak, N.; Torun, H.; Gruz, J.; Strnad, M.; Hermosín-Gutiérrez, I.; Hayirlioglu-Ayaz, S.; Ayaz, F.A. *Bog Bilberry Phenolics, Antioxidant Capacity and Nutrient Profile*; Elsevier Ltd.: Amsterdam, The Netherlands, 2016; Volume 201, ISBN 9046237737.
160. Ji, R.; Wu, J.; Zhang, J.; Wang, T.; Zhang, X.; Shao, L.; Chen, D.; Wang, J. Extending Viability of *Bifidobacterium longumin* Chitosan-Coated Alginate Microcapsules Using Emulsification and Internal Gelation Encapsulation Technology. *Front. Microbiol.* **2019**, *10*, 1389. [CrossRef] [PubMed]
161. Chavarri, M.; Maranon, I.; Carmen, M. Encapsulation Technology to Protect Probiotic Bacteria. In *Probiotics*; IntechOpen: London, UK, 2012; pp. 501–540. [CrossRef]
162. Sosnik, A.; Seremeta, K.P. Advantages and Challenges of the Spray-Drying Technology for the Production of Pure Drug Particles and Drug-Loaded Polymeric Carriers. *Adv. Colloid Interface Sci.* **2015**, *223*, 40–54. [CrossRef] [PubMed]
163. Eckert, C.; Serpa, V.G.; Felipe dos Santos, A.C.; Marinês da Costa, S.; Dalpubel, V.; Lehn, D.N.; Volken de Souza, C.F. Microencapsulation of *Lactobacillus plantarum* ATCC 8014 through Spray Drying and Using Dairy Whey as Wall Materials. *LWT-Food Sci. Technol.* **2017**, *82*, 176–183. [CrossRef]
164. John, R.P.; Tyagi, R.D.; Brar, S.K.; Prévost, D.; Surampalli, R.Y. Effect of emulsion formulation of *Sinorhizobium meliloti* and pre-inoculated seeds on alfalfa nodulation and growth: A pouch study. *J. Plant Nutr.* **2012**, *36*, 231–242. [CrossRef]

applied biosciences

Communication

Bacterial Community Analysis of Biofilm Formed on Metal Joint

Hironaga Akita [1], Yoshiki Shinto [2] and Zen-ichiro Kimura [2,*]

1 Research Institute for Sustainable Chemistry, National Institute of Advanced Industrial Science and Technology (AIST), 3-11-32 Kagamiyama, Higashi-Hiroshima 739-0046, Japan
2 Department of Civil and Environmental Engineering, National Institute of Technology, Kure College, 2-2-11 Aga-Minami, Kure 737-8506, Japan
* Correspondence: z-kimura@kure-nct.ac.jp; Tel.: +81-(0)-823-73-8486

Abstract: Microbiologically influenced corrosion (MIC) is caused by biofilms formed on metal surfaces, and MIC of metal alloys on marine infrastructure leads to severe accidents and great economic losses. Although bacterial community analyses of the biofilms collected from corroded metal have been studied, the analyses of biofilms collected from uncorroded metal are rarely reported. In this study, a biofilm formed on an uncorroded metal joint attached to a metal dock mooring at Akitsu Port was used as a model for bacterial community analysis. The bacterial community was analyzed by high-throughput sequencing of the V3–V4 variable regions of the 16S rRNA gene. Bacterial species contained in the biofilms were identified at the genus level, and *Alkanindiges* bacteria were the dominant species, which have been not reported as the dominant species in previous research on MIC. The genome sequences of known *Alkanindiges* bacteria do not have conserved gene clusters required to cause metal corrosion, which suggests that *Alkanindiges* bacteria do not corrode metals but act on the formation of biofilms. Those findings indicated that the bacterial community may change significantly during the process from biofilm formation to the occurrence of metal corrosion.

Keywords: 16S rRNA gene; bacterial community analysis; microbiologically influenced corrosion; biofilm; *Alkanindiges*

Citation: Akita, H.; Shinto, Y.; Kimura, Z.-i. Bacterial Community Analysis of Biofilm Formed on Metal Joint. *Appl. Biosci.* **2022**, *1*, 221–228. https://doi.org/10.3390/applbiosci1020014

Academic Editor: Robert Henry

Received: 30 June 2022
Accepted: 2 September 2022
Published: 6 September 2022

1. Introduction

Metal alloys are widely used as raw materials for the construction of vessels, bridges, and factories using established molding technology to create products intended for long-term use in marine environments. In marine environments, biofilms are formed when microorganisms adhere to metal surfaces. Biofilms are complexes of microorganisms with a three-dimensional structure, which is enclosed in an extracellular polymeric substance (EPS) matrix [1]. Inside the biofilm, extracellular matrix proteins and exogenous genes are transported through the EPS matrix, which encourages microbial growth [1]. Moreover, a well-developed biofilm tightly adheres to metal surfaces by extracellular polysaccharides of the EPS matrix, which is difficult to remove completely. The microorganisms inside the biofilm are protected from stresses caused by environmental changes and chemical substances. As such, a favorable environment for microbial growth is formed inside the biofilm, and the microbial community develops to a high density over time. In biofilms formed on metal surfaces, microbiologically influenced corrosion (MIC) is caused by various microbial factors, such as metal oxidation by electron transfer from the metal surface to bacteria [2] and accumulation of corrosive metabolites [3]. MIC causes serious problems for metal infrastructure in the marine environment, and economic losses from MIC were estimated to be $2.5 trillion in 2013, which is equivalent to about 3.4% of gross world product [4]. MIC assessments in the marine environment have been reported for carbon steel [5], stainless steel [6], and low-alloy steel [7]. In those reports, *Bacillus*, *Desulfovibrio*, and *Desulfobacter* have been detected as the dominant bacteria, and the developmental mechanisms of those bacteria are discussed. By contrast, bacterial community analyses of biofilms collected from uncorroded metal have rarely been reported, and our understanding of the community structure is limited. Thus, changes in bacterial community from development of biofilm to occurrence of metal corrosion have not been elucidated in detail. We consider that microbial community analyses of biofilms collected from uncorroded metal in the marine environment are important to understand the mechanism of MIC, and elucidation of the mechanism may lead to the establishment of measures to prevent the occurrence of metal corrosion [8].

High-throughput amplicon sequencing is a powerful method for the identification of target genes in large amounts of unexamined genetic material from the environmental microbiome [9]. In this method, microbial DNA is extracted from the environmental sample, and target genes are then amplified by polymerase chain reaction (PCR) using the extracted DNA as a template and primers designed to bind to the template. Subsequently, the amplified regions (amplicons) are sequenced by a next-generation sequencer. Since hundreds to thousands of different amplicons can be sequenced together by a next-generation sequencer, this method can efficiently identify target genes. In bacterial genomic DNA, several housekeeping genes are conserved to ensure cell maintenance and growth. In particular, the 16S rRNA gene encodes the 30S ribosomal subunit that is essential for the translation of mRNA and protein synthesis, but it has genus-specific variable regions, such as V1–V9 [10]. Almost all bacterial taxa can theoretically be identified based on the 16S rRNA gene sequence. Thus, bacterial community analysis based on high-throughput sequencing of 16S rRNA gene fragments can characterize the bacterial microbiome [10].

In this study, to investigate the bacterial community that contributes to MIC, a biofilm formed on an uncorroded metal joint attached to a metal dock mooring was used as a model. Using bacterial genomic DNA prepared from the biofilm, the variable region V3–V4 of the 16S rRNA gene was amplified by PCR, and high-throughput amplicon sequencing was carried out. The bacterial community was identified at the genus level based on the operation taxonomic units (OTUs).

2. Materials and Methods

2.1. Sampling

A biofilm that formed on an uncorroded metal joint attached to a metal dock mooring was collected in Akitsu Port in Hiroshima prefecture, Japan (Figure 1). After collection, the samples were put into sterile tubes and immediately placed in a cool box at 4 °C, then transported to the lab within a few hours.

Figure 1. Image of the sampling point. The sampling point for the collection of the biofilm is indicated by a black arrow.

2.2. Genomic DNA Extraction and PCR Conditions

Genomic DNA was extracted from the biofilm using an illustra™ bacteria genomicPrep Mini Spin Kit (GE Healthcare, Chicago, IL, USA) according to the manufacturer's instructions, and the concentration and purity were measured using a NanoDrop ND-1000 spectrophotometer (Thermo Fisher Scientific, Waltham, MA, USA). Using the genomic DNA as a template, the V3–V4 variable regions of the 16S rRNA genes were amplified by KOD -Plus- (TOYOBO, Osaka, Japan) using bacterial domain-specific primers 341F (5′-TCGTCGGCAGCGTCAGATGTGTATAAGAGACAGCCTACGGGNGGCWGCAG-3′) [11] and 805R (5′-GTCTCGTGGGCTCGGAGATGTGTATAAGAGACAGGACTACHVGGGTAT CTAATCC-3′) [12]. For amplification of the 16S rRNA genes, samples were subjected to initial denaturation at 94 °C for 2 min, followed by 35 cycles of denaturation at 98 °C for 10 s, annealing at 55 °C for 30 s, and extension at 68 °C for 30 s, and the final extension at 68 °C for 5 min.

2.3. Sequencing Library Preparation, Sequencing, and Bioinformatics Analysis

The length of the PCR products was confirmed using 1.0% agarose gel electrophoresis, and the samples (approximately 450 bp) were then purified using a Wizard SV Gel and PCR Clean-up System (Promega, Madison, WI, USA).

The microbiome sequence libraries were prepared using a Nextera XT DNA Library Preparation Kit (Illumina, San Diego, CA, USA) according to the manufacturer's instructions. The concentrations of the sequence libraries were measured using a Quanti Fluor™ dsDNASystem (Promega). The libraries were sequenced using a MiSeq sequencer (Illumina) with a MiSeq Reagent Kit v3 (Illumina).

The bacterial community was analyzed using the QIIME 2 microbiome bioinformatics platform [13]. After the quality of the sequence data was confirmed using a FastQC ver.0.11.9 [14], the amplicon sequence variants (ASVs) were prepared based on the filtered data using the DADA2 algorithm [15]. The ASVs were grouped into OTUs at 99% sequence identity using the taxonomic databases of Sliva ver. 138 [16,17]. To construct a phylogenetic tree based on the variable regions of 16S rRNA gene sequences, the OTU dataset was also subjected to phylogenetic analysis with MEGA 11 software [18]. The consensus bootstrap phylogenetic tree was constructed using the neighbor-joining method.

3. Results and Discussion

Among the OTUs, a total of 18 genera of bacteria were observed, and the proportion of unclassified bacteria was less than 0.1% (Figure 2). *Alkanindiges* bacteria were observed as the dominant species, and their relative abundance was more than 95%. Apart from *Alkanindiges*, several bacteria such as *Sphingomonas* and *Pseudomonas* were also observed as OTUs, but the relative abundances of those bacteria were less than 3%. To confirm the phylogenetic relationship of the bacteria comprising more than 1% of the OTUs, a neighbor-joining tree based on the V3–V4 variable regions of the 16S rRNA gene was constructed. In the neighbor-joining tree, nine kinds of *Alkanindiges* bacteria formed a cluster with *A. hongkongensis* HKU9^T, *A. illinoisensis* MVAB Hex1^T, and *A. hydrocarboniclasticus* H1^T (Figure 3). *Alkanindiges* bacteria belong to a class of Gammaproteobacteria and can grow using several kinds of hydrocarbons degraded from petroleum [19], so *Alkanindiges* bacteria are a major contributor to the bioremediation of petroleum-contaminated soil [20–22]. To our knowledge, there is a report of Gammaproteobacteria being observed as more than 90% of OTUs in a biofilm collected from a metal surface [6], but there are no reports of *Alkanindiges* bacteria accounting for more than 90% of the total.

Lichen is a complex organism of algae, fungi, and bacteria that grows on a wide range of substrates, from natural surfaces to artificial materials such as metal, glass, rubber, and plastic [23]. For example, *Stereocaulon japonicum* and *Cladonia humilis* grow in Cu-polluted sites around Cu-roofed temple buildings by their Cu-accumulation capacity [24]. Fernández-Brime et al. reported the bacterial communities associated with a lichen symbiosis, and more than 20 orders of bacteria were observed in several lichenized samples [25].

Moreover, 11 genera of bacteria observed in this study, such as *Alkanindiges*, *Sphingomonas*, *Pseudomonas*, *Roseateles*, *Allorhizobium/Neorhizobium/Pararhizobium/Rhizobium*, *Brevundimonas*, *Reyranella*, *Massilia*, *Pantoea*, *Mucilaginibacter*, *Asinibacterium*, *Roseomonas*, *Hymenobacter*, *Paracoccus*, and *Bradyrhizobium* were also observed in the report by Fernández-Brime et al. [25]. Thus, the occurrence of metal corrosion may be induced by bacteria adhered to metal surfaces via lichen. To improve the accuracy of the measures to prevent the occurrence of metal corrosion, it is necessary to consider bacterial community analysis including lichen in the future.

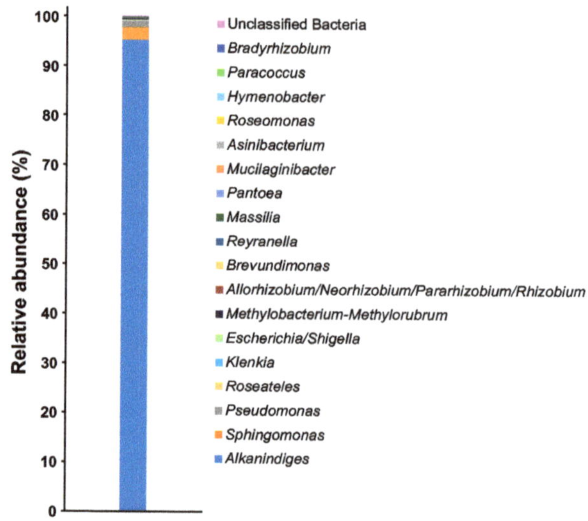

Figure 2. Relative abundances of the bacterial community in the biofilm at the genus level.

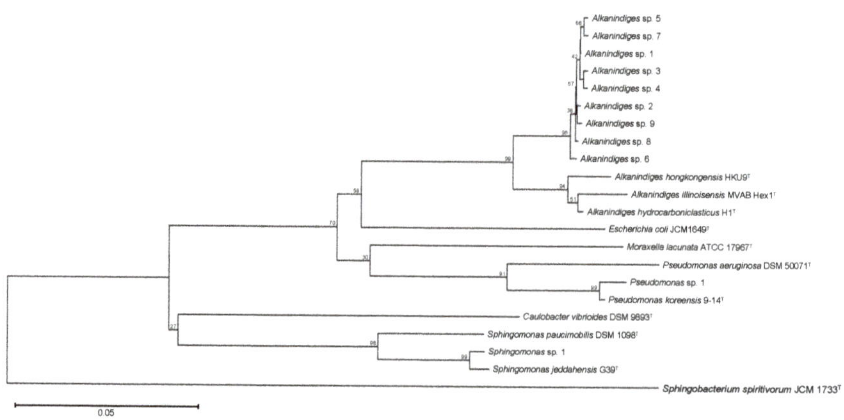

Figure 3. Consensus bootstrap phylogenetic tree of OTUs comprising more than 1% of the total 16S rRNA gene sequences. Each OTU is denoted by its representative strain type. The numbers at the nodes are percentages indicating the levels of bootstrap support based on neighbor-joining analysis of 1000 resampled datasets. The tree was rooted using *Sphingobacterium spiritivorum* JCM 1733^T as the outgroup. The bar indicates a 0.05% nucleotide substitution rate.

The detailed molecular mechanisms of MIC are still unclear, but electron transfer from the metal surface to MIC-causing bacteria has been demonstrated in a few reports. For example, *Pseudomonas aeruginosa* MCCC 1A00099 produces water-soluble electron transfer mediators, such as phenazine-1-carboxamide and pyocyanin, by phenazine-modifying enzyme, methyl transferase, and flavin-containing monooxygenase, which are encoded by *phzH*, *phzM*, and *phzS* genes. Those mediators activate the extracellular electron transfer pathway [26,27]. In the genome of *Methanococcus maripaludis* OS7, the genes that encode the small and large subunits of a [NiFe] hydrogenase (MMOS7_11590 and MMOS7_11600), the membrane proteins for secretion of the [NiFe] hydrogenase (MMOS7_11620 and MMOS7_11630), and a hydrogenase maturation protease (MMOS7_11610) are clustered [28]. Although *M. maripaludis* OS7 promotes MIC of iron, a deletion mutant strain lacking the gene cluster does not corrode iron. In *Alkanindiges* bacteria, the genome has been sequenced for two species, including *A. hydrocarboniclasticus* H1^{T} and *A. illinoisensis* DSM 15370^{T}. In those genome sequences, with the exception of MMOS7_11620 and MMOS7_11630, genes related to MIC are not present. This suggests that *Alkanindiges* bacteria do not cause MIC but act in the formation of biofilms before MIC occurrence.

In the KEGG PATHWAY Database [29], the biofilm formation pathway is registered for three species of Gammaproteobacteria, including *Escherichia coli*, *P. aeruginosa*, and *Vibrio cholerae*. To investigate the biofilm-forming capacity of *Alkanindiges* bacteria, the biofilm formation pathways of *A. hydrocarboniclasticus* H1^{T} and *A. illinoisensis* DSM 15370^{T} were compared using that of *P. aeruginosa*, which is the closest of the three species of Gammaproteobacteria to *Alkanindiges* bacteria (Figure 4). In *P. aeruginosa*, cAMP/Vfr signaling, Quorum sensing, and Gac/Rsm and c-di-GMP signaling pathways have been identified, and those pathways except the Quorum sensing pathway seem to be conserved in both strains. In particular, the function of c-di-GMP signaling pathway is enhanced by Psl, which is a neutral exopolysaccharide typically comprising D-glucose, D-mannose, and L-rhamnose. This exopolysaccharide is a crucial adhesive component of the biofilm matrix that promotes both cell-to-cell interactions and cell-to-metal surface attachment [30]. Thus, *Alkanindiges* bacteria may adhere to a metal surface by the function of the c-di-GMP signaling pathway, resulting in the formation of a biofilm. In addition, the development of biofilms can migrate to bacterial communities composed of MIC-causing bacteria and eventually cause MIC on the surface of metal joints. *Pseudomonas* species have been isolated and identified from various sources, such as soil, fresh water, and metal surfaces, and several species are suggested to be MIC-causing strains because those species can obtain energy by transferring electrons to extracellular solid substances [31]. In fact, we also observed *Pseudomonas* sp. 1, which shows high identity with the MIC-causing strain *P. koreensis* Ps 9-14^{T} [32]. Thus, the relative abundance of *Pseudomonas* sp. 1 in the biofilm may increase as time proceeds, which then encourages MIC on the surface of the metal joint.

Figure 4. Presence/absence matrix of biofilm formation pathway genes in *A. hydrocarboniclasticus* $H1^T$ and *A. illinoisensis* DSM 15370^T. Strains: 1, *A. hydrocarboniclasticus* $H1^T$; 2, *A. illinoisensis* DSM 15370^T. Green, orange, and gray squares indicate cases where the genes are conserved, similar genes are conserved, and genes are not conserved, respectively.

4. Conclusions

Bacterial community analyses of metal corrosion-free biofilms have rarely been reported, though analyses within different environments are required to understand the mechanism of MIC. Unlike previous studies on MIC, *Alkanindiges* bacteria were observed as the dominant species in this study. Moreover, the gene clusters required to cause metal corrosion were not conserved in the genome sequences of known *Alkanindiges* bacteria. Those findings indicated that the bacterial community changes significantly during the process from development of biofilm to occurrence of metal corrosion. For more detailed investigations, it will be necessary to analyze changes in the bacterial community over time.

Author Contributions: Conceptualization, H.A.; methodology, H.A. and Y.S.; validation, H.A. and Y.S.; formal analysis, H.A. and Y.S.; investigation, H.A. and Y.S.; resources, H.A.; data curation, H.A. and Y.S.; writing—original draft preparation, H.A.; writing—review and editing, H.A., Y.S., and Z.-i.K.; visualization, H.A.; supervision, H.A. and Z.-i.K.; project administration, H.A.; funding acquisition, H.A. and Z.-i.K. All authors have read and agreed to the published version of the manuscript.

Funding: This work was supported by a grant from The Salt Science Research Foundation (No. 2101).

Institutional Review Board Statement: Not applicable.

Informed Consent Statement: Not applicable.

Data Availability Statement: The 16S rRNA gene sequences of *A. hongkongensis* HKU9^T, *A. hydrocarboniclasticus* H1^T, *A. illinoisensis* MVAB Hex1^T, *C. vibrioides* DSM 9893^T, *E. coli* JCM 1649^T, *M. lacunata* ATCC 17967^T, *P. aeruginosa* DSM 50071^T, *P. koreensis* 9-14^T, *S. jeddahensis* G39^T, *S. paucimobilis* DSM 1098^T, and *S. spiritivorum* JCM 1733^T are available in the GenBank/EMBL/DDBJ databases under accession numbers NR_114676, LT827059, NR_025254, NR_037099, NR_024570, NR_036825, AF094713, NR_025228, NR_158139, LN681566, and LC060921, respectively. Draft genome sequences of *A. hydrocarboniclasticus* H1^T and *A. illinoisensis* DSM 15370^T were deposited in DDBJ/EMBL/GenBank databases under accession numbers MLCN01000001–MLCN01000092 and JHUX01000001–JHUX01000012, respectively.

Acknowledgments: We are grateful to all members of the Bio-conversion Research Group at our Institute (Research Institute for Sustainable Chemistry, National Institute of Advanced Industrial Science and Technology (AIST)) for their technical assistance and valuable discussion.

Conflicts of Interest: The authors declare no conflict of interest.

References

1. Karygianni, L.; Ren, Z.; Koo, H.; Thurnheer, T. Biofilm matrixome: Extracellular components in structured microbial communities. *Trends Microbiol.* **2020**, *28*, 668–681. [CrossRef] [PubMed]
2. Meiying, L.; Min, D. A review: Microbiologically influenced corrosion and the effect of cathodic polarization on typical bacteria. *Rev. Environ. Sci. Biotechnol.* **2018**, *17*, 431–446.
3. Jia, R.; Unsal, T.; Xu, D.; Lekbach, Y.; Gu, T. Microbiologically influenced corrosion and current mitigation strategies: A state of the art review. *Int. Biodeterior. Biodegradation* **2019**, *137*, 42–58. [CrossRef]
4. Koch, G.; Varney, J.; Thompson, N.; Moghissi, O.; Gould, M.; Payer, J. *International Measures of Prevention, Application, and Economics of Corrosion Technologies Study*; Jacobson, G., Ed.; NACE International: Houston, TX, USA, 2016; pp. 1–216.
5. Procópio, L. Microbial community profiles grown on 1020 carbon steel surfaces in seawater-isolated microcosm. *Ann. Microbiol.* **2020**, *70*, 13. [CrossRef]
6. Capão, A.; Moreira-Filho, P.; Garcia, M.; Bitati, S.; Procópio, L. Marine bacterial community analysis on 316L stainless steel coupons by Illumina MiSeq sequencing. *Biotechnol. Lett.* **2020**, *42*, 1431–1448. [CrossRef]
7. Lv, M.; Du, M.; Li, Z. Investigation of mixed species biofilm on corrosion of X65 steel in seawater environment. *Bioelectrochemistry* **2022**, *143*, 107951. [CrossRef] [PubMed]
8. Procópio, L. The role of biofilms in the corrosion of steel in marine environments. *World J. Microbiol. Biotechnol.* **2019**, *35*, 73. [CrossRef] [PubMed]
9. Gołębiewski, M.; Tretyn, A. Generating amplicon reads for microbial community assessment with next-generation sequencing. *J. Appl. Microbiol.* **2020**, *128*, 330–354. [CrossRef]

10. Johnson, J.S.; Spakowicz, D.J.; Hong, B.Y.; Petersen, L.M.; Demkowicz, P.; Chen, L.; Leopold, S.R.; Hanson, B.M.; Agresta, H.O.; Gerstein, M.; et al. Evaluation of 16S rRNA gene sequencing for species and strain-level microbiome analysis. *Nat. Commun.* **2019**, *10*, 5029. [CrossRef]
11. Muyzer, G.; de Waal, E.C.; Uitterlinden, A.G. Profiling of complex microbial populations by denaturing gel electrophoresis analysis of polymerase chain reaction amplified genes coding for 16S rRNA. *Appl. Environ. Microbiol.* **1993**, *59*, 695–700. [CrossRef]
12. Herlemann, D.P.; Labrenz, M.; Jürgens, K.; Bertilsson, S.; Waniek, J.J.; Andersson, A.F. Transitions in bacterial communities along the 2000 km salinity gradient of the Baltic Sea. *ISME J.* **2011**, *5*, 1571–1679. [CrossRef] [PubMed]
13. Bolyen, E.; Rideout, J.R.; Dillon, M.R.; Bokulich, N.A.; Abnet, C.C.; Al-Ghalith, G.A.; Alexander, H.; Alm, E.J.; Arumugam, M.; Asnicar, F.; et al. Reproducible, interactive, scalable and extensible microbiome data science using QIIME 2. *Nat. Biotechnol.* **2019**, *37*, 852–857. [CrossRef] [PubMed]
14. Andrews, S. FastQC: A Quality Control Tool for High Throughput Sequence Data. 2010. Available online: http://www.bioinformatics.babraham.ac.uk/projects/fastqc (accessed on 30 November 2021).
15. Callahan, B.J.; McMurdie, P.J.; Rosen, M.J.; Han, A.W.; Johnson, A.J.; Holmes, S.P. DADA2: High-resolution sample inference from Illumina amplicon data. *Nat. Methods* **2016**, *13*, 581–583. [CrossRef] [PubMed]
16. Quast, C.; Pruesse, E.; Yilmaz, P.; Gerken, J.; Schweer, T.; Yarza, P.; Peplies, J.; Glöckner, F.O. The SILVA ribosomal RNA gene database project: Improved data processing and web-based tools. *Nucleic Acids Res.* **2013**, *41*, D590–D596. [CrossRef]
17. Bokulich, N.A.; Kaehler, B.D.; Rideout, J.R.; Dillon, M.; Bolyen, E.; Knight, R.; Huttley, G.A.; Caporaso, J.G. Optimizing taxonomic classification of marker-gene amplicon sequences with QIIME 2's q2-feature-classifier plugin. *Microbiome* **2018**, *6*, 90. [CrossRef]
18. Tamura, K.; Stecher, G.; Kumar, S. MEGA11: Molecular evolutionary genetics analysis version 11. *Mol. Biol. Evol.* **2021**, *38*, 3022–3027. [CrossRef]
19. Bogan, B.W.; Sullivan, W.R.; Kayser, K.J.; Derr, K.D.; Aldrich, H.C.; Paterek, J.R. *Alkanindiges illinoisensis* gen. nov., sp. nov., an obligately hydrocarbonoclastic, aerobic squalane-degrading bacterium isolated from oilfield soils. *Int. J. Syst. Evol. Microbiol.* **2003**, *53*, 1389–1395. [CrossRef]
20. Zheng, J.; Feng, J.Q.; Zhou, L.; Mbadinga, S.M.; Gu, J.D.; Mu, B.Z. Characterization of bacterial composition and diversity in a long-term petroleum contaminated soil and isolation of high-efficiency alkane-degrading strains using an improved medium. *World J. Microbiol. Biotechnol.* **2018**, *34*, 34. [CrossRef]
21. Vergeynst, L.; Greer, C.W.; Mosbech, A.; Gustavson, K.; Meire, L.; Poulsen, K.G.; Christensen, J.H. Biodegradation, photo-oxidation, and dissolution of petroleum compounds in an arctic fjord during summer. *Environ. Sci. Technol.* **2019**, *53*, 12197–12206. [CrossRef]
22. Liang, J.; Gao, S.; Wu, Z.; Rijnaarts, H.H.M.; Grotenhuis, T. DNA-SIP identification of phenanthrene-degrading bacteria undergoing bioaugmentation and natural attenuation in petroleum-contaminated soil. *Chemosphere* **2021**, *266*, 128984. [CrossRef]
23. Fernández-Brime, S.; Muggia, L.; Maier, S.; Grube, M.; Wedin, M. Bacterial communities in an optional lichen symbiosis are determined by substrate, not algal photobionts. *FEMS Microbiol. Ecol.* **2019**, *95*, fiz012. [CrossRef] [PubMed]
24. Sierra, M.A.; Danko, D.C.; Sandoval, T.A.; Pishchany, G.; Moncada, B.; Kolter, R.; Mason, C.E.; Zambrano, M.M. The Microbiomes of seven lichen genera reveal host specificity, a reduced core community and potential as source of antimicrobials. *Front. Microbiol.* **2020**, *11*, 398. [CrossRef] [PubMed]
25. Nakajima, H.; Yamamoto, Y.; Yoshitani, A.; Itoh, K. Effect of metal stress on photosynthetic pigments in the Cu-hyperaccumulating lichens *Cladonia humilis* and *Stereocaulon japonicum* growing in Cu-polluted sites in Japan. *Ecotoxicol. Environ. Saf.* **2013**, *97*, 154–159. [CrossRef]
26. Huang, Y.; Zhou, E.; Jiang, C.; Jia, R.; Liu, S.; Xu, D.; Gu, T.; Wang, F. Endogenous phenazine-1-carboxamide encoding gene *PhzH* regulated the extracellular electron transfer in biocorrosion of stainless steel by marine *Pseudomonas aeruginosa*. *Electrochem. Commun.* **2018**, *94*, 9–13. [CrossRef]
27. Huang, L.; Huang, Y.; Lou, Y.; Qian, H.; Xu, D.; Ma, L.; Jiang, C.; Zhang, D. Pyocyanin-modifying genes *phzM* and *phzS* regulated the extracellular T electron transfer in microbiologically-influenced corrosion of X80 carbon steel by *Pseudomonas aeruginosa*. *Corros. Sci.* **2020**, *164*, 108355. [CrossRef]
28. Tsurumaru, H.; Ito, N.; Mori, K.; Wakai, S.; Uchiyama, T.; Iino, T.; Hosoyama, A.; Ataku, H.; Nishijima, K.; Mise, M.; et al. An extracellular [NiFe] hydrogenase mediating iron corrosion is encoded in a genetically unstable genomic island in Methanococcus maripaludis. *Sci. Rep.* **2018**, *8*, 15149. [CrossRef]
29. Kanehisa, M.; Goto, S. KEGG: Kyoto encyclopedia of genes and genomes. *Nucleic Acids Res.* **2000**, *28*, 27–30. [CrossRef]
30. Thi, M.T.T.; Wibowo, D.; Rehm, B.H.A. *Pseudomonas aeruginosa* Biofilms. *Int. J. Mol. Sci.* **2020**, *21*, 8671. [CrossRef]
31. Chugh, B.; Sheetal; Singh, M.; Thakur, S.; Pani, B.; Singh, A.K.; Sajil, V.S. Extracellular electron transfer by *Pseudomonas aeruginosa* in biocorrosion: A review. *ACS Biomater. Sci. Eng.* **2022**, *8*, 1049–1059. [CrossRef]
32. Raut, I.; Calin, M.; Oancea, F.; Arsene, M.L.; Jecu, L. Isolation and identification of microbial strains involved in industrial systems materials corrosion in aquatic environment. *Rev. Roum. Chim.* **2013**, *58*, 59–64.

applied biosciences

Opinion

Does Intestine Morphology Still Have Secrets to Reveal? A Proposal about the "Ghost" Layer of the Bowel

Francesco Cappello [1,2,*], Dario Saguto [1], Stefano Burgio [1], Letizia Paladino [1,2] and Fabio Bucchieri [1]

1 Department of Biomedicine, Neuroscience and Advanced Diagnostics, University of Palermo, 90127 Palermo, Italy; dariosaguto@gmail.com (D.S.); stefano.burgio94@gmail.com (S.B.); letizia.paladino91@gmail.com (L.P.); fabio.bucchieri@unipa.it (F.B.)
2 Euro-Mediterranean Institute of Science and Technology, 90139 Palermo, Italy
* Correspondence: francesco.cappello@unipa.it

Abstract: In this brief Opinion paper, the term "muco-microbiotic layer" is introduced to describe the innermost layer of the intestinal wall. This layer may contribute not only to the overall health of the bowel, but also to that of extraintestinal organs. Its constituents, in terms of soluble molecules and nanovesicles, need to be studied further. Moreover, one can hypothesize the existence of an analogous layer in other organs, such as the airways or some parts of the genital tracts. Further studies on it are needed.

Keywords: microbiota; bowel; muco-microbiotic layer; cell differentiation; tissue homeostasis; organ remodeling; human anatomy; histology; pathophysiology; nanovesicles; exosomes

Citation: Cappello, F.; Saguto, D.; Burgio, S.; Paladino, L.; Bucchieri, F. Does Intestine Morphology Still Have Secrets to Reveal? A Proposal about the "Ghost" Layer of the Bowel. *Appl. Biosci.* **2022**, *1*, 95–100. https://doi.org/10.3390/applbiosci1010006

Academic Editor: Robert Henry

Received: 13 May 2022
Accepted: 6 June 2022
Published: 7 June 2022

1. An Anatomical Introduction

In Anatomy and Histology treatises, within the chapters on the structure of the intestine, little space is left to uncertainty and the readers' imagination, whether they be young students or expert scientists: the intestinal wall—just like the other organs of the alimentary canal—consists of four overlapping tunicae as if they were four soft cylinders inserted one inside the other, which have been named—from the inner to the outer—"mucosa", "submucosa", "muscularis propria" and "adventitia" (the latter incorporated in the submesothelial connective tissue of the serosa, the peritoneum, when present) [1].

Particular attention is given to the mucosa, usually defined as the innermost tunic, which is, in turn, composed of three layers, all of the equal importance: the epithelium (the layer facing the lumen), whose cellular elements morpho-functionally better characterize the two portions into which it is divided the intestine (i.e., small and large); the lamina propria, containing not only the blood vessels for tissue trophism but also lymphatics, immune cells, undifferentiated mesenchymal cells, nerve fibers, and other cells and structures that harbor—in some parts—glandular introflexions derived from the lining epithelium; and, last but not least, the muscularis mucosae, a thin bundle of smooth muscle fiber cells, which has the dual purpose of protecting the submucosa (more richly vascularized and innervated than the lamina propria) from chemical or traumatic insults that can affect the more abluminal layers on one hand, and to physically separate the two connective layers of the lamina propria and submucosa, thus actively participating in the creation of two different microenvironments, on the other [1].

The concept of the microenvironment becomes more relevant as our knowledge of the constituents of the human body increases, since it is now clear that cells can differentiate, perform metabolic functions, communicate with each other, and cease (more or less voluntarily) their existence based on the chemical–physical stimuli provided by the environment [2].

2. The Microbiota and Its Microenvironment

In the last few decades, a further constituent of the intestine structure has emerged in the scientific literature: the microbiota, that is, the array of all living elements, bacterial and not, that colonize our intestine, not having been generated together with it—therefore not sharing their genetic heritage with the cells of the individual who hosts them—but having reached a saprophytic balance with it [3]. The constituents of microbiotic flora are presented in Figure 1 and Tables 1–3.

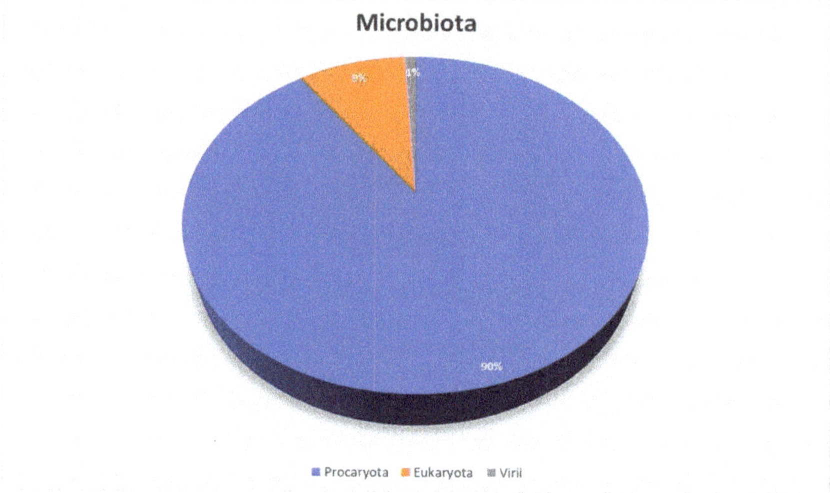

Figure 1. The microbiotic flora. The microbiota is mainly made up of prokaryotic elements (90%) and, in a lesser extent, of eukaryotic (9%) and virii (1%) species. See Tables 1–3 for more information about its composition.

The microbiota is defined by some—improperly, from a strictly morphological point of view—as "the organ within the organ"; instead, we would like to redefine and reframe it in a morpho-functionally more appropriate way in the short text that follows. To do this, we have to start by correlating the "morphology" with the "pathophysiology". Thus, please forgive us for the slightly pedantic nature of the following paragraphs.

The human body can be subdivided into apparatuses (or systems), which can be split into organs, each formed of tissues, in turn composed of cells, each specialized for their function (just remember how different they are, morpho-functionally, not just a neuron from a muscle cell, but also an absorbing epithelial cell from a secreting one) [4].

The alteration of cellular differentiation triggers the disorganization of the tissue, which, in turn, causes organ malfunction, causing the organism to display signs of illness. Physicians, through the knowledge of topographical anatomy, are called to identify the organ and/or the anatomical area affected by the disease, using semiotic tools (physical and instrumental); the knowledge of the structures of the human body permits them to identify where those natural mechanisms regulating cell differentiation, tissue homeostasis and organ remodeling (all phenomena that continue throughout life, both pre- and postnatally, including not only infancy and adolescence but also tissue senescence and organ meiopragia later in life) are malfunctioning [5].

Table 1. The Prokaryotic kingdom and their phyla in the human microbiota. The percentages derive from literature data.

Procaryota				
Kingdoms	**Phylum**	**Genus**	**(%)**	**References**
Bacteria	Firmicutes	Lactobacillus	5%	[6–12]
		Faecalibacterium	5%	[12]
		Halomonas	3%	[12]
		Clostridium	3%	[12]
		Ruminococcus	7%	[12]
		Roseburia	4%	[12]
		Streptococcus	2%	[12]
		Others	21%	[12]
	Bacteroidetes	Bacteroides	19%	[6–13]
		Others	2%	[12]
	Actinobacteria	Bifidobacterium	4%	[9,12,14]
		Others	1%	[12]
	Proteobacteria	Escherichia	5%	[10,13]
		Shigella	2%	[10]
		Others	9%	[13]
	Spirochaetes	Brachyspira	3%	[12]
		Others	2%	[12]
	Others	Others	2%	[11]
Archea	Euryarchaeota	Halobacteria	2%	[15]

Table 2. The Eukaryotic kingdom and their phyla in the human microbiota. The percentages derive from literature data.

Eukaryota			
Phylum	**Genus**	**(%)**	**References**
Ascomycota	Debaryomycetaceae	13%	[16–18]
	Dipodascaceae	14%	[17]
	Saccharomycetaceae	13%	[17]
Basidiomycota	Tremellomycetes	20%	[16]
	Malasseziaceae	20%	[19]
Mucorimycota	Fusicatenibacter	5%	[17]
	Aspergiullus	5%	[19]

Table 3. The Virii kingdom in the human microbiota. The percentages derive from literature data.

Virii			
Phylum	**Genus**	**(%)**	**References**
Caudovirales	Siphoviridae	20%	[11,20–23]
	Myoviridae	20%	[17,22,23]
	Podoviridae	20%	[17,22,23]
	Others	40%	[11,22,23]

The identification of "breakdown" at the cellular level can lead to targeted therapy (sometimes just lifestyle changes are enough; other times, drug therapy may be necessary; and, in extremis, surgical intervention could be inevitable for the demolition and eventual reconstruction of the irremediably sick part), and the efforts of the biomedical and bioengineering community are concentrated above all on developing such therapy protocols.

It now seems certain that the microbiota is involved in many intestinal pathologies, therefore implying that it can transform itself from a combination of commensal germs to a mix of pathogens under certain conditions and for causes that are not yet fully understood [24–27]. What mainly affects the health of the microbiota is certainly the microenvironment in which we live and, in the first place, our diet: what passes through the intestinal lumen throughout our existence. However, to better understand how and why the microbiota is altered and, above all, how and why this alteration affects not only intestinal homeostasis but also that of other tissues physically distant from it, the context of its microenvironment, i.e., the mucus that lines the epithelial surface of the intestine and in which the microbiota is contained, must be contemplated [28].

3. The Muco-Microbiotic Layer

Mucus and microbiota form a morpho-functional layer, the biological elements of the microbiota being comparable to a "cellular component", and the mucus to the "matrix" in which not only these cells reside, but also in which their products—both soluble molecules (proteins, lipids, etc.) and nanovesicles (for example, the "outer membrane vesicles", OMV, of prokaryotic origin, equivalent to the exosomes produced by eukaryotic cells)—are released [28,29].

Figure 2 shows, schematically, the composition of the muco-microbiotic (MuMi) layer with its main constituents. Why has the MuMi layer been ignored until now, and why has it not yet been described in anatomy and histology books? This is probably due to the fact that widely used histochemical stains contain alcohol, which solubilizes the mucus by sweeping it away along with the microbiota. Therefore, in histological sections, neither of these two components are visible, if not in minute traces, and this has led scientists to dismiss it as a physiological constituent of our bodies.

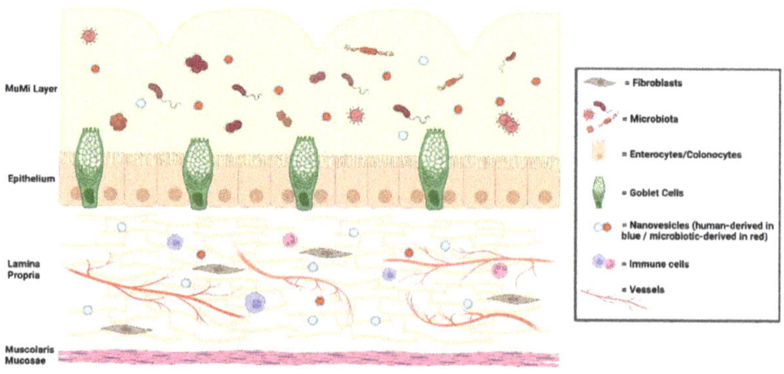

Figure 2. The muco-microbiotic (MuMi) layer. This figure shows, schematically, the MuMi layer, its main constituents (in terms of microbiotic elements and nanovesicles) and its relationships with the underlying mucosal components, i.e., epithelial cells and lamina propria. Whereas nanovesicles, either human- and microbiota-derived ones, can easily reach the lamina propria (in turn contributing to determine its microenvironment), the muscularis mucosae represents a wall between the connective tissue of the lamina propria and that of the submucosa (not shown).

The MuMi layer is therefore located in the bowel internally to the mucosa, representing the innermost "fifth layer" (or the first one) of the intestinal wall; the alterations of the mucous matrix can cause a qualitative and quantitative modification of the populations of germs that housed in it, and vice versa [28,29]. In addition, the intense traffic of soluble products and nano-vesicles mutually influence human and microbiotic cells; the increase in pathogenic bacteria generates, for example, a disturbance in the microenvironment of the lamina propria, triggering or perpetuating inflammatory processes; and since the nanovesicles (both human and bacterial) can easily reach both the bloodstream and the lymphatic one, the well-being of our intestinal MuMi layer can affect the homeostasis of extra-intestinal tissues, thus compromising the function of other organs, including the liver, lung, heart, and brain, just to mention the most important ones considering the anatomical progression of the blood mass originating from the abdominal splanchnic district [28].

4. Conclusions

Based on the consideration reported above, the existence of the MuMi layer, in our modest opinion, should not only no longer be ignored, but must also be investigated (also in other anatomical districts, e.g., the airways or some genital tract organs) in order to characterize, from a morpho-molecular point of view, its constituents and the mechanisms that fine-tune all the related physiological and pathophysiological events.

Limiting our conclusion to the bowel, if the Hippocratic axiom "we are what we eat" is still true, and if it is correct to think that the health of our body depends, in the first instance, on the health of our intestine, omitting the observations of molecular phenomena that occur in this layer from scientific reasoning—as well as from scientific protocols—would undermine the achievement of better knowledge on human health.

Author Contributions: Conceptualization, F.C. and F.B.; methodology, S.B. and L.P.; software, S.B. and D.S.; writing—original draft preparation, F.C., S.B. and L.P.; writing—review and editing, D.S. and F.B. All authors have read and agreed to the published version of the manuscript.

Funding: This research received no external funding.

Institutional Review Board Statement: Not applicable.

Informed Consent Statement: Not applicable.

Data Availability Statement: Not applicable.

Conflicts of Interest: The authors declare no conflict of interest.

References

1. Gartner, L. *Textbook of Histology*, 5th ed.; Elsevier: Amsterdam, The Netherlands, 2020.
2. Jiang, T.; Yang, T.; Chen, Y.; Miao, Y.; Xu, Y.; Jiang, H.; Yang, M.; Mao, C. Emulating interactions between microorganisms and tumor microenvironment to develop cancer theranostics. *Theranostics* **2022**, *12*, 2833–2859. [CrossRef] [PubMed]
3. Suzuki, T.A.; Ley, R.E. The role of the microbiota in human genetic adaptation. *Science* **2020**, *370*, eaaz6827. [CrossRef] [PubMed]
4. Standring, S. *Gray's Anatomy. The Anatomical Basis of Clinical Practice*, 42nd ed.; Elsevier: Amsterdam, The Netherlands, 2020.
5. Campanella, C.; Caruso Bavisotto, C.; Logozzi, M.; Marino Gammazza, A.; Mizzoni, D.; Cappello, F.; Fais, S. On the Choice of the Extracellular Vesicles for Therapeutic Purposes. *Int. J. Mol. Sci.* **2019**, *20*, 236. [CrossRef]
6. Bäckhed, F.; Ley, R.E.; Sonnenburg, J.L.; Peterson, D.A.; Gordon, J.I. Host-bacterial mutualism in the human intestine. *Science* **2005**, *307*, 1915–1920. [CrossRef]
7. Eckburg, P.B.; Bik, E.M.; Bernstein, C.N.; Purdom, E.; Dethlefsen, L.; Sargent, M.; Gill, S.R.; Nelson, K.E.; Relman, D.A. Diversity of the Human Intestinal Microbial Flora. *Science* **2005**, *308*, 1635–1638. [CrossRef]
8. Kim, J.Y.; Whon, T.W.; Lim, M.Y.; Kim, Y.B.; Kim, N.; Kwon, M.S.; Kim, J.; Lee, S.H.; Choi, H.J.; Nam, I.H.; et al. The human gut archaeome: Identification of diverse haloarchaea in Korean subjects. *Microbiome* **2020**, *8*, 114. [CrossRef] [PubMed]
9. King, C.H.; Desai, H.; Sylvetsky, A.C.; LoTempio, J.; Ayanyan, S.; Carrie, J.; Crandall, K.A.; Fochtman, B.C.; Gasparyan, L.; Gulzar, N.; et al. Baseline human gut microbiota profile in healthy people and standard reporting template. *PLoS ONE* **2019**, *14*, e0206484. [CrossRef]
10. Lin, A.; Bik, E.M.; Costello, E.K.; Dethlefsen, L.; Haque, R.; Relman, D.A.; Singh, U. Distinct distal gut microbiome diversity and composition in healthy children from Bangladesh and the United States. *PLoS ONE* **2013**, *8*, e53838. [CrossRef] [PubMed]
11. Nayfach, S.; Páez-Espino, D.; Call, L.; Low, S.J.; Sberro, H.; Ivanova, N.N.; Proal, A.D.; Fischbach, M.A.; Bhatt, A.S.; Hugenholtz, P.; et al. Metagenomic compendium of 189,680 DNA viruses from the human gut microbiome. *Nat. Microbiol.* **2021**, *6*, 960–970. [CrossRef]

12. Zhang, Q.; Zhao, H.; Wu, D.; Cao, D.; Ma, W. A comprehensive analysis of the microbiota composition and gene expression in colorectal cancer. *BMC Microbiol.* **2020**, *20*, 308. [CrossRef] [PubMed]
13. Manrique, P.; Dills, M.; Young, M.J. The Human Gut Phage Community and Its Implications for Health and Disease. *Viruses* **2017**, *9*, 141. [CrossRef]
14. Miller, T.L.; Wolin, M.J. Methanogens in human and animal intestinal Tracts. *Syst. Appl. Microbiol.* **1986**, *7*, 223–229. [CrossRef]
15. Hoffmann, C.; Dollive, S.; Grunberg, S.; Chen, J.; Li, H.; Wu, G.D.; Lewis, J.D.; Bushman, F.D. Archaea and fungi of the human gut microbiome: Correlations with diet and bacterial residents. *PLoS ONE* **2013**, *8*, e66019. [CrossRef]
16. Li, J.; Chen, D.; Yu, B.; He, J.; Zheng, P.; Mao, X.; Yu, J.; Luo, J.; Tian, G.; Huang, Z.; et al. Fungi in Gastrointestinal Tracts of Human and Mice: From Community to Functions. *Microb. Ecol.* **2018**, *75*, 821–829. [CrossRef]
17. Raimondi, S.; Amaretti, A.; Gozzoli, C.; Simone, M.; Righini, L.; Candeliere, F.; Brun, P.; Ardizzoni, A.; Colombari, B.; Paulone, S.; et al. Longitudinal Survey of Fungi in the Human Gut: ITS Profiling, Phenotyping, and Colonization. *Front. Microbiol.* **2019**, *10*, 1575. [CrossRef]
18. Kühbacher, T.; Ott, S.J.; Helwig, U.; Mimura, T.; Rizzello, F.; Kleessen, B.; Gionchetti, P.; Blaut, M.; Campieri, M.; Fölsch, U.R.; et al. Bacterial and fungal microbiota in relation to probiotic therapy (VSL#3) in pouchitis. *Gut* **2006**, *55*, 833–841.
19. Underhill, D.M.; Iliev, I.D. The mycobiota: Interactions between commensal fungi and the host immune system. *Nat. Rev. Immunol.* **2014**, *14*, 405–416. [CrossRef]
20. Abeles, S.R.; Pride, D.T. Molecular bases and role of viruses in the human microbiome. *J. Mol. Biol.* **2014**, *426*, 3892–3906. [CrossRef]
21. Mayneris-Perxachs, J.; Castells-Nobau, A.; Arnoriaga-Rodríguez, M.; Garre-Olmo, J.; Puig, J.; Ramos, R.; Martínez-Hernández, F.; Burokas, A.; Coll, C.; Moreno-Navarrete, J.M.; et al. Caudovirales bacteriophages are associated with improved executive function and memory in flies, mice, and humans. *Cell Host Microbe* **2022**, *30*, 340–356.e348. [CrossRef]
22. Zhang, T.; Breitbart, M.; Lee, W.H.; Run, J.Q.; Wei, C.L.; Soh, S.W.; Hibberd, M.L.; Liu, E.T.; Rohwer, F.; Ruan, Y. RNA viral community in human feces: Prevalence of plant pathogenic viruses. *PLoS Biol.* **2006**, *4*, e3. [CrossRef]
23. Scarpellini, E.; Ianiro, G.; Attili, F.; Bassanelli, C.; De Santis, A.; Gasbarrini, A. The human gut microbiota and virome: Potential therapeutic implications. *Dig. Liver Dis. Off. J. Ital. Soc. Gastroenterol. Ital. Assoc. Study Liver* **2015**, *47*, 1007–1012. [CrossRef]
24. Bellavia, M.; Tomasello, G.; Romeo, M.; Damiani, P.; Lo Monte, A.I.; Lozio, L.; Campanella, C.; Marino Gammazza, A.; Rappa, F.; Zummo, G.; et al. Gut microbiota imbalance and chaperoning system malfunction are central to ulcerative colitis pathogenesis and can be counteracted with specifically designed probiotics: A working hypothesis. *Med. Microbiol. Immunol.* **2013**, *202*, 393–406. [CrossRef]
25. Untersmayr, E.; Brandt, A.; Koidl, L.; Bergheim, I. The Intestinal Barrier Dysfunction as Driving Factor of Inflammaging. *Nutrients* **2022**, *14*, 949. [CrossRef] [PubMed]
26. Stolfi, C.; Maresca, C.; Monteleone, G.; Laudisi, F. Implication of Intestinal Barrier Dysfunction in Gut Dysbiosis and Diseases. *Biomedicines* **2022**, *10*, 289. [CrossRef] [PubMed]
27. Nandwana, V.; Nandwana, N.K.; Das, Y.; Saito, M.; Panda, T.; Das, S.; Almaguel, F.; Hosmane, N.S.; Das, B.C. The Role of Microbiome in Brain Development and Neurodegenerative Diseases. *Molecules* **2022**, *27*, 3402. [CrossRef]
28. Cappello, F.; Rappa, F.; Canepa, F.; Carini, F.; Mazzola, M.; Tomasello, G.; Bonaventura, G.; Giuliana, G.; Leone, A.; Saguto, D.; et al. Probiotics Can Cure Oral Aphthous-Like Ulcers in Inflammatory Bowel Disease Patients: A Review of the Literature and a Working Hypothesis. *Int. J. Mol. Sci.* **2019**, *20*, 5026. [CrossRef]
29. Cappello, F.; Mazzola, M.; Jurjus, A.; Zeenny, M.N.; Jurjus, R.; Carini, F.; Leone, A.; Bonaventura, G.; Tomasello, G.; Bucchieri, F.; et al. Hsp60 as a Novel Target in IBD Management: A Prospect. *Front. Pharmacol.* **2019**, *10*, 26. [CrossRef] [PubMed]

applied biosciences

MDPI

Communication

Boronium Salt as an Antiviral Agent against Enveloped Viruses Influenza A and SARS-CoV-2

Terrence J. Ravine [1,*], Jonathan O. Rayner [2], Rosemary W. Roberts [2], James H. Davis, Jr. [3] and Mohammad Soltani [3]

1 Department of Biomedical Sciences, University of South Alabama, Mobile, AL 36688, USA
2 Department of Microbiology & Immunology, University of South Alabama, Mobile, AL 36688, USA
3 Department of Chemistry, University of South Alabama, Mobile, AL 36688, USA
* Correspondence: travine@southalabama.edu

Abstract: Quaternary ammonium compounds (QACs) are routinely used as disinfectants in a variety of settings. They are generally effective against a wide range of microbes but often exhibit undesirable toxicity. Consequently, companies are constantly seeking alternatives to QACs that are just as effective but with reduced health and environmental hazards. Two boronium salt derivatives were tested against influenza A and SARS-CoV-2 viruses. One salt possessed a terminal benzyl group, while the other lacked the same terminal benzyl group. Both salts demonstrated virus inactivation similar to a commercial QAC disinfectant. The non-benzylated form exhibited the same cell toxicity profile as the QAC. However, the benzylated form displayed less cell toxicity than both the non-benzylated form and QAC. These results suggest that the boronium salts may be suitable for use as a disinfecting agent against enveloped viruses in lieu of using a QAC. Continued evaluation of the boronium salts is warranted to determine the lowest effective concentration capable of effectively controlling influenza A and SARS-CoV-2 viruses that also demonstrates low cytotoxicity.

Keywords: boronium ion; disinfectants; quaternary ammonium compound; cytotoxicity; antiviral

Citation: Ravine, T.J.; Rayner, J.O.; Roberts, R.W.; Davis, J.H., Jr.; Soltani, M. Boronium Salt as an Antiviral Agent against Enveloped Viruses Influenza A and SARS-CoV-2. *Appl. Biosci.* **2022**, *1*, 289–298. https://doi.org/10.3390/applbiosci1030018

Academic Editor: Robert Henry

Received: 20 October 2022
Accepted: 25 November 2022
Published: 1 December 2022

1. Introduction

Quaternary ammonium compounds (R_4N^+), also known simply as quats or QACs, are effective for low-level disinfection of many surfaces. Most quats are cationic compounds containing the organic salt benzalkonium chloride. QACs are generally soluble in water and alcohol as long as their chain length does not exceed C14 [1]. They act by disrupting cell membranes, allowing intracellular contents to exit the cell. They are mainly bacteriostatic when used at low concentrations and are bactericidal at higher concentrations. Commercial QACs are usually seen as a mixture of compounds with differing alkyl chain lengths, ranging from C8 to C18. Greater biocide activity is seen with C12 and C14 lengths [2].

QACs are widely used as antiseptics, disinfectants, preservatives, and sterilization agents in a variety of settings, such as homes, medical facilities, and water treatment facilities. They are also extensively used in both the textile and food industries. QACs are also employed as herbicides and pesticides, although environmental safety concerns are associated with their use. For instance, the popular QAC herbicide paraquat dichloride (paraquat) has been banned for use in several countries, including the United States, due to toxicity [1].

Toxicity and biosafety profiles are both top priorities for chemical agents with possible human contact that will also eventually make their way into the environment. In this regard, the safety profile for boronium salts has yet to be established. However, it is well known that QACs such as those found in Bacdown disinfectant pose both health and environmental hazards. The Agency for Toxic Substances and Disease Registry (ATSDR) states that Category 1 agents pose an "Urgent public health hazard", while Category 2 agents carry a "Public health hazard" [3]. The Safety Data Sheet (SDS) [4] indicates the following health and environmental hazard levels:

- Skin corrosion/irritation (Category 2)
- Serious eye damage/eye irritation (Category 1)
- Hazardous to the aquatic environment, acute hazard (Category 1)
- Hazardous to the aquatic environment, long-term hazard (Category 2)

Moreover, increased global production and use of synthetic QACs carries with it similar environmental and health hazards [1]. Most notably:

- General toxicity
- Accumulation of QACs in soil, sludge/sewage, water, and plants
- Bacterial and antibiotic resistance and co-/cross-resistance
- Risk of asthma

While QACs are effective against several bacterial types, some bacteria, such as pseudomonads, are not adequately controlled by these compounds. In fact, some *Pseudomonas* strains are known to grow (multiply) in QACs. Furthermore, viruses, bacterial spores, and some fungi are recognized to be more resistant to the action of QACs and require a different type of intermediate-level disinfection [5].

Two boronium salts with neutral charges were synthesized, each containing a boronium ion attached to a C16 alkyl tail, as previously characterized [6]. One salt possessed a terminal benzyl group (B) while the other salt lacked a benzyl group (A), as is seen in Figure 1. Both salts were subsequently tested against three different bacteria. The benzylated form was demonstrated to be four times more effective than the commercial QAC Bacdown disinfectant detergent (D + E), as well as BDD (Decon Labs, King Of Prussia, PA), at inhibiting the growth of both methicillin-sensitive *Staphylococcus aureus* (MSSA) and *Escherichia coli*. Conversely, the non-benzylated form did not display growth inhibition in any test bacterium at the highest concentration tested. The most noteworthy finding was that the benzylated form was eight times more effective against *Pseudomonas aeruginosa* than the Bacdown QAC [6].

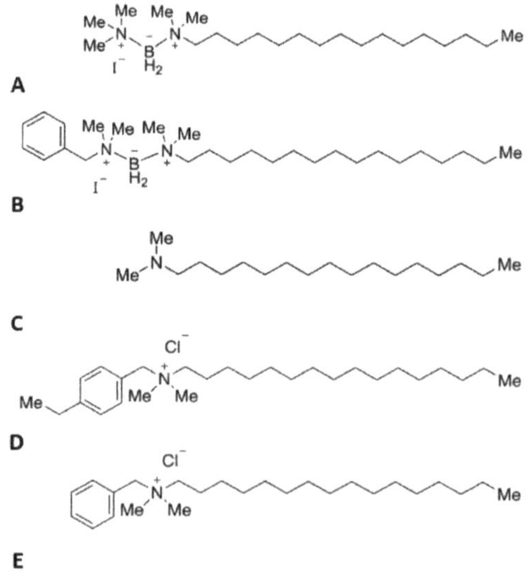

Figure 1. Test articles. Chemical structures of (**A**)—boronium salt without benzyl group ($C_{21}H_{50}BIN_2$), (**B**)—boronium salt with benzyl group ($C_{27}H_{54}BIN_2$), (**C**)—N,N-dimethylhexadecylamine, a negative control without boronium ion ($C_{18}H_{39}N$), (**D**)—alkyl dimethyl ethylbenzyl ammonium chloride ($C_{27}H_{50}ClN$), (**E**)—alkyl dimethyl benzyl ammonium chloride ($C_{25}H_{46}ClN$). Structures D and E are similar to those found in the Bacdown disinfectant, which acted as a positive control (C16 structures are shown).

Subsequently, these same two boronium salts were tested against two opportunistic fungal pathogens, namely *Candida albicans* yeast and *Aspergillus fumigatus* mold. Yeasts grows at higher temperatures closer to the human body temperature of 37 °C, whereas molds tend to favor a lower, more ambient (room) temperature of 25 °C [7]. Here, the non-benzylated form showed better control of *C. albicans* yeast growth and equal control of *A. fumigatus* mold growth when compared to the Bacdown QAC disinfectant. In contrast, the benzylated form was less effective than either the non-benzylated form or Bacdown QAC against either fungus type [8].

Differences seen in the boronium salts' antibacterial and antifungal activity are most likely attributed to variations in bacterial and fungal membranes. The underlying plasma membrane is the target region, where the bornium salts are thought to have a disruptive effect. Bacteria are classified as prokaryotes, while fungi are eukaryotes. One such membrane difference is sterols, like ergosterol present in fungal membranes, which are absent in prokaryotes. Instead, bacterial membranes contain hopanoids, a sterol-like lipid [6–8].

During maturation and prior to leaving the cell, enveloped viruses acquire their envelope from host cell membrane, such as the plasma membrane or other intracellular membranes, depending on the virus. Disruption of the outer virus envelope has been demonstrated by QACs [9]. Plasma membrane disruption is also the suspected mechanism for boronium salts' activity against bacteria and fungi [6,8]. This suggested an additional assessment of boronium salts for their ability to similarly inactivate infectivity of enveloped RNA viruses, as a potential alternative to QACs.

To accomplish this, the first step was to determine if either boronium salt exhibited any antiviral activity against two notable enveloped RNA viral pathogens, namely influenza A and SARS-CoV-2. Serial dilutions of each test article had to be prepared, but it was unknown what a reasonable starting concentration should be that would not negatively impact the health of culture cells in permissible host tissue. Results from prior bacterial and fungal studies could not be used as a predictor of what concentration would be needed to show antiviral activity. An additional study facet was to determine what effect each boronium salt had on these tissue culture cells. An estimation of cell toxicity caused by each test article was also of great interest.

The purpose of this investigation was to establish initial concentration breakpoints of each test article to be used during later definitive testing. The authors wanted to see how each boronium salt compared to an approved virucidal QAC disinfectant, such as Bacdown, when adjusted to an equal concentration and not to individual Bacdown components. During this process, both antiviral effectiveness and cell cytotoxicity were examined. We now present the results of preliminary testing using two frequently encountered viral pathogens, namely influenza A (INFA) and SARS-CoV-2.

2. Results

2.1. Virus Inactivation

The degree to which a specific test compound inactivated INFA and SARS-CoV-2 viruses was evaluated after a 1 h co-incubation, as described in the Section 4. INFA virus titers decreased in comparison to the untreated control virus when incubated with as little as 9.6 μg/mL of Bacdown QAC, the boronium salt with benzyl group, and the aliphatic compound lacking the boronium group (Table 1).

A minimum of 48 μg/mL of the boronium salt without benzyl group was required to decrease INFA titers in comparison to the untreated control. In contrast, the boronium salt w/benzyl group, boronium salt without (w/o) benzyl group, and Bacdown QAC all reduced INFA virus titers below the limit of detection (LOD) at concentrations $\geq$240 μg/mL, while the aliphatic compound w/o boronium group required concentrations $\geq$1200 μg/mL to achieve inactivation of INFA virus below the LOD.

For SARS-CoV-2, a concentration > 48 μg/mL of the boronium salt w/benzyl group, boronium salt w/o benzyl group, and Bacdown QAC was required to decrease infectious virus titers in comparison to the untreated control: however, all three demonstrated virus inactivation below the LOD at concentrations ≥ 240 µg/mL like that seen with INFA (Table 1). These results suggest that while SARS-CoV-2 virus is perhaps more stable than INFA virus when exposed to lower concentrations of these test compounds, the optimal effective concentration for eliminating both viruses is similarly between 48 and 240 µg/mL. In contrast, the aliphatic compound w/o boronium required a concentration > 240 µg/mL to decrease SARS-CoV-2 virus titers, and inactivation below the LOD required concentrations ≥ 6000 µg/mL. Therefore, the aliphatic compound w/o boronium is less effective at inactivating SARS-CoV-2 virus as compared to INFA.

Table 1. Results of test article virus inactivation capability.

Test Article	Concentration at which Virus Titer Fell below the LOD (µg/mL)	
	INFA	**SARS-CoV-2**
Quaternary ammonium compound	≥240	≥240
Boronium ion w/o benzyl group	≥240	≥240
Boronium ion w/benzyl group	≥240	≥240
Aliphatic compound w/o boronium ion	≥1200	≥6000

2.2. *Cell Toxicity*

Each of the test compounds was expected to be toxic to the cells, potentially confounding the assessment of infectious virus titers in the tissue culture assay. Thus, the degree by which each test compound alone caused cell death at each concentration was assessed in both MDCK and Vero cells. The boronium salt without benzyl group and Bacdown QAC both demonstrated MDCK cell toxicity at concentrations ≥ 240 µg/mL. In contrast, the boronium salt with benzyl group and aliphatic compound lacking the boronium group were less toxic in MDCK cells, requiring concentrations ≥ 1200 µg /mL to cause toxicity (Table 2). Vero cells in general were less subject to the toxic effects of all test compounds when compared to MDCK cells. Specifically, the boronium salt without benzyl group and Bacdown QAC both caused cytotoxicity in Vero cells at concentrations ≥ 1200 µg/mL, while the boronium salt with benzyl group required concentrations ≥ 6000 µg/mL to cause cytotoxicity. Yet, the aliphatic compound lacking the boronium group had no toxic effect on Vero cells at any of the concentrations tested (Table 2).

Table 2. Results of test article cell toxicity.

Test Article	Concentration at which Cell Toxicity Was First Observed (µg/mL)	
	MDCK	**Vero**
Quaternary ammonium compound	≥240	≥1200
Boronium ion w/o benzyl group	≥240	≥1200
Boronium ion w/benzyl group	≥1200	≥6000
Aliphatic compound w/o boronium ion	≥1200	ND

ND = none detected.

3. Discussion

Overall Assessment

Based on all assay results, the boronium salt with a terminal benzyl group (B) displayed the best combination of both virus inactivation and decreased cell toxicity when compared to Bacdown QAC (D + E). Both compounds were equally effective at inactivating INFA and SARS-CoV-2 virus, with virus titers falling below the LOD at ≥240 ug/mL. However, the benzylated boronium salt was significantly less toxic to the cells, requiring concentrations ≥ 11200 µg/mL and ≥6000 µg/mL to cause cytotoxicity in MDCK and Vero cells, respectively. Bacdown QAC caused MDCK cell toxicity at concentrations ≥ 240 µg/mL and Vero cell toxicity at concentrations ≥ 1200 µg/mL (Table 1). Importantly, the concentration at which MDCK cell toxicity was first observed for Bacdown QAC is the same as the concentration at which INFA titers fell below the LOD, supporting concerns regarding the toxicity of this compound at effective concentrations. In contrast, the boronium salt with benzyl group required greater concentrations to cause toxicity in both MDCK and Vero cells when compared to the concentration required to decrease INFA and SARS-CoV-2 titers below the LOD, suggesting that it would be safer to use as a disinfectant when compared to Bacdown QAC.

The boronium salt without the benzyl group (A) was also equally effective as Bacdown QAC (D + E) at inactivating both INFA and SARS-CoV-2 (≥240 µg/mL); however, it exhibited the same cell toxicity profile as Bacdown QAC in both MDCK (≥240 µg/mL) and Vero cells (≥1200 µg/mL), making it less attractive as an alternative to Bacdown QAC (Table 1). The aliphatic compound without boronium ion (C), in contrast, demonstrated MDCK cell toxicity at concentrations ≥ 1200 µg/mL and no toxicity on Vero cells, which is similar to or better than the boronium salt with benzyl group. However, higher concentration of the aliphatic compound without boronium ion was required for INFA (≥1200 µg/mL) and SARS-CoV-2 (≥6000 µg/mL) virus inactivation as compared to the boronium salt with benzyl group (B) or Bacdown QAC (≥240 µg/mL) negating its utility as an effective disinfectant.

With respect to the antiviral mechanism of action, QACs are well characterized as cationic surfactants containing a positively charged/hydrophilic nitrogen head and a long chain hydrophobic tail [1]. They are known to interact with key virus elements of DNA/RNA, proteins, and lipids [9]. QACs are recognized to be effective against hydrophobic enveloped viruses, such as human immunodeficiency virus, hepatitis B, herpes simplex, influenza, and severe acute respiratory syndrome (SARS-CoV). Their effectiveness is a result of interacting primarily with hydrophobic outer viral envelope phospholipids.

Both influenza and SARS-CoV-2 outer membrane phospholipids share similarity with bacterial membranes, with the exception of some variation in bacterial phosphate heads [10]. The QAC benzalkonium chloride interdigitates within bacterial and yeast plasma membranes via its alkyl chains, thereby disrupting cell membrane function [11]. Likewise, QACs are thought to disrupt viral envelopes by this same mechanism. QACs are less effective against hydrophilic nonenveloped viruses, such as adenovirus, enterovirus, rhinovirus, and rotavirus [12].

The amphiphilic character of the boronium salts (A, B) may allow them to act in a similar fashion. They could likewise act to disrupt the outer virus lipid envelope during the initial co-incubation phase prior to exposure of permissible culture cells (Figure 2). In doing so, the hydrophobic 16C alkyl chain of the boronium ion inserts itself between adjacent envelope phospholipids, thereby altering the structural integrity of the viral outer membrane. In theory, disruption of the outer membrane should result in loss of embedded INFA hemagglutinin glycoproteins to adversely affect virus attachment. This would result in decreased virus adherence to host cell membranes, preventing subsequent host cell entry.

Computational modeling had previously revealed a noticeable difference in head group dipole movement between the benzylated and non-benzylated boronium salts. This result suggested that each salt head group would interact differently with negatively-charged regions of microbe membranes [8]. The greater antiviral activity exhibited by the boronium salt with benzyl group may be related to a better ability to interact with virus envelope outer membrane phospholipids than the non-benzylated form. In this regard, the chemical nature of each boronium salt's head group is thought to account for these differences. Here, the benzylated form demonstrates more interaction with negatively-charged membrane regions of tested bacterial and fungal microbes, leading to greater membrane disruption [4,6]. Moreover, it is entirely possible that the terminal-located benzyl group may also be interfering with a different viral component, such as viral RNA and/or proteins, as previously indicated for QACs. Similar testing of rhinovirus is being considered to establish the boronium salt's effectiveness against a nonenveloped virus, where QACs are less effective. It is interesting to note that the alkyl dimethyl benzyl ammonium chloride (C12-18) in the Bacdown disinfectant also has a terminal benzyl group. The presence of the benzyl group alone did not account for the difference seen in the concentration at which cell toxicity was first observed in MDCK and Vero cells.

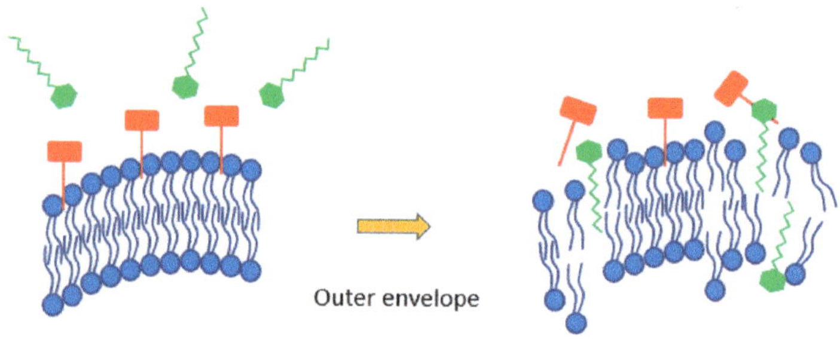

Figure 2. Schematic indicating benzylated boronium salt (green) disrupting outer viral envelope phospholipids (blue) causing a release of INFA hemagglutinin glycoproteins (red) that are required for host cell attachment.

Study limitations include testing of a low number of compound concentrations, along with the inherent limit of detection (LOD) in performed virus inactivation and cell cytotoxicity assays. Further testing will be conducted to determine the lowest effective concentration of each test article. This can be accomplished now that targets for effective test article concentrations have been established. Replicate assays assessing virus inactivation are also indicated, although the reported cell toxicity values represent the average of triplicate cultures. This will allow a statistical analysis to determine if differences in effective concentrations are significant. The use of a shorter contact period (e.g., 10–30 min) will also be examined.

The suggestion that the terminal benzyl group bound to one boronium salt (B) exerts an additional antiviral effect is purely speculative. Further testing is required to establish the exact nature of its interaction with non-enveloped virus components or DNA-containing viruses. In this regard, an examination of virus and host cell membrane integrity, pre- and post-agent exposure, may be helpful in revealing any additional virus inactivation mechanism(s). This could be accomplished using a combination of specialized techniques, such as cryo-electron microscopy of vitreous sections (CEMOVIS), electron tomography (ET), and lipid analysis by mass spectrometry [13,14].

Of particular interest was that both boronium salts (A, B) demonstrated equal or reduced cell toxicity to Bacdown QAC (D + E) in Vero cells of primate origin (African green monkey kidney) and MDKC cells of canine origin (normal kidney epithelium). These findings suggest a comparable or decreased cytotoxicity potential of the benzylated salt in human cell lines when compared to Bacdown QAC. Validation of reduced cell toxicity could be accomplished using an MTT assay. The MTT (3-[4,5-dimethylthiazol-2-yl]-2,5 diphenyl tetrazolium bromide) assay is routinely used to assess cell metabolic activity, viability, membrane integrity, and proliferation, all of which are indicators of overall cell health [15]. Here, several different human cell lines could be tested to determine the effects of either boronium salt on both cell viability and cell proliferation.

4. Materials and Methods

4.1. Cells and Viruses

Madin-Darby canine kidney cells (MDCK cells) [London Line, FR-58] were obtained from International Reagent Resource (Manassas, VA, USA). After in-house adaptation, cells were maintained at 37 °C with 5% CO_2 in serum-free media [OptiPRO SFM (Gibco, Waltham,

MA, USA) supplemented with 2X GlutaMAX (Gibco) and 1% Penicillin/Streptomycin (1% Pen/Strep) (100 units/mL Penicillin and 100 μg/mL Streptomycin, Gibco)].

Shell Vial Vero E6 Cells (SV-VeroE6 cells) [Cercopithecus aethiops kidney clone from Vero 76] (Quidel, San Diego, CA, USA) were recovered from glass coverslips and maintained at 37 °C with 5% CO_2 in complete media [Dulbecco's Modified Eagle Media (DMEM) (Lonza, Basel, Switzerland) supplemented with 10% Fetal Bovine Serum (FBS) (Millipore, Burlington, Massachusetts), 1% L-glutamine (2 mM, Lonza), and 1% Pen/Strep (100 units/mL Penicillin and 100 μg/mL Streptomycin, Lonza)].

Influenza virus strain A/PR8/34 (Identifier H1N1, Lot # 111519JHK, 4.1×10^8 Focus Forming Units (FFU)/mL) was rescued using reverse genetics methods by transfection with a 12-plasmid system of influenza A (IAV) A/PR8/34 (H1N1) strain in 293T cells [ATCC, CRL-11268], as described previously [16]. The stock of IAV was prepared by infecting rescued IAV to confluent MDCK cells in T75 flasks for 1 h at 37 °C. The inoculum was removed and MDCK cells were washed twice with Dulbecco's Phosphate Buffered Saline (DPBS) (Gibco). Fresh growth medium [(OptiPRO SFM supplemented with 2% GlutaMAX-1 and 0.5 μg/mL trypsin treated with N-tosyl-L-phenylalanine chloromethyl ketone (TPCK-trypsin)] was added, and the infected MDCK cells were incubated at 37 °C for 3 days. Harvested IAVs were filtered with a 0.45 μm polyethersulfone filter (Fisherbrand, Waltham, MA, USA), divided into aliquots, and stored at −80 °C. The stock of IAV was sequenced after PCR amplification of all 8 segments to ensure that there were no unwanted mutations (GENEWIZ, South Plainfield, New Jersey).

SARS-CoV-2 strain MOBAL-05 (Identifier MOBAL-05, Lot # 110420JOR, 2.92×10^6 Plaque Forming Units (PFU)/mL) was isolated from a clinical sample in Mobile, Alabama. Emergency Department (ED) swabs in viral media were received from University Hospital (Mobile, Alabama) and were vortexed for 30 s then allowed to settle prior to aliquoting into 1.7 mL cryovials for storage at −80 °C. Samples were thawed and 0.5 mL was inoculated per well onto approximately 80% confluent SV-VeroE6 cells in 12-well tissue culture plates (Corning, Corning, NY, USA). Plates were centrifuged at 1900× *g* for 1 h at room temperature to promote interaction between the virus and cells. Plates were incubated at 37 °C with 5% CO_2 and monitored daily for Cytopathic Effects (CPE). After 72 h, 0.2 mL of culture supernatant was transferred to a well of ~80% confluent SV-VeroE6 cells in a 6-well plate. The remaining media in the 12-well plate was removed and replaced with 1 mL maintenance media (DMEM supplemented with 2% FBS, 1% L-glutamine, and 1% Pen/Strep); a 1 mL pipette tip was used to scrape cells, pipetting to mix, and 0.2 mL was added to the corresponding supernatant in the 6-well plate. Plates were incubated at 37 °C with 5% CO_2 and monitored for CPE. After 72 h culture, supernatants were filtered with a 0.45 μm PVDF syringe filter and aliquoted into cryovials for storage at −80 °C.

Master and Working stocks were confirmed to be free from mycoplasma contamination and had endotoxin levels below 0.5 endotoxin units (EU)/mL. Influenza virus titers were determined via focus forming assay in MDCK cells. At 1 h after inoculation at 37 °C, infected MDCK cells were overlayed with a 2:1 mixture of growth medium and microcrystalline cellulose (3.6% Avicel, RC-591, DuPont Pharma, Wilmington, DE, USA) including TPCK-treated trypsin (0.5 μg mL^{-1}) and incubated at 37 °C. At 3 days post infection, MDCK cells were fixed with 10% buffered formalin for 1 h. After removing the overlay, MDCK cells were permeabilized with PBS containing 0.5% Triton X-100 and 20 mM glycine for 20 min. The plates were washed 3 times with PBS containing 0.05% Tween 20 (PBST), and plaques were immunostained with a 1:3000 dilution of anti-IAV-NP antibodies (MAB8257 and MAB8258, 1:1 mixture, EMD Millipore, Saint Louis, MO, USA) and peroxidase-labeled goat anti-mouse IgG (SeraCare Life Sciences, Milford, MA, USA). Immunostained plaques were treated with peroxidase substrate (TrueBlue, SeraCare) according to the manufacturer's protocol and counted manually using an inverted light microscope (Laxco, Mill Creek, WA, USA). SARS-CoV-2 titers were determined by standard plaque assay on SV-VeroE6 cells.

4.2. Antiviral Assessment

Bacdown QAC (D + E) and test articles (A, B, C) were serially diluted 1:5 in DPBS from 1:5 out to 1:3125 (Figure 3). Virus stocks were thawed and diluted to 5.0×10^5 FFU/mL (H1N1) or PFU/mL (MOBAL-05) in DPBS. An aliquot of diluted virus was mixed with an equal volume of each QAC or test article dilution, or an equal volume of DPBS as a control, and incubated at room temperature for 1 h. To assess cytotoxicity of the compounds, each QAC or test article dilution was mixed with an equal volume of DPBS and incubated at room temperature for 1 h.

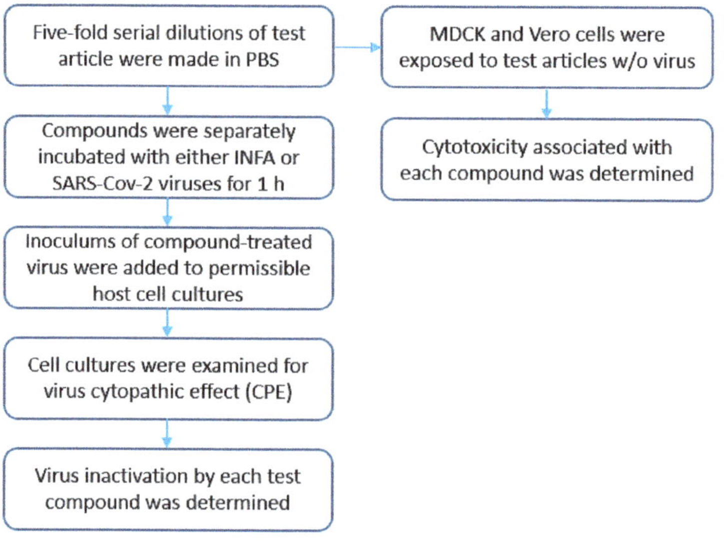

Figure 3. Testing schematic. Simplified diagram outlining general steps accomplished during the testing process.

After incubation, the 50% tissue culture infectious dose ($TCID_{50}$) was assessed. For H1N1, MDCK cells were plated in 96-well tissue culture plates (Alkali Scientific, Ft. Lauderdale, FL, USA) at a density of 1.25×10^4 cells/well in serum-free media and incubated at 37 °C with 5% CO_2 for 18–24 h to reach a confluency of ~60–70%. Samples were added to wells of 96-well dilution plates (CELLTREAT, Pepperell, MA, USA) and serially diluted 1:10 in serum-free media containing 0.5 µg/mL TPCK-trypsin from 1:10 out to 1:10,000. Media was removed from cells and replaced with 0.1 mL of sample dilutions. Cells were then incubated at 37 °C with 5% CO_2 and monitored daily for CPE. Final CPE scores were recorded at 48 h post-infection. $TCID_{50}$ titers were calculated according to the methods of Reed and Muench [16]. The limit of detection (LOD) for this assay was 55 $TCID_{50}$/mL.

For $TCID_{50}$ assessment of MOBAL-05, SV-VeroE6 cells were plated in 96-well tissue culture plates at a density of 2.0×10^4 cells/well in complete media and incubated at 37 °C with 5% CO_2 for 18–24 h to reach a confluency of ~60–70%. After incubation, samples were added to wells of 96-well dilution plates and serially diluted 1:10 in maintenance media from 1:10 out to 1:10,000. Media was removed from cells and replaced with 0.1 mL of sample dilutions, and cells were incubated at 37 °C with 5% CO_2 and monitored daily for CPE. Final CPE scores were recorded at 72 h post-infection.

4.3. Cytotoxicity Assessment

Cell cytotoxicity associated with each test article was also assessed in MDCK cells and SV-VeroE6 cells by exposing cells in 96-well plates to dilutions of each test article at each concentration and observing cells under the microscope for cell death after 48 and 72 h incubation, respectively. The 50% tissue culture cytotoxic dose ($TCCD_{50}$) of each compound at each concentration was calculated using the method of Reed and Muench as above. The LOD for this assay was 55 $TCCD_{50}$/mL.

5. Conclusions

Initial testing of enveloped RNA viruses INFA and SARS-CoV-2 accomplished in the current study suggests that the boronium salts may be effective in controlling and/or eliminating enveloped viruses. Additionally, the benzylated salt form (B) demonstrated decreased cell toxicity, when compared to Bacdown QAC (D+E). If true, boronium salts would represent a viable option to using QAC disinfectants, which have known health and environmental consequences. Given the positive results from additional studies, a defined mixture of both boronium salts (A, B) containing varying alkyl tail lengths may result in an extremely effective disinfectant formulation that controls a broad spectrum of bacterial, fungal, and viral pathogens, the outcome being a new class of boronium ion-based disinfectants with fewer health hazards and environmental concerns than QACs.

Author Contributions: Conceptualization, J.H.D.J., T.J.R. and M.S.; Methodology, T.J.R., J.O.R. and R.W.R.; Validation, J.O.R. and R.W.R.; Formal Analysis, T.J.R. and J.O.R.; Investigation, T.J.R., J.O.R. and R.W.R.; Resources, J.H.D.J.; Writing—Original Draft Preparation, T.J.R. and J.O.R.; Writing—Review & Editing, J.H.D.J., T.J.R., J.O.R., R.W.R. and M.S.; Visualization, T.J.R. and J.O.R.; Supervision, T.J.R. and J.O.R.; Project Administration, T.J.R. and J.O.R.; Funding Acquisition, J.H.D.J. All authors have read and agreed to the published version of the manuscript.

Funding: This research was funded in part (synthesis and characterization of boronium salts) by National Science Foundation, grant number CHE-2102978.

Institutional Review Board Statement: Not applicable.

Informed Consent Statement: No human or animal subjects were used in this study.

Data Availability Statement: Available upon request.

Acknowledgments: We thank the National Science Foundation for supporting this work.

Conflicts of Interest: The authors declare no conflict of interest.

References

1. Bureš, F. Quaternary Ammonium Compounds: Simple in Structure, Complex in Application. *Top Curr Chem.* **2019**, *377*, 14. [CrossRef] [PubMed]
2. Merchel Piovesan Pereira, B.; Tagkopoulos, I. Benzalkonium Chlorides: Uses, Regulatory Status, and Microbial Resistance. *Appl Environ Microbiol.* **2019**, *85*, e00377-19. [CrossRef] [PubMed]
3. Public Health Assessment Guidance Manual (2005 Update): Appendix H: Public Health Conclusion Categories. Agency for Toxic Substances and Disease Registry (ATSDR), Centers for Disease Control and Prevention (CDC). Available online: https://www.atsdr.cdc.gov/hac/phamanual/apph.html (accessed on 17 March 2022).
4. Murray, P.R.; Rosenthal, K.S.; Pfaller, M.A. *Medical Microbiology: Sterilization, Disinfection, and Antisepsis*, 9th ed.; Elsevier: Amterdam, The Netherlands, 2021.
5. Soltani, M.; Ravine, T.J.; Davis, J.H., Jr. Novel boronium salt exhibits substantial antibacterial activity when compared to a commercial quaternary ammonium disinfectant. *Bioorganic Med. Chem. Lett.* **2021**, *36*, 127808. [CrossRef] [PubMed]
6. Tortora, G.J.; Funke, B.R.; Case, C.L. (Eds.) *Microbiology: An Introduction*, 13th ed.; Pearson Education: Boston, MA, USA, 2019; p. 334.
7. Ravine, T.J.; Soltani, M.; Davis, J.H., Jr.; Salter, E.A.; Wierzbicki, A. Unusual Boronium Salt Shows Antifungal Activity Comparable to a Commercial Quaternary Ammonium Disinfectant. *ChemistrySelect* **2022**, *7*, e202104344. [CrossRef]
8. Simon, M.; Veit, M.; Osterrieder, K.; Gradzielski, M. Surfactants—Compounds for inactivation of SARS-CoV-2 and other enveloped viruses. *Curr. Opin. Colloid Interface Sci.* **2021**, *55*, 101479. [CrossRef] [PubMed]
9. Schrank, C.L.; Minbiole, K.P.C.; Wuest, W.M. Are Quaternary Ammonium Compounds, the Workhorse Disinfectants, Effective against Severe Acute Respiratory Syndrome-Coronavirus-2? *ACS Infect. Dis.* **2020**, *6*, 1553–1557. [CrossRef] [PubMed]
10. Julien, C. (Novo Nordisk Pharmatech A/S). Quaternary Ammonium Compounds (Quats) against Viruses, Pharma Focus Asia. Available online: https://www.pharmafocusasia.com/advertorials/quaternary-ammonium-compounds-against-viruses (accessed on 31 March 2022).
11. Wan, W.Y.; Thoon, K.C.; Loo, L.H.; Chan, K.S.; Oon, L.L.E.; Ramasamy, A.; Maiwald, M. Trends in Respiratory Virus Infections During the COVID-19 Pandemic in Singapore, 2020. *JAMA Netw. Open* **2021**, *4*, e2115973. [CrossRef] [PubMed]

12. Krijnse Locker, J.; Chlanda, P.; Sachsenheimer, T.; Brügger, B. Poxvirus membrane biogenesis: Rupture not disruption. *Cell. Microbiol.* **2013**, *15*, 190–199. [CrossRef] [PubMed]
13. Sachse, M.; Tenorio, R.; Fernández de Castro, I.; Muñoz-Basagoiti, J.; Perez-Zsolt, D.; Raïch-Regué, D.; Rodon, J.; Losada, A.; Avilés, P.; Cuevas, C.; et al. Unraveling the antiviral activity of plitidepsin against SARS-CoV-2 by subcellular and morphological analysis. *Antivir. Res.* **2022**, *200*, 105270. [CrossRef] [PubMed]
14. Tolosa, L.; Donato, M.T.; Gómez-Lechón, M.J. General Cytotoxicity Assessment by Means of the MTT Assay. *Methods Mol. Biol.* **2015**, *1250*, 333–348. [CrossRef] [PubMed]
15. Kim, J.H.; Bryant, H.; Fiedler, E.; Cao, T.; Rayner, J.O. Real-time tracking of bioluminescent influenza A virus infection in mice. *Sci. Rep.* **2022**, *12*, 3152. [CrossRef] [PubMed]
16. Reed, L.J.; Muench, H. A Simple Method of Estimating Fifty Per Cent Endpoints. *Am. J. Epidemiol.* **1938**, *27*, 493–497. [CrossRef]

applied biosciences

MDPI

Review

Citric Acid Production by *Aspergillus niger* Using Solid-State Fermentation of Agricultural Processing Coproducts

Thomas P. West

Department of Chemistry, Texas A&M University-Commerce, Commerce, TX 75429, USA; thomas.west@tamuc.edu; Tel.: +1-903-886-5399

Abstract: The ability of *Aspergillus niger* strains to support citric acid production using solid-state fermentation of agricultural processing coproducts was examined in this review. Citric acid has been shown to have a number of commercial applications in the food and beverage industries. The *A. niger* strains capable of elevated citric acid production are known to contain genetic mutations that stimulate overproduction of the organic acid likely involving citric acid cycle reactions. The agricultural processing coproducts previously examined for their ability to support citric acid production by *A. niger* solid-state fermentation include fruit processing wastes, sugarcane bagasse, starch vegetable processing wastes and cereal grain processing coproducts. A comparison of citric acid production by *A. niger* strains using solid-state fermentation demonstrated that certain agricultural processing coproducts were more effective in supporting a high level of acid synthesis. In particular, fruit processing wastes, such as apple pomace, banana peels, grape pomace and orange peels, supported high levels of citric acid by the fungal strains following solid-state fermentation. On the other hand, processing coproducts of cereal grains, such as brans and ethanol processing coproducts, supported low levels of citric acid production by the *A. niger* strains using solid-state fermentation. It appeared that the cereal processing coproducts provided less available sugar content to support citric acid production by the fungal strains. It was concluded that the level of citric acid produced by the *A. niger* strains during solid-state fermentation was dependent on the sugar content of the agricultural processing coproduct utilized.

Keywords: citric acid; solid-state; *Aspergillus niger*; fermentation; processing coproducts

Citation: West, T.P. Citric Acid Production by *Aspergillus niger* Using Solid-State Fermentation of Agricultural Processing Coproducts. *Appl. Biosci.* **2023**, 2, 1–13. https://doi.org/10.3390/applbiosci2010001

Academic Editor: Robert Henry

Received: 3 November 2022
Revised: 23 December 2022
Accepted: 29 December 2022
Published: 4 January 2023

1. Introduction

Citric acid is a water-soluble, specialty organic acid, which exists as a white, crystalline powder and exhibits a significant buffering capacity in water (Figure 1). The global citric acid market in 2023 is projected increase to $3.2 billion [1]. It is estimated that over a million tons of citric acid are produced globally every year [2]. Citric acid has been designated as Safe by the World Health Organization [1]. The applications of citric acid include its use as a flavor enhancer, a pH regulator, a preservative, a chelating agent, a stabilizer and an antioxidant [1]. Citric acid has a number of commercial applications in foods and beverages. It is used as an acidifier in foods plus it leaves little aftertaste [1]. In beverages, citric acid is used to balance sweetness since it imparts a tart taste [1]. Healthwise, citric acid is known to support digestion and kidney function [1]. Citric acid is used in pharmaceuticals to modify pH as well as an antioxidant in preserving vitamins. In the textile industry, citrate is used as a foaming agent to improve textile softness. Citrate is also used in the detergent industry as a replacement for phosphate in detergents to make them more environmentally friendly. It should also be noted that an organic acid, such as citric acid, could be used as a building block in the production of commodity chemicals [2].

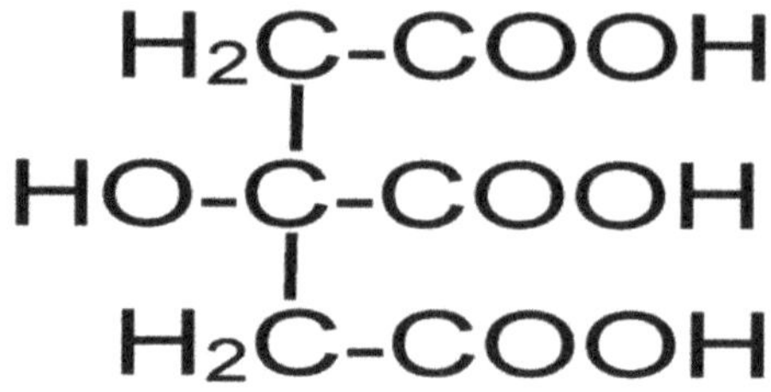

Figure 1. The structure of citric acid.

Citric acid can be isolated from citrus fruit [3–5]. Originally, citric acid was extracted from lime juice but the process was not cost effective. Similarly, chemical synthesis of citric acid using glycerol was not feasible economically. It was replaced by the microbial fermentation of citric acid, which was far more economically feasible. The majority of citric acid that is produced commercially is synthesized by microbial fermentation. Although a number of bacteria and yeast synthesize citric acid, the majority of the citric acid produced commercially is synthesized by the fungus *Aspergillus niger*. Biochemically, the synthesis of citric acid by *A. niger* occurs in its mitochondria (Figure 2). The synthesis of citric acid occurs in the citric acid cycle located in the mitochondria [3–6]. Pyruvate is formed from fermentable sugars by glycolysis. The pyruvate crosses the mitochondrial membrane and is converted to oxaloacetate by the enzyme pyruvate carboxylase. Citrate synthase catalyzes the formation of citric acid from oxaloacetate and acetyl CoA. It should be noted that not all strains of *A. niger* strains are capable of citric acid production. The citric acid-producing *A. niger* strains have been shown to be genetically modified so that they are capable of excreting high levels of citric acid [3–6]. Fermentation of citric acid is considered a green chemistry process since it does not generate any toxic products unlike during the chemical synthesis of citric acid. Most commonly, moist agricultural processing coproducts are used as substrate for a citric acid-producing *A. niger* strain to synthesize the organic acid. For example, it has been shown that the fungal stain *A. niger* ATCC 10577 was capable of producing citric acid from spent grain liquor using surface fermentation [7]. It has also been shown that the addition of phosphate or methanol to *A. niger* cultures can stimulate citric acid production by the fungus [8]. This review will focus on citric acid production by *A. niger* cultures using solid-state fermentation rather than surface or submerged fermentation. With solid-state fermentation, low value biomass from agricultural processing can be transformed into the high value commodity chemical citric acid but the type of biomass utilized is important [9].

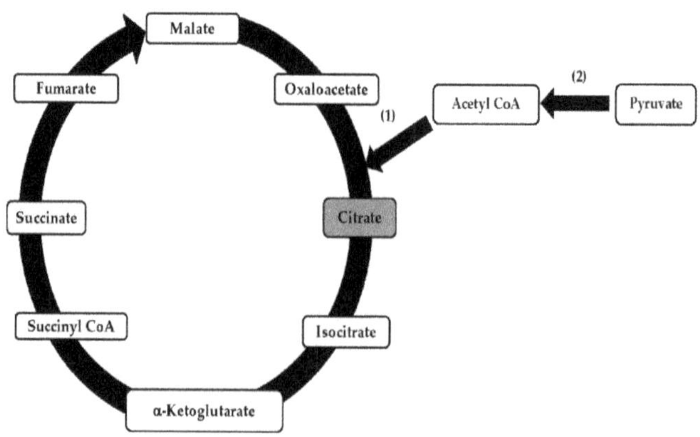

Figure 2. Citric acid synthesis by the citric acid cycle in *Aspergillus niger*. (1) Citrate synthase; (2) Pyruvate carboxylase.

2. Fruit Processing Wastes

2.1. Apple Processing Wastes

The ability of *A. niger* to synthesize citric acid using solid-state fermentation of apple processing wastes has been investigated (Table 1). Peels of African star apples (*Chrysophyllum albidum*) were used for solid-state fermentation (initial pH 3.5) by *A. niger* strains [10]. It was determined that *A. niger* strain DTO: 133-E8 produced a low citric acid concentration after 192 h at 28 °C as did *A. niger* strain DTO: 131-H5 of solid-state fermentation of the peels [10]. In the presence of 2% methanol, solid-state fermentation of the peels by strain DTO: 133-E8 or DTO: 131-H5 increased citric acid production to 1.4-fold or 1.3-fold, respectively, after 192 h at 28 °C [10]. Several studies have used apple pomace as a substrate for the fungal solid-substrate fermentation of citric acid (Table 1). An early study examined solid-state fermentation of apple pomace by *A. niger* strains ATCC 13794, ATCC 11414, ATCC 9142, ATCC 1015 and ATCC 12846 [11]. It was observed that ATCC 9142 produced the highest citric acid level following solid-state fermentation of apple pomace for 120 h at 30 °C compared to the other strains tested [11]. When 4% methanol was added to the apple pomace, solid-state fermentation by ATCC 12846 produced the highest citric acid level after 120 h at 30 °C [11]. It was 1.8-fold higher than the citric acid level produced by ATCC 9142 on the untreated apple pomace [11]. A later study found that *A. niger* MTCC 281 produced a low level of citric acid when used for solid-state fermentation of apple pomace in flasks in the presence of 4% methanol after 120 h at 30 °C [12]. Another study found that solid-state fermentation of apple pomace by *A. niger* ATCC 12846 produced a slightly higher citric acid concentration than did *A. niger* ATCC 13794 after 30 °C for 72 h when the substrate (55% initial moisture) was inoculated with 10^7 spores in flasks [13]. The addition of 3% ethanol or 4% methanol to the apple pomace resulted in about a doubling of the citric acid concentration produced by *A. niger* ATCC 12846 when the solid-state fermentation of apple pomace for 96 h at 30 °C [13]. A similar study using *A. niger* ATCC 12846 for the solid-state fermentation of apple pomace supplemented with rice husks (75% moisture and initial pH of 3.4) for 168 h at 30 °C resulted in the production of nearly 350 g citric acid/kg substrate [14]. Solid-state fermentation by ATCC 12846 of the apple pomace supplemented with rice husks increased citric acid production when either methanol or ethanol was added [14]. When solid-state fermentation of apple pomace (75% initial moisture and initial pH 3.0) by ATCC 12846 occurred in a 12-L rotating-drum-type bioreactor in the presence of 3% methanol after 120 h, a higher level of citric acid was synthesized by ATCC 12846 than if solid-state fermentation of the apple pomace occurred in a rotating-type bioreactor in the presence of 3% ethanol [15]. It should be mentioned that the use of a rotating-type bioreactor would be considered large-scale production. It is clear that the majority of solid-state fermentation studies of agricultural processing coproducts by *A. niger* have been performed using small-scale, shake flask fermentations. Apple pomace supplemented with rice husk (70% moisture) has also been used as a substrate for solid-state fermentation by *A. niger* ATCC 13794 (10^7 spores/g pomace) in 500 mL Erlenmeyer flasks for 72 h at 30 °C. Less citric acid was produced by ATCC 13794 than if solid-state fermentation of the apple pomace supplemented with rice husks was done in trays for 120 h at 30 °C [16]. It appeared that the use of apple pomace for fungal citric acid production using solid-state fermentation was highly feasible.

2.2. Banana and Plantain Peels

The solid-state fermentation of banana and plantain peels have examined as a potential substrate for citric acid production (Table 1). It was noted that solid-state fermentation of banana peels (supplemented with methanol and copper ions) by *A. niger* UABN210 produced a higher level of citric acid after 96 h at 30 °C than that level produced by *A. niger* ATCC 9142 solid-state fermentation of banana peels (initial pH 8.0) for 120 h at 30 °C [17,18]. Solid-state fermentation of banana peels (initial pH 4) by *A. niger* EU440768.1 for 240 h at 28 °C in the presence of 3% methanol resulted in citric acid production being higher than the levels synthesized by *A. niger* UABN210 or ATCC 9142 [17–19]. Solid-state fermentation of banana peels (pH 5.0) in flasks by *A. niger* ATCC 16888 produced less citric acid after 72 h of incubation at 30 °C than when solid-state fermentation of banana peels (pH 5.0) by ATCC 16888 was performed in an aerated, air-jacketed 2 L glass column for 48 h at 30 °C [20]. Solid-state fermentation of plantain peels by *A. niger* ATCC 6275 was investigated [21]. The fermentation of plantains by ATCC 6275 resulted in a 1.5-fold higher citric acid concentration being synthesized after 96 h at 30 °C compared to its citric acid production after 144 h at 30 °C [21]. It was concluded that solid-state fungal fermentation of banana peels or plantains peels was an excellent substrate for citric acid production.

Table 1. Growth conditions and maximum citric acid production by *Aspergillus niger* strains on processing wastes of apples, bananas and plantains using solid-state fermentation.

Substrate	Strain	Sugar Addition	Growth Conditions	Methanol Addition	Citric Acid (g/kg)	Reference
Apple peels	DTO: 133-E8	15% sucrose	192 h, 28 °C	None	16	[10]
Apple peels	DTO: 133-E8	15% sucrose	192 h, 28 °C	2%	21	[10]
Apple peels	DTO: 131-H5	15% sucrose	192 h, 28 °C	None	10	[10]
Apple peels	DTO: 131-H5	15% sucrose	192 h, 28 °C	2%	13	[10]
Apple pomace	ATCC 1015	None	120 h, 30 °C	None	314	[11]
Apple pomace	ATCC 1015	None	120 h, 30 °C	4%	771	[11]
Apple pomace	ATCC 9142	None	120 h, 30 °C	None	482	[11]
Apple pomace	ATCC 9142	None	120 h, 30 °C	4%	791	[11]
Apple pomace	ATCC 11414	None	120 h, 30 °C	None	455	[11]
Apple pomace	ATCC 11414	None	120 h, 30 °C	4%	816	[11]
Apple pomace	ATCC 12846	None	120 h, 30 °C	None	259	[11]
Apple pomace	ATCC 12846	None	120 h, 30 °C	4%	883	[11]
Apple pomace	ATCC 13794	None	120 h, 30 °C	None	359	[11]
Apple pomace	ATCC 13794	None	120 h, 30 °C	4%	766	[11]
Apple pomace	MTCC 281	None	120 h, 30 °C	4%	46	[12]
Apple pomace	ATCC 12846	None	72 h, 30 °C	None	66	[13]
Apple pomace	ATCC 12896	None	96 h, 30 °C	4%	116	[13]
Apple pomace	ATCC 13794	None	72 h, 30 °C	None	61	[13]
Apple pomace	ATCC 12846	None	144 h, 30 °C	3%	342	[14]
Apple pomace	ATCC 12846	None	120 h, 30 °C	3%	221	[15]
Apple pomace	ATCC 13794	None	72 h, 30 °C	None	61	[16]
Apple pomace	ATCC 13794	None	120 h, 30 °C	3%	364	[16]
Banana peels	UABN 210	None	96 h, 30 °C	1%	82	[17]
Banana peels	ATCC 9142	None	120 h, 30 °C	None	2495	[18]
Banana peels	EU440768.1	5% glucose	240 h, 28 °C	3%	98	[19]
Banana peels	ATCC 16888	None	48 h, 30 °C	None	124	[20]
Plantains	ATCC 6275	None	96 h, 30 °C	None	29	[21]

2.3. Cocoa Pod and Coffee Husk Processing Wastes

Solid-state fermentation by *A. niger* of cocoa pods and coffee husk processing wastes to synthesize citric acid has been explored in prior studies (Table 2). Citric acid production by *A. niger* LPB BC-CCT 7717 using solid-state fermentation of cocoa pod husks was examined when 4% methanol was supplemented [22]. The strain produced the highest citric acid concentration after growth on the husks (65% moisture, initial pH 5.5) at 30 °C for 72 h with relatively high yields [22]. It was determined that the level of citric acid synthesized by the strain on the husks after growth at 30 °C for 24 or 48 h was reduced by 31% or 72%, respectively, relative to citric acid production by the strain on the husks after 72 h at 30 °C [22]. Coffee husk processing wastes was used as a substrate for citric acid production by an isolate of *A. niger* using solid-state fermentation [23]. The addition of 3% methanol to the husks (45% moisture, initial pH 4.5) increased citric acid production and the yield by strain RCNM17 after 72 h at 30 °C by 1.6-fold or 1.3-fold, respectively, compared to when the strain was grown on unsupplemented coffee husks [23]. It appeared that both types of processing wastes were excellent substrates capable of supporting fungal citric acid production.

Table 2. Growth conditions and maximum citric acid production by *Aspergillus niger* strains on processing wastes of cocoa pods, coffee husks, figs, kiwifruit, grapes, pineapples, oranges and pomegranates using solid-state fermentation.

Substrate	Strain	Sugar Addition	Growth Conditions	Methanol Addition	Citric Acid (g/kg)	Reference
Cocoa pod husks	LPB B6–CCT 7717	None	72 h, 30 °C	4%	979	[22]
Coffee husks	RCNM 17	None	72 h, 30 °C	None	82	[23]
Coffee husks	RCNM 17	None	120 h, 30 °C	3%	188	[23]
Figs	ATCC 10577	None	360 h, 30 °C	6%	96	[24]
Kiwifruit peels	ATCC 12846	None	120 h, 30 °C	None	70	[25]
Kiwifruit peels	ATCC 12846	None	120 h, 30 °C	2%	100	[25]
Grape pomace	ATCC 1015	None	96 h, 30 °C	None	112	[26]
Grape pomace	ATCC 1015	None	96 h, 30 °C	3%	413	[26]
Grape pomace	ATCC 9142	None	96 h, 30 °C	None	304	[26]
Grape pomace	ATCC 9142	None	96 h, 30 °C	3%	498	[26]
Grape pomace	ATCC 11414	None	96 h, 30 °C	None	160	[26]
Grape pomace	ATCC 11414	None	96 h, 30 °C	3%	511	[26]
Grape pomace	ATCC 12846	None	96 h, 30 °C	None	275	[26]
Grape pomace	ATCC 12846	None	96 h, 30 °C	3%	600	[26]
Grape pomace	ATCC 13794	None	96 h, 30 °C	None	112	[26]
Grape pomace	ATCC 13794	None	96 h, 30 °C	3%	413	[26]
Grape pomace	Isolate	1.5%	24 h, 30 °C	4%	34	[27]
Pineapple peel	ATCC 9142	None	96 h, 30 °C	None	164	[28]
Pineapple peel	ATCC 9142	None	96 h, 30 °C	3%	178	[28]
Pineapple peel	ATCC 10577	None	96 h, 30 °C	None	100	[28]
Pineapple peel	ATCC 12846	None	96 h, 30 °C	None	105	[28]
Pineapple pulp	KS-7	None	120 h, 30 °C	None	17	[29]
Pineapple pulp	KS-7	None	120 h, 30 °C	3%	61	[29]
Orange peels	ATCC 9142	None	86 h, 30 °C	None	171	[30]
Orange peels	Isolate	None	144 h	None	86	[31]
Orange pulp	LPB BC	None	96 h, 30 °C	4%	261	[32]
Pomegranate peel	B60	None	192 h, 25 °C	3%	352	[33]

2.4. Figs and Kiwifruit Peels

Previous investigations have explored solid-state fermentation by *A. niger* of figs and kiwifruit peels for citric acid production (Table 2). Solid-state fermentation of dried figs (75% moisture content with initial pH of pH 7.0) by *A. niger* ATCC 10577 produced 1.5-fold less citric acid after 360 h at 30 °C than if ATCC 10577 was grown on the dried figs supplemented with 6% methanol [24]. Similarly, solid-state fermentation of kiwifruit peels by *A. niger* ATCC 12846 after 96 h at 30 °C resulted in a 1.4-lower citric acid concentration being produced compared to when the strain was grown on 2% methanol-supplemented kiwifruit peels for 96 h at 30 °C [25].

2.5. Grape Pomace

It has been reported that solid-state fermentation of grape pomace supports citric acid production by *A. niger* (Table 2). The solid-state fermentation of grape pomace by *A. niger* strains ATCC 13794, ATCC 11414, ATCC 9142, ATCC 1015 and ATCC 12846 was investigated [26]. Of the strains tested, ATCC 9142 produced the highest citric acid level following solid-state fermentation of apple pomace (moisture content 65.4% and initial pH 3.8) for 96 h at 30 °C compared to the other strains tested [26]. The addition of 3% methanol stimulated citric acid production by the strains using solid-state fermentation on grape pomace compared to no addition [26]. Methanol addition was noted to increase citric acid production by all the strains by at least 1.6-fold relative to citric acid production of grape pomace alone [26]. Solid-state fermentation by ATCC 12846 on the grape pomace supplemented with 3% methanol produced the highest citric acid level after 96 h at 30 °C compared to citric acid production by the other strains screened [26]. Although methanol addition to the grape pomace affected citric acid by ATCC 9142 using solid-state fermentation slightly, its presence increased citric acid production by ATCC 13794 by 3.7-fold higher than the citric acid level produced by ATCC 13794 in the absence of methanol [26]. In another study [27], it was shown that an *A. niger* isolate produced a significant concentration of citric acid following solid-state-fermentation of grape pomace for 24 h at 30 °C.

2.6. Pineapple Processing Waste

Pineapple processing waste has been investigated for its ability to support solid-state fermentation by *A. niger* strains (Table 2). Several *A. niger* strains, including ATCC 10577, ATCC 9142 and ATCC 12846, were examined for their ability to utilize pineapple peels for solid-state fermentation of citric acid [28]. Citric acid production by ATCC 9142 following solid-state fermentation of the pineapple waste for 96 h at 30 °C was 1.6-fold higher than citric acid production by ATCC 10577 or ATCC 12846 solid-state fermentation of pineapple waste for 96 h at 30 °C [28]. When 3% methanol was added to the pineapple waste during solid-state fermentation by all three strains, citric acid production was stimulated [28]. It was reported that solid-state fermentation by *A. niger* KS-7 of pineapple waste (65% moisture) supplemented with 2% methanol produced a 3.5-fold higher citric acid concentration after 120 h at 30 °C compared to the citric acid level synthesized by the strain grown on the unsupplemented pineapple waste for 120 h at 30 °C [29].

2.7. Orange Processing Wastes and Pomegranate Peels

Solid-state fermentation of processing wastes of oranges and pomegranates by *A. niger* have been studied for their ability to support citric acid production (Table 2). Orange peels (initial pH 4.5 and 75% moisture) were used as a substrate for solid-state fermentation by *A. niger* ATCC 9142 [30]. After 86 h at 30 °C using an inoculum of about 10^6 spores/g of dry orange peels, ATCC 9142 produced a high level of citric acid production [30]. Both the spore concentration used as the inoculum as well as the moisture content of the orange peels were critical factors in the level of citric acid produced by ATCC 9142 during solid-state fermentation. The addition of methanol to the orange peels did not enhance citric acid production by ATCC 9142 [30]. Another study explored citric acid production by an isolate of *A. niger* grown on orange peels using solid-state fermentation [31]. When the isolate was grown on the orange peels (60% moisture content) for 144 h of growth at 35 °C more than 100 g citric acid/kg peels was synthesized [31]. Orange juice processing waste was utilized as a substrate for solid-state fermentation by *A. niger* strain LPBBC [32]. After 96 h at 28 °C of solid-state fermentation of the orange juice processing waste (initial pH 2.70 and 65% moisture content), strain LPBBC produced nearly 450 g citric acid/kg waste [32]. Pomegranate peel waste (75% moisture and initial pH 8.0) was also explored as a possible substrate for solid-state fermentation of *A. niger* [33]. Solid-state fermentation of wet and dried pomegranate peel wastes by strain B60 was investigated for citric acid production at 25 °C after 192 h [33]. Higher amounts of citric acid was produced by the strain when it was grown on the wet peel wastes than the dried peel wastes. Methanol (3%) addition to the pomegranate peel waste stimulated citric acid production by strain B60 grown at 25 °C after 192 h [33].

3. Sugarcane Bagasse

In several studies [27,32,34–41], sugarcane bagasse has been used as a substrate for solid-state fermentation by *A. niger* to produce citric acid production (Table 3). After 216 h at 30 °C, *A. niger* strain DS1 produced a 2.2-fold higher level of citric acid on sugarcane bagasse (65% moisture) in a sucrose-containing medium supplemented with 4% methanol than if the strain was grown on the unsupplemented sugarcane bagasse-containing medium [34]. Citric acid production by *A. niger* ATCC 9142 on sugarcane bagasse pretreated with urea increased citric acid production by 1.5-fold after 120 h at 30 °C compared to citric acid production by ATCC 9142 on the untreated bagasse [35]. In another investigation, it was observed that citric acid production by *A. niger* strain 318 following solid-state fermentation of sugarcane bagasse in a sucrose-containing medium for 264 h at 30 °C was about a g citric acid/kg dry bagasse [36]. A prior study showed that solid-state fermentation of sugarcane bagasse by *A. niger* strain DS1 produced a 1.4-fold higher citric acid concentration in a bioreactor (representing large-scale production) compared to the citric acid levels synthesized in shake flasks (representing small-scale production) in a sucrose-containing medium supplemented with 4% methanol for 192 h at 30 °C [37]. An isolate of *A. niger* used for the solid-state fermentation of untreated sugarcane bagasse in a sucrose-containing medium (pH 4.0) supplemented with 4% methanol in 250 mL Erlenmeyer flasks produced a relatively low citric acid concentration after growth for 24 h at 30 °C [27]. Untreated and acid-treated sugarcane bagasse in a sucrose medium containing 4% methanol was investigated for citric acid production by *A. niger* MTCC [38].

Citric acid production by *A. niger* MTCC was 2.3-fold higher on the acid-treated bagasse compared to the level of production on the untreated or treated bagasse after 240 h at 25 °C [38]. When acid-treated sugarcane bagasse supplemented with 4% methanol and urea was used as a substrate for solid-state fermentation in 250 mL Erlenmeyer flasks, *A. niger* ATCC 9142 produced a 1.2-fold higher level of citric acid after 96 h at 25 °C compared to citric acid production by the strain grown on the untreated bagasse after 120 h at 30 °C [39]. Solid-state fermentation of sodium hydroxide-treated sugarcane bagasse supplemented with 4% methanol and urea resulted in a 1.3-fold elevation of citric acid concentration relative to citric acid production by the strain grown on the untreated bagasse for 120 h at 30 °C [39]. Citric acid production by *A. niger* ATCC 9142 on acid-treated sugarcane bagasse (75% moisture content) in a sucrose medium supplemented with 4% methanol was studied and was found to produce a low level of citric acid after 144 h at 30 °C [40]. Using solid-state fermentation of untreated sugarcane bagasse (65% moisture), *A. niger* strain MCCB0201 produced a very low level of citric acid after 120 h at 30 °C [41]. If the bagasse was treated with acid, urea or heat, citric acid production by the strain increased by 3.1-fold, 2.1-fold or 1.6-fold, respectively, after 120 h at 30 °C [41]. The highest level of citric acid was produced by the strain when grown on acid-treated bagasse that was supplemented with ammonium sulfate as a nitrogen source [41].

Table 3. Growth conditions and maximum citric acid production by *Aspergillus niger* strains on sugarcane bagasse and starchy vegetable processing wastes using solid-state fermentation.

Substrate	Strain	Sugar Addition	Growth Conditions	Methanol Addition	Citric Acid (g/kg)	Reference
Sugarcane bagasse	Soil isolate	1.5% sucrose	24 h, 30 °C	4%	25	[27]
Sugarcane bagasse	DS1	15% sucrose	216 h, 30 °C	None	122	[34]
Sugarcane bagasse	DS1	15% sucrose	216 h, 30 °C	4%	206	[34]
Sugarcane bagasse	ATCC 9142	None	120 h, 30 °C	None	95	[35]
Sugarcane bagasse	318	14% sucrose	264 h, 30 °C	None	1	[36]
Sugarcane bagasse	14/20	14% sucrose	264 h, 30 °C	None	1	[36]
Sugarcane bagasse	DS1	31% sucrose	192 h, 30 °C	4%	124	[37]
Sugarcane bagasse	MTCC	31% sucrose	240 h, 25 °C	4%	3	[38]
Sugarcane bagasse	ATCC 9142	None	120 h, 30 °C	4%	76	[39]
Sugarcane bagasse	ATCC 9142	31% sucrose	144 h, 30 °C	4%	233	[40]
Sugarcane bagasse	MCCB0201	None	120 h, 30 °C	None	1	[41]
Kumara	Yang 2	None	144 h, 30 °C	None	69	[42]
Potato	Yang 2	None	144 h, 30 °C	None	3	[42]
Taro	Yang 2	None	144 h, 30 °C	None	66	[42]
Cassava bagasse	LPB 21	None	144 h, 28 °C	None	220	[43]
Cassava peels	CRL isolate	None	72 h, 32 °C	None	25	[44]

4. Starchy Vegetable Processing Wastes

4.1. Sweet Potato and Potato Processing Wastes

As shown in Table 3, processing wastes of different types of potatoes (71% moisture content) have been examined as a substrate for solid-state fermentation by *A. niger* strain Yang No. 2 [42]. Citric acid production by strain Yang No. 2 was measured following solid-state fermentation of potato waste for 144 h at 30 °C and a very low citric acid level was synthesized with a very low yield of 0.03 g citric acid/g starch used. A 23-fold higher level of citric acid was produced by strain Yang No. 2 on kumara after 144 h at 30 °C (yield of 0.58 g citric acid/g starch consumed) compared to the strain grown on the potato waste [42]. Strain Yang No. 2 produced a 22-fold higher citric acid level on taro after 144 h at 30 °C using solid-state fermentation (yield of 0.79 g citric acid/g starch consumed) compared to the strain grown on the potato waste [42]. The addition of the metal ions iron, copper, zinc or manganese was found to slightly increase citric acid production by strain Yang No. 2 on the starch substrates after 144 h at 30 °C [42]. Fungal biomass production by strain Yang No. 2 was also affected by the addition of different combinations of metal ions to the starch substrates during solid-state fermentation [42]. It was clear that the fungal citric acid production varied according to the type of starchy processing waste used during solid-state fermentation.

4.2. Cassava Processing Waste

Processing cassava waste has been explored as a substrate for solid-state fermentation by *A. niger* for the production of citric acid (Table 3). Solid-state fermentation by *A. niger* strain LPB21 of gelatinized cassava bagasse (70% moisture) after 144 h at 28 °C produced a high concentration of citric acid in 250 mL Erlenmeyer flasks [43]. In a column bioreactor (60 mL/min), strain LPB21 grown on bagasse for 144 h at 28 °C under optimized conditions produced a 1.6-fold higher citric acid concentration than the strain did under non-optimized conditions [43]. In a tray bioreactor, strain LPB21 produced about the same citric acid level as the strain produced on bagasse after 144 h at 28 °C under optimized column bioreactor conditions [43]. In a horizontal drum bioreactor (representing large-scale production conditions), strain LPB21 produced a slightly higher concentration of citric acid on the bagasse after 120 h at 28 °C than the level of citric acid produced by the strain grown on the bagasse for 144 h at 28 °C under optimized conditions [43]. Peels of cassava were tested as a substrate for *A. niger* solid-state fermentation [44]. It was determined that solid-state fermentation of cassava peels (60% moisture) by an *A. niger* strain at 35 °C for 96 h produced relatively little citric acid [44]. In the presence of 3% ethanol, solid-state fermentation of cassava peels by the strain at 35 °C for 96 h only slightly increased its citric acid production [44].

5. Cereal Grain Processing Products

5.1. Oat and Wheat Bran

Brans have been utilized as substrates for solid-state fermentation by *A. niger* for citric acid production (Table 4). Oat bran was utilized as a substrate for solid-state fermentation by isolate C of *A. niger* [45]. When the isolate was grown on the sucrose-containing medium containing 3% methanol and oat bran, the strain produced a relatively low citric acid concentration after 72 h at 28 °C [45]. When wheat bran was utilized as a substrate, *A. niger* strain DS1 produced a 4.8-fold higher citric acid level after 72 h at 28 °C compared to the citric acid level synthesized by *A. niger* strain N1 after solid-state fermentation of wheat bran for 72 h at 30 °C [34,46]. It appeared that either bran did not sustain a high level of citric acid production by *A. niger*.

Table 4. Growth conditions and maximum citric acid production by *Aspergillus niger* strains on cereal grain processing coproducts utilizing solid-state fermentation.

Substrate	Strain	Sugar Addition	Growth Conditions	Methanol Addition	Citric Acid (g/kg)	Reference
Oat bran	Isolate C	0.1% sucrose	72 h, 28 °C	3%	70	[45]
Wheat bran	DS1	0.15% sucrose	216 h, 30 °C	None	122	[34]
Wheat bran	N1	None	72 h, 28 °C	None	4	[46]
Corncob	ATCC 10549	None	192 h, 30 °C	None	48	[47]
Corncob	KA88	None	192 h, 28 °C	None	19	[48]
Corn distillers grains & solubles	ATCC 9029	None	240 h, 25 °C	None	4	[49]
Corn distillers grains & solubles	ATCC 9142	None	240 h, 25 °C	None	10	[49]
Corn distillers grains & solubles	ATCC 10577	None	240 h, 25 °C	None	4	[49]
Corn distillers grains & solubles	ATCC 11414	None	240 h, 25 °C	None	7	[49]
Corn distillers grains & solubles	ATCC 12846	None	240 h, 25 °C	None	7	[50]
Corn distillers grains & solubles	ATCC 26550	None	240 h, 25 °C	None	5	[50]
Corn distillers grains & solubles	ATCC 201122	None	240 h, 25 °C	None	5	[50]

5.2. Corncob

Corncobs were examined as a possible substrate for solid-state fermentation by *A. niger* (Table 4). Solid-state fermentation by *A. niger* ATCC 10549 on corncobs was capable of synthesizing a 2.6-fold higher citric acid concentration after 192 h at 30 °C compared to the level produced by *A. niger* strain KA88 after 144 h at 28 °C [47,48]. The addition of sucrose or glucose to the corncobs stimulated citric acid production by strain KA88 [48]. The addition of a nitrogen source to a medium containing sucrose and corncobs further stimulated citric acid production by strain KA88 by about double [48].

5.3. *Coproducts from Corn-Based Ethanol Production*

When corn-based ethanol is produced, com distillers grains and condensed com solubles are the primary coproducts that result from the process. For each bushel of com processed at ethanol plants, approximately 18 pounds of 90% corn distillers grains with solubles are produced. The corn distillers grains with solubles was explored as a substrate for solid-state fermentation by *A. niger*. Citric acid production by *A. niger* strains were analyzed on the untreated grains or autoclaved grains and was found to occur [49–52]. The highest citric acid level on the untreated grains was produced by *A. niger* ATCC 9142. It was determined that ATCC 9029 and ATCC 10577 produced about 3-fold less citric acid after 240 h relative to ATCC 9142. Similarly, citric acid production by ATCC 26550 and ATCC 201122 was noted to be half the level produced by ATCC 9142 after 240 h at 25 °C. Citric acid production by ATCC 201122 on the autoclaved grains was found to be highest after 240 h at 25 °C at a level that was 61% lower than on the untreated grains. Less citric acid was synthesized by the majority of strains screened for citric acid production on the autoclaved grains. Citric acid production by only ATCC 10577 or ATCC 12846 following solid-state fermentation of autoclaved grains after 240 h at 25 °C was more than 1.5-fold higher compared to untreated grains [49,50]. A prior study found that methanol addition to the corn distillers grains diminished citric acid production by *A. niger* ATCC 9142 using solid-state fermentation after 240 h at 25 °C while phosphate addition to the grains slightly increased citric acid production by ATCC 9142 [8]. The addition of 3% methanol to the untreated corn distillers grain with solubles using solid-state fermentation by ATCC 11414 or ATCC 26550 increased citric acid production after 240 h at 25 °C [51]. With respect to solid-state fermentation by *A. niger* ATCC 12846 or ATCC 26550 on 0.5% sulfuric acid-treated corn distillers grains for 240 h at 25 °C, it was noted that there was a 1.4-fold or 2.5-fold increase, respectively, in citric acid production by the strain compared to citric acid production on the untreated grains [52]. When the grains were treated with 1% sulfuric acid, *A. niger* ATCC 10577 produced double the concentration of citric acid on the acid-treated grains compared to the untreated grains using solid-state fermentation after 240 h at 25 °C [49]. Solid-state fermentation of acid-treated grains was found to reduce citric acid production by ATCC 11414 or ATCC 201122 compared to the untreated grains after 240 h at 25 °C [52].

6. Conclusions

The objective of the review was to specifically examine solid-state fermentation of agricultural processing coproducts as substrates to synthesize citric acid. It is clear that a variety of studies have been conducted on the solid-state fermentation of a wide range of agricultural processing coproducts to produce citric acid with varying success. It was observed that citric acid production by *A. niger* strains was dependent on the type of agricultural processing coproduct that was utilized for solid-state fermentation. It appeared that solid-state fermentation of fruit processing wastes (such as apple and grape pomace) by *A. niger* strains resulted in the highest levels of citric acid production. The lowest level of citric acid production was produced by *A. niger* strains following solid-state fermentation of potato or cereal grain processing coproducts as substrates. It was concluded that the sugar content of the substrate being utilized by the *A. niger* strain for solid-state fermentation was critical to the level of citric acid that was produced by the strain. Despite the number of *A. niger* solid-state fermentation studies investigating citric acid synthesis from agricultural processing coproducts, commercial synthesis of citric acid has relied on submerged or surface fermentation. Although this review has focused on solid-state fermentation on low value biomass by *A. niger*, any interest in comparing citric acid production by solid-state, submerged or surface fermentation from a systems efficiency has been addressed in recent reviews [53,54].

Funding: Funding of this work was supported by Welch Foundation Grant T-0014.

Institutional Review Board Statement: Not applicable.

Informed Consent Statement: Not applicable.

Data Availability Statement: Not applicable.

Conflicts of Interest: The author declares no conflict of interest.

References

1. Behera, B.C.; Mishra, R.; Mohapatra, S. Microbial citric acid: Production, properties, application, and future perspectives. *Food Front.* **2021**, *2*, 62–76. [CrossRef]
2. Hu, W.; Li, W.; Yang, H.; Chen, J. Current strategies and future prospects for enhancing microbial production of citric acid. *Appl. Microbiol. Biotechnol.* **2019**, *103*, 201–209. [CrossRef] [PubMed]
3. Angumeenal, A.R.; Venkappayya, D. An overview of citric acid production. *LWT—Food Sci. Technol.* **2013**, *50*, 367–370. [CrossRef]
4. Show, P.L.; Oladele, K.O.; Siew, Q.Y.; Zakry, F.A.A.; Lan, J.C.-W.; Ling, T.C. Overview of citric acid production from *Aspergillus niger*. *Front. Life Sci.* **2015**, *8*, 271–283. [CrossRef]
5. Lende, A.V.; Karemore, H.; Umekar, M.J. Review on production of citric acid by fermentation technology. *GSC Biol. Pharm. Sci.* **2021**, *17*, 85–93. [CrossRef]
6. Karaffa, L.; Kubicek, C.P. *Aspergillus niger* citric acid accumulation: Do we understand this well working black box? *Appl. Microbiol. Biotechnol.* **2003**, *61*, 189–196. [CrossRef]
7. Roukas, T.; Kotzekidou, P. Production of citric acid from brewery wastes by surface fermentation using *Aspergillus niger*. *J. Food Sci.* **1986**, *51*, 225–226. [CrossRef]
8. Xie, G.; West, T.P. Citric acid production by *Aspergillus niger* ATCC 9142 from a treated ethanol fermentation coproduct using solid-state fermentation. *Lett. Appl. Microbiol.* **2009**, *48*, 639–644. [CrossRef]
9. Chilakamarry, C.R.; Sakinah, A.M.M.; Zularisam, A.W.; Sirohi, R.; Khilji, I.A.; Ahmad, N.; Pandey, A. Advances in solid-state fermentation for bioconversion of agricultural wastes to value-added products: Opportunities and challenges. *Bioresour. Technol.* **2022**, *343*, 126065. [CrossRef]
10. Dienye, B.N.; Ahaotu, I.; Agwa, O.K.; Odu, N.N. Citric acid production potential of *Aspergillus niger* using *Chrysophyllum albidum* peel. *Adv. Biosci. Biotechnol.* **2018**, *9*, 190–203. [CrossRef]
11. Hang, Y.D.; Woodams, E.E. Apple pomace: A potential substrate for citric acid production by *Aspergillus niger*. *Biotechnol. Lett.* **1984**, *6*, 763–764. [CrossRef]
12. Kumar, D.; Verma, R.; Bhalla, T.C. Citric acid production by *Aspergillus niger* van. Tieghem MTCC 281 using waste apple pomace as a substrate. *J. Food Sci. Technol.* **2010**, *47*, 458–460. [CrossRef] [PubMed]
13. Dhillon, G.S.; Brar, S.K.; Verma, M.; Tyagi, R.D. Utilization of different agro-industrial wastes for sustainable bioproduction of citric acid by *Aspergillus niger*. *Biochem. Eng. J.* **2011**, *54*, 83–92. [CrossRef]
14. Dhillon, G.S.; Brar, S.K.; Verma, M.; Tyagi, R.D. Enhanced solid-state citric acid bio-production using apple pomace waste through surface response methodology. *J. Appl. Microbiol.* **2011**, *110*, 1045–1055. [CrossRef] [PubMed]
15. Dhillon, G.S.; Brar, S.K.; Kaur, S.; Verma, M. Bioproduction and extraction optimization of citric acid from *Aspergillus niger* by rotating drum type solid-state bioreactor. *Ind. Crops Prod.* **2013**, *41*, 78–84. [CrossRef]
16. Dhillon, G.S.; Brar, S.K.; Kaur, S.; Verma, M. Screening of agro-industrial wastes for citric acid bioproduction by *Aspergillus niger* NRRL 2001 through solid state fermentation. *J. Sci. Food Agric.* **2013**, *93*, 1560–1567. [CrossRef]
17. Kareem, S.O.; Rahman, R.A. Utilization of banana peels for citric acid by *Aspergillus niger*. *Agric. Biol. J. N. Am.* **2011**, *4*, 384–387. [CrossRef]
18. Amenaghawon, N.A.; Oiwoh, O.; Ebewele, O.E. Optimisation of solid state fermentation of banana peels for citric acid production. *Niger. J. Technol.* **2015**, *34*, 716–723. [CrossRef]
19. Odu, N.N.; Uzah, G.A.; Akani, N.P. Optimization of citric acid production by *Aspergillus niger* and *Candida tropicalis* for solid state fermentation using banana peel substrate. *J. Life Bio-Sci. Res.* **2020**, *1*, 51–60. [CrossRef]
20. Chysirichote, T. Valorization of banana peel for citric acid production under solid state fermentation with *Aspergillus niger*. *Chem. Biochem. Eng. Q.* **2020**, *34*, 49–57. [CrossRef]
21. Monrroy, M.; Rueda, L.; Aparicio, A.L.; Garcia, J.R. Fermentation of *Musa paradisiaca* peels to produce citric acid. *J. Chem.* **2019**, *2019*, 8356712. [CrossRef]
22. Oliveira, P.Z.; Vandenberghe, L.P.S.; Rodriques, C.; Pereira, V.M.; Soccol, C.R. Exploring cocoa pod husks as a potential substrate for citric acid production by solid-state fermentation using *Aspergillus niger* mutant strain. *Process Biochem.* **2022**, *113*, 107–112. [CrossRef]
23. Ramachandra, Y.L.; Narayanamurthy, G.; Jois, S.; Chavan, A.; Satwadi, P.R. Production of citric acid in basal coffee husk medium by *Aspergillus niger* under solid state fermentation. *Adv. Biol. Res.* **2013**, *7*, 234–240.

24. Roukas, T. Citric and gluconic acid production from fig by *Aspergillus niger* using solid-state fermentation. *J. Ind. Microbiol. Biotechnol.* **2000**, *25*, 298–304. [CrossRef] [PubMed]
25. Hang, Y.D.; Luh, B.S.; Woodams, E.E. Microbial production of citric acid by solid state fermentation of kiwifruit peel. *J. Food Sci.* **1987**, *52*, 226–227. [CrossRef]
26. Hang, Y.D.; Woodams, E.E. Grape pomace: A novel substrate for microbial production of citric acid. *Biotechnol. Lett.* **1985**, *7*, 253–254. [CrossRef]
27. Goud, K.H.; Srilakshmi, A.; Kumar, A.P.; Narasimha, G. Citric acid production by *Aspergillus niger* through solid state fermentation using fruit wastes. *Biotechnol. Indian J.* **2012**, *6*, 93–96.
28. Tran, C.T.; Sly, L.I.; Mitchell, D.A. Selection of a strain of *Aspergillus* for the production of citric acid from pineapple waste in solid-state fermentation. *World J. Microbiol. Biotechnol.* **1998**, *14*, 399–404. [CrossRef]
29. Kareem, S.O.; Akpan, I.; Alebiowu, O.O. Production of citric acid by *Aspergillus niger* using pineapple waste. *Malays. J. Microbiol.* **2010**, *6*, 161–165.
30. Torrado, A.M.; Cortes, S.; Salgado, J.M.; Max, B.; Rodriquez, N.; Bibbins, B.P.; Converti, A.; Dominguez, J.M. Citric acid production from orange peel wastes by solid state fermentation. *Braz. J. Microbiol.* **2011**, *42*, 394–409. [CrossRef]
31. Zafar, M.; Bano, H.S.; Anwar, Z. Orange peels valorization for citric acid production through single and co-culture fermentation. *Jordan J. Biol. Sci.* **2021**, *14*, 261–266.
32. Rodrigues, C.; Vandenberghe, L.P.S.; Teodoro, J.; Pandey, A.; Soccol, C.R. Improvement on citric acid production in solid-state fermentation by *Aspergillus niger* LPB BC mutant using citric pulp. *Appl. Biochem. Biotechnol.* **2009**, *158*, 72–87. [CrossRef] [PubMed]
33. Roukas, T.; Kotzekidou, P. Pomegranate peel waste: A new substrate for citric acid production by *Aspergillus niger* in solid-state fermentation under non-aseptic conditions. *Environ. Sci. Pollut. Res.* **2020**, *27*, 13105–13113. [CrossRef] [PubMed]
34. Kumar, D.; Jain, V.K.; Shanker, G.; Srivastava, A. Citric acid production by solid state fermentation using sugarcane bagasse. *Process Biochem.* **2003**, *38*, 1731–1738. [CrossRef]
35. Khosravi-Darani, K.; Zoghi, A. Comparison of pretreatment strategies of sugarcane baggase: Experimental design for citric acid production. *Bioresour. Technol.* **2008**, *99*, 6986–6993. [CrossRef]
36. Al-Mahin, A.; Hasan, S.M.; Khan, M.H.; Begum, R. Citric acid production by *Aspergillus niger* through solid-state fermentation on sugarcane bagasse. *Bangladesh J. Microbiol.* **2008**, *25*, 9–12. [CrossRef]
37. Kumar, A.; Jain, V.K. Solid state fermentation studies of citric acid production. *Afr. J. Biotechnol.* **2008**, *7*, 644–650.
38. Vaishnavi, R.; Chairman, K.; Singh, A.J.A.R.; Ramesh, S.; Viswanathan, S. Screening the effect of bagasse—An agro waste for the production of citric acid using *Aspergillus niger* through solid state fermentation. *Biosci. Meth.* **2012**, *3*, 48–54. [CrossRef]
39. Yadegary, M.; Hamidi, A.; Alavi, S.A.; Khodaverdi, E.; Yahaghi, H.; Sattari, S.; Bagherpour, G.; Yahaghi, E. Citric acid production from sugarcane bagasse through solid state fermentation method using *Aspergillus niger* mold and optimization of citric acid production by Taguchi method. *Jundishapur J. Microbiol.* **2013**, *6*, e7625. [CrossRef]
40. Amenaghawon, N.A.; Areguamen, S.O.; Agbroko, N.T.; Ogbeide, S.E.; Okieimen, C.O. Modelling and statistical optimisation of citric acid production from solid state fermentation of sugar cane bagasse using *Aspergillus niger*. *Int. J. Sci.* **2013**, *2*, 56–62.
41. Sharan, A.; Charan, A.A.; Bind, A.; Tiwara, S.B. Citric acid production from pre-treated sugarcane bagasse by *Aspergillus niger* under solid state fermentation. *Asian J. Biosci.* **2015**, *10*, 162–166. [CrossRef]
42. Lu, M.Y.; Maddox, I.S.; Brooks, J.D. Citric acid production by *Aspergillus niger* in solid-substrate fermentation. *Bioresour. Technol.* **1995**, *54*, 235–239. [CrossRef]
43. Vandenberghe, L.P.S.; Soccol, C.R.; Prado, F.C.; Pandey, A. Comparison of citric acid production by solid-state fermentation in flask, column, tray, and drum bioreactors. *Appl. Biochem. Biotechnol.* **2004**, *118*, 293–303. [CrossRef]
44. Ajala, A.S.; Adeoye, A.O.; Olaniyan, S.A.; Fasonyin, O.T. A study on effect of fermentation conditions on citric acid production from cassava peels. *Sci. Afr.* **2020**, *8*, e00396. [CrossRef]
45. Rao, P.R.; Reddy, M.K. Production of citric acid by *Aspergillus niger* using oat bran as a substrate. *Int. J. Chem. Eng.* **2013**, *3*, 181–190.
46. Hussain, A.M. Citric acid production using wheat bran by *Aspergillus niger*. *Indian J. Public Health Res. Dev.* **2019**, *10*, 1213–1217. [CrossRef]
47. Ashour, A.; El-Sharkawy, S.; Amer, M.; Marzouk, A.; Zaki, A.; Kishikawa, A.; Ohzono, M.; Kondo, R.; Shimizu, K. Production of citric acid from corncobs with its biological evaluation. *J. Cosmet. Dermatol. Sci. Appl.* **2014**, *4*, 141–149. [CrossRef]
48. Addo, M.G.; Kusi, A.; Andoh, L.A.; Obiri-Danso, K. Citric Acid Production by *Aspergillus niger* on a corn cob solid substrate using one-factor-at-a-time optimisation method. *Int. Adv. Res. J. Sci. Eng. Technol.* **2016**, *3*, 95–99.
49. Xie, G.; West, T.P. Citric acid production by *Aspergillus niger* on wet corn distillers grains. *Lett. Appl. Microbiol.* **2006**, *43*, 269–273. [CrossRef]
50. West, T.P.; Reed, B.; Xie, G. Fungal citric acid production on distillers grains. *Curr. Trends Microbiol.* **2004**, *1*, 97–102.
51. Xie, G.; West, T.P. Comparison of citric acid production by *Aspergillus niger* ATCC 9029 and ATCC 12846 on corn distillers' grains with solubles. *Res. J. Microbiol.* **2006**, *1*, 540–545.
52. Xie, G.; West, T.P. Citric acid production by *Aspergillus niger* on corn distillers' grains with solubles. *Res. J. Microbiol.* **2006**, *1*, 228–233.

53. Tong, Z.; Zheng, X.; Tong, Y.; Shi, Y.C.; Sun, J. Systems metabolic engineering for citric acid production by *Aspergillus niger* in the post-genomic era. *Microb. Cell Fact.* **2019**, *18*, 28. [CrossRef]
54. Behera, B.C. Citric acid from *Aspergillus niger*: A comprehensive overview. *Crit. Rev. Microbiol.* **2020**, *46*, 727–749. [CrossRef]

applied biosciences

MDPI

Communication

Sedimentation Rate of *Dunaliella salina* in Dark Conditions

Angelica Naka * and Midori Kurahashi

Graduate School of Agricultural and Life Sciences, The University of Tokyo, Tokyo 113-0032, Japan
* Correspondence: angelica.naka@mail.ecc.u-tokyo.ac.jp

Abstract: Microalgae are a source of carbohydrates, proteins and lipids. Thus, they can be considered as raw material to transition from current fossil fuel-based refineries to biorefineries. Microalgae harvesting is considered a major challenge in biomass production. There are several harvesting techniques, but the majority of them are either expensive or not effective. The harvesting method that we propose is sedimentation-induced by light blockage, taking advantage of the motility characteristics of certain microalgae. In this research, the halophilic microalgae *Dunaliella salina* was selected. Experiments were conducted under light and dark conditions to compare the sedimentation rates. Sedimentation behavior was measured by collecting data on the optical density and cell count under both light and dark conditions. The results showed that, under light conditions, the cell count in the middle of the flask decreased from 1×10^6 cell/mL to 5×10^4 cell/mL after 50 days. Under dark conditions sedimentation took less than 10 days for complete settlement. Leaving *Dunaliella salina* under dark conditions may constitute a promising harvest method as this provides a high recovery rate and requires low energy.

Keywords: *Dunaliella salina*; microalgae harvesting; gravity sedimentation

Citation: Naka, A.; Kurahashi, M. Sedimentation Rate of *Dunaliella salina* in Dark Conditions. *Appl. Biosci.* **2023**, *2*, 14–20. https://doi.org/10.3390/applbiosci2010002

Academic Editor: Robert Henry

Received: 13 November 2022
Revised: 21 December 2022
Accepted: 4 January 2023
Published: 10 January 2023

1. Introduction

Microalgae are photosynthetic microorganisms that can be found in marine or freshwater environments. They convert sunlight, water, nutrients and carbon dioxide to produce microalgal biomass that consists of carbohydrates, proteins, lipids and other compounds over short periods of time. Microalgal biomass can be considered a renewable and sustainable source of raw material for different industries such as food, feed, chemical, fermentation, pharmaceutical, etc. [1–5]. They do not compete with agriculture for arable land as they can be cultivated in areas that are arid or affected by drought or salinity [6–8]. Some of them grow in saline water or wastewater, which do not represent any threat to freshwater supply [9]. They also have a short period of harvesting (1 to 10 days), and can be harvested practically year round [6]. Therefore, microalgae can help make the transition from current fossil fuel-based refineries to biorefineries in order to tackle climate change by creating a carbon neutral society. To reduce CO_2 emissions from the current fossil fuel use and reach carbon neutrality, massive amounts of microalgae become necessary.

Except for large species such as *Arthrospira*, microalgae are too small (2 to 20 μm) for conventional filtration systems. Therefore, a major challenge in biomass production lies in their harvesting [10–17], mainly because of the high cost and energy demand [18]. For instance, a research team reported that harvesting was 20–30% of the total cost of biomass production [19]. Microalgae harvesting results in 50 to 200 fold concentration of algae [20] and can be performed by mechanical, chemical, biological or electrical methods [21,22]. Some of the harvesting methods include gravity sedimentation, flocculation, flotation, filtration and screening, centrifugation, electricity-based techniques, ultrasound and immobilization, including combinations [17,22–26]. The majority of them are expensive and not very efficient, requiring long extraction periods or the use of chemicals [23,27]. Thus, there is a need to seek eco-friendly, cost-efficient, and low-energy systems.

Sedimentation appears to be one of the pre-dewatering biomass preconcentration methods with the lowest cost and energy consumption [11,28]. The settlement can be described by Stokes' Law (Equation (1)) which describes the velocity of a sphere falling in a fluid in terms of the radius of the cells and the difference in density between the microalgae cells and the medium:

$$v = \frac{2}{9} g \frac{r^2}{\eta} (\rho_s - \rho_l) \tag{1}$$

where v is the settlement velocity, g is the acceleration due to gravity (9.81 m/s^2), r is the spherical particle radius, η is the dynamic viscosity and ρ_s and ρ_l are the solid (particle) and liquid (medium) densities, respectively.

In this research, the halophilic microalgae *Dunaliella salina* was selected because of the high salinity that is required for its growth (up to 5 M NaCl), making it suitable for open culture systems while reducing the risk of contamination from other microalgae and/or protozoa. Outdoor systems reduce the production cost because of freely available sunlight, while being suitable only for microalgae that can live in extreme conditions such as high acidity, alkalinity or salinity. *Dunaliella salina* is well known as a source of beta carotene that is produced when exposed to harsh conditions especially in terms of light, salinity, temperature and nutrients. It also produces a wide range of compounds with different potential applications. We propose the massive and open culture of this microalgae on non-arable land to use as raw materials in different industries and be able to make the transition to a biorefinery. Considering that *Dunaliella salina* is a motile microalgae, it can probably be assumed that light blockage may affect its motility which may lead to settlement. There is research on the effect of carbon dioxide and pH on motility of flagellated microalgae, including *Dunalliela salina* [29]. However, to the best knowledge of the authors, there is no research that studies sedimentation speed and compares settlement rates in light and dark conditions. Therefore, this research was conducted to monitor *Dunaliella salina* sedimentation with time, using optical density sensors in both light and dark conditions.

2. Materials and Methods

The microalgae *Dunaliella salina* (strain CS-744/01, from the Australian National Algae Culture Collection, isolated from a saline lake in Western Australia) was used in the settlement experiments. The microalgae were cultivated in f/2 medium [30,31] with a salinity of 12% (2 M NaCl). The f/2 medium was prepared by mixing all the chemicals in sea salt (Red Sea Salt used in aquariums) solution. A pre-culture of 400 mL was inoculated into 3600 mL of fresh medium in a 5 L beaker and incubated at 25 °C under 200 $\mu mol/m^2/s$ of continuous illumination (white LED) for 10 days without agitation. The initial cell concentration was 5.0×10^4 cell/mL, and the pH was 7.14. The 5 L beaker was lightly covered with a plastic film to prevent evaporation. The pH was measured using a portable pH meter (Hanna Instruments HI98130), and the salinity was measured with a portable salt refractometer (Tekcoplus Refractometer ATC). The settlement experiments were conducted when the cell concentration was around 1×10^6 cell/mL. *Dunaliella salina* was placed in 25 cm^2 culture flasks with double sealed caps (canted neck, slim type, Iwaki). Flasks were filled to the top (approximately 73 mL) just before overflowing and then they were closed with the caps. The settlement of *Dunaliella salina* was tested in bright and dark conditions. To monitor the settlement speed of the microalgae, optical density (OD_{600}) equipment (Taitec OD-Monitor A&S connected to OD sensor-S) with light transmission (horizontal projection, 0.00 to 2.55) was used. The data were taken every minute and automatically stored on a USB attached to the system. The cell concentration was periodically monitored by taking a 25 μL sample from the middle of the flask and measured in the Invitrogen Tali Image Based Cytometer. Experiments were conducted until settlement was complete or OD values were close to zero. Thus, tests performed in dark conditions were monitored for 10 days, whereas experiments conducted in light conditions were monitored for 70 days. To obtain a consistent result, experiments were repeated four times for dark conditions and three times for light conditions.

3. Results

The settlement experiments were conducted when the average concentration of *Dunaliella salina* was 1.28×10^6 cell/mL, the cell size was 12.5 μm and the pH was 8.03. The settlement or sinking of *Dunaliella salina* to the bottom of the flask took place gradually (Figure 1).

Figure 1. Picture of the settlement process under dark condition: (**a**) initial condition, (**b**) during settlement (on the 6th day), and (**c**) after settlement (on the 10th day).

Optical density (OD_{600}) change under dark and light conditions are presented in Figure 2. Optical density is generally used to measure culture growth, but in this study, we used it to determine the microalgae sinking rate. When *Dunaliella salina* was in suspension, the OD was between 0.2 and 1. However, when the microalgae settled, the OD was below 0.2. The OD change graphs that we obtained were very similar, and thus, only one result for each case was presented. Experiments conducted under dark conditions showed that *Dunaliella salina* settlement was completed after 8 days, while experiments conducted in light conditions showed that sinking took around 50 days.

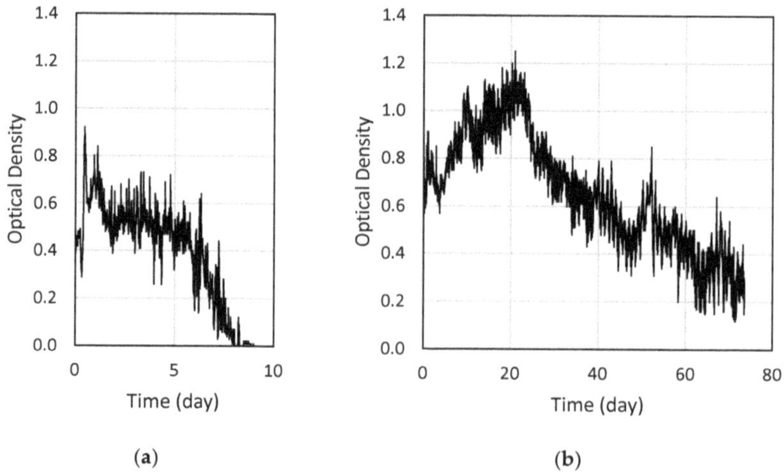

Figure 2. Optical density (OD) changes in (**a**) dark and (**b**) light conditions with time.

In order to study the cell concentration change with time, total cell concentration was measured periodically by taking 25 µL samples from the middle of the flask for both dark and light conditions. The results are presented in Figure 3. The results showed that under dark conditions, the microalgae concentration decreased to 2×10^4 cell/mL after 8 days, which correlates with biomass settlement. On the other hand, for light conditions, after 50 days, the cell concentration decreased to 5×10^4 cell/mL. Considering that the initial cell concentration of microalgae was 1.28×10^6 cell/mL, for dark conditions, it decreased by 98% after 8 days, and with light conditions, it took 50 days to reach 96%. When comparing both results, it can be said that gravity sedimentation of *Dunaliella salina* occurs faster in darker conditions rather than light available conditions. This is probably because blocking light decreases microalgae movement, making it sink. Gravity sedimentation is a simple harvesting method, but if carried out under dark conditions, it also reduces time from an accelerated sinking rate. Therefore, in terms of time and cost reduction, it is recommended to harvest *Dunaliella salina* under dark conditions.

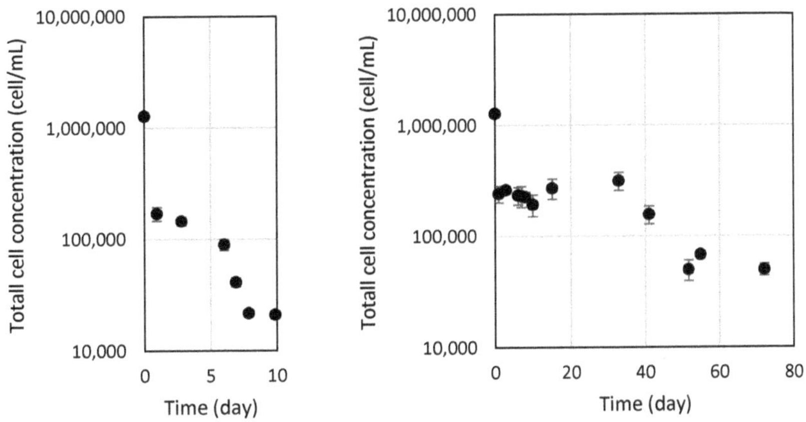

(**a**) (**b**)

Figure 3. *Dunaliella salina* total cell concentration in (**a**) dark and (**b**) light conditions with time.

Stokes' law (Equation (1)) was used to predict the settlement velocity of *Dunaliella salina*. The density and the viscosity of the medium, obtained from the literature for seawater with 12% salinity, were considered to be 1090.25 kg/m^3 and 1320.6 µPa·s, respectively [32]. According to previous research, the density of the cytoplasm of marine microalgae ranges between 1030 and 1100 kg/m^3 [33]. Another study considered the density of *Dunaliella salina* as 1260 kg/m^3, the density of glycerol [34]. Due to the small density difference between the microalgae and the medium, it can be said that density contribution to settlement is small compared to the size of the microalgae. Stoke's law holds for spherical shaped microalgae, so *Dunaliella salina* was assumed to be spherical. The average size of *Dunaliella salina* was 12.5 µm, so for a radius of 6.25 µm, the settlement velocity is expected to be around 0.05 m/day. This value is not in accordance with experimental results probably because *Dunaliella salina* is a motile microalgae, which makes it more difficult to predict its settlement. Flagella are considered a dynamic morphological means to suspension [33], and thus, complex hydromechanical [35] and biophysical [36] parameters might be considered for settlement prediction models.

4. Discussion

Microalgae can potentially be a great source of raw materials such as protein, carbohydrates and lipids. However, its elevated price compared with raw materials from fossil fuel is still a great barrier to overcome. Therefore, reducing biomass production and processing costs are crucial for large scale culture and establishing a microalgae-based biorefinery. For instance, cost effective cultivation and harvesting methods may drastically reduce their cost. In this study, we proposed the gravity sedimentation technique of the halophilic microalgae *Dunaliella salina* under dark conditions taking advantage of its motility characteristic. In our previous study, we presented the settlement degree of *Dunaliella salina* at different temperatures and found a sedimentation rate of 79 to 96% between 20 to 30 degrees Celsius [37]. Thus, in the present study, we conducted experiments at 25 degrees Celsius and monitored the speed of settlement when experiments were conducted both in dark and light conditions, using OD and cell concentration data. Motile phytoplankton sank to the bottom of the flask when they lost their flagella, when stressed physiologically or with increasing age (senescent cells) [33]. In this study, light blockage might have triggered *Dunaliella salina*'s loss of movement and subsequent gravity settlement, but further research should be conducted to determine the mechanism involved. Light, photoperiod and nutrients may affect suspension and sinking, but these factors have not been well studied, so their level of influence are, to a certain degree, unknown [33]. Another hypothesis that can potentially explain the sinking of this microalgae is a possible aggregation between cells [33], especially when exposed to stressful conditions. However, further research should be conducted to test this hypothesis.

Costs can be kept low as the energy required for this harvesting method is low, the recovery rate is high, and no consumables are required. As such, it can probably constitute a cost-effective recovery method. In addition, chemical substances such as flocculants and pH adjusters are not necessary, which avoid chemical contamination. Physical and mechanical methods such as centrifugation, filtration (filter), evaporation to dryness, voltage application, etc. are not required, and equipment operation or maintenance are not necessary. This method is simple, suppressing costs.

For high volumes of microalgae or open and large-scale cultivation, we suggest temporarily transferring the culture to a collection container, then blocking the light for several days. The culture water can be removed by decantation or other methods, and the supernatant can be reused. This research was conducted in a laboratory under controlled conditions. As part of further research, we need to conduct experiments at a larger scale, performing the experiments outdoors to assure the efficacy of this method under a broad range of temperatures.

5. Conclusions

Sedimentation under dark conditions of *Dunaliella salina* was completed in less than 10 days, whereas settlement under light conditions took more than 50 days. The cell count in the middle of the flask decreased from 1×10^6 cell/mL to 5×10^4 cell/mL after 50 days (96% recovery rate). Under dark conditions, sedimentation took less than 10 days for almost complete settlement (98% recovery rate). Leaving *Dunaliella salina* under dark conditions may constitute a promising harvest method as this provides a high recovery rate, requires low energy and has high efficiency. A large-scale pilot is necessary before its application.

Author Contributions: Conceptualization, A.N. and M.K.; methodology, A.N. and M.K.; validation, A.N. and M.K.; formal analysis, A.N. and M.K.; investigation, A.N. and M.K.; resources, M.K.; data curation, A.N. and M.K.; writing—original draft preparation, A.N. and M.K.; writing—review and editing, A.N. and M.K.; visualization, A.N. and M.K.; supervision, M.K.; project administration, M.K. All authors have read and agreed to the published version of the manuscript.

Funding: This research received no external funding.

Institutional Review Board Statement: Not applicable.

Informed Consent Statement: Not applicable.

Data Availability Statement: Data is contained within the article.

Acknowledgments: We want to express our deepest appreciation to Mitsubishi Gas Chemical Company for donating the Taitec OD Monitor A&S and OD sensor-S that were used for all experiments.

Conflicts of Interest: The authors declare no conflict of interest.

References

1. Chisti, Y. Society and microalgae: Understanding the past and present. In *Microalgae in Health and Disease Prevention*; Elsevier: Amsterdam, The Netherlands, 2018; pp. 11–21.
2. Zhuang, D.; He, N.; Khoo, K.S.; Ng, E.-P.; Chew, K.W.; Ling, T.C. Application progress of bioactive compounds in microalgae on pharmaceutical and cosmetics. *Chemosphere* **2021**, *291*, 132932. [CrossRef] [PubMed]
3. Yarkent, Ç.; Gürlek, C.; Oncel, S.S. Potential of microalgal compounds in trending natural cosmetics: A review. *Sustain. Chem. Pharm.* **2020**, *17*, 100304. [CrossRef]
4. Harun, R.; Danquah, M.K.; Forde, G.M. Microalgal biomass as a fermentation feedstock for bioethanol production. *J. Chem. Technol. Biotechnol.* **2010**, *85*, 199–203. [CrossRef]
5. Nagarajan, D.; Varjani, S.; Lee, D.-J.; Chang, J.-S. Sustainable aquaculture and animal feed from microalgae–Nutritive value and techno-functional components. *Renew. Sustain. Energy Rev.* **2021**, *150*, 111549. [CrossRef]
6. Schenk, P.M.; Thomas-Hall, S.R.; Stephens, E.; Marx, U.C.; Mussgnug, J.H.; Posten, C.; Kruse, O.; Hankamer, B. Second generation biofuels: High-efficiency microalgae for biodiesel production. *Bioenergy Res.* **2008**, *1*, 20–43. [CrossRef]
7. Vanthoor-Koopmans, M.; Wijffels, R.H.; Barbosa, M.J.; Eppink, M.H. Biorefinery of microalgae for food and fuel. *Bioresour. Technol.* **2013**, *135*, 142–149. [CrossRef]
8. Chew, K.W.; Yap, J.Y.; Show, P.L.; Suan, N.H.; Juan, J.C.; Ling, T.C.; Lee, D.-J.; Chang, J.-S. Microalgae biorefinery: High value products perspectives. *Bioresour. Technol.* **2017**, *229*, 53–62. [CrossRef] [PubMed]
9. Pessôa, L.C.; Deamici, K.M.; Pontes, L.A.M.; Druzian, J.I.; de Jesus Assis, D. Technological prospection of microalgae-based biorefinery approach for effluent treatment. *Algal Res.* **2021**, *60*, 102504. [CrossRef]
10. Schlesinger, A.; Eisenstadt, D.; Bar-Gil, A.; Carmely, H.; Einbinder, S.; Gressel, J. Inexpensive non-toxic flocculation of microalgae contradicts theories; overcoming a major hurdle to bulk algal production. *Biotechnol. Adv.* **2012**, *30*, 1023–1030. [CrossRef]
11. Vandamme, D.; Foubert, I.; Muylaert, K. Flocculation as a low-cost method for harvesting microalgae for bulk biomass production. *Trends Biotechnol.* **2013**, *31*, 233–239. [CrossRef]
12. Talukder, M.M.R.; Das, P.; Wu, J.C. Immobilization of microalgae on exogenous fungal mycelium: A promising separation method to harvest both marine and freshwater microalgae. *Biochem. Eng. J.* **2014**, *91*, 53–57. [CrossRef]
13. Estime, B.; Ren, D.; Sureshkumar, R. Cultivation and energy efficient harvesting of microalgae using thermoreversible sol-gel transition. *Sci. Rep.* **2017**, *7*, 1–9. [CrossRef]
14. Poelman, E.; De Pauw, N.; Jeurissen, B. Potential of electrolytic flocculation for recovery of micro-algae. *Resour. Conserv. Recycl.* **1997**, *19*, 1–10. [CrossRef]
15. Chen, C.-Y.; Yeh, K.-L.; Aisyah, R.; Lee, D.-J.; Chang, J.-S. Cultivation, photobioreactor design and harvesting of microalgae for biodiesel production: A critical review. *Bioresour. Technol.* **2011**, *102*, 71–81. [CrossRef]
16. Najjar, Y.S.H.; .Abu-Shamleh, A. Harvesting of microalgae by centrifugation for biodiesel production: A review. *Algal Res.* **2020**, *51*, 102046. [CrossRef]
17. Li, S.; Hu, T.; Xu, Y.; Wang, J.; Chu, R.; Yin, Z.; Mo, F.; Zhu, L. A review on flocculation as an efficient method to harvest energy microalgae: Mechanisms, performances, influencing factors and perspectives. *Renew. Sustain. Energy Rev.* **2020**, *131*, 110005. [CrossRef]
18. Brennan, L.; Owende, P. Biofuels from microalgae—A review of technologies for production, processing, and extractions of biofuels and co-products. *Renew. Sustain. Energy Rev.* **2010**, *14*, 557–577. [CrossRef]
19. Gudin, C.; Thepenier, C. Bioconversion of solar energy into organic chemicals by microalgae. *Adv. Biotechnol. Process. (USA)* **1986**, *6*, 73–110.
20. Molina Grima, E.; Belarbi, E.-H.; Acién Fernández, F.G.; Robles Medina, A.; Chisti, Y. Recovery of microalgal biomass and metabolites: Process options and economics. *Biotechnol. Adv.* **2003**, *20*, 491–515. [CrossRef]
21. Morais Junior, W.G.; Gorgich, M.; Corrêa, P.S.; Martins, A.A.; Mata, T.M.; Caetano, N.S. Microalgae for biotechnological applications: Cultivation, harvesting and biomass processing. *Aquaculture* **2020**, *528*, 735562. [CrossRef]
22. Singh, G.; Patidar, S. Microalgae harvesting techniques: A review. *J. Environ. Manag.* **2018**, *217*, 499–508. [CrossRef]
23. Goh, B.H.H.; Ong, H.C.; Cheah, M.Y.; Chen, W.-H.; Yu, K.L.; Mahlia, T.M.I. Sustainability of direct biodiesel synthesis from microalgae biomass: A critical review. *Renew. Sustain. Energy Rev.* **2019**, *107*, 59–74. [CrossRef]
24. Ho, S.-H.; Chen, C.-Y.; Lee, D.-J.; Chang, J.-S. Perspectives on microalgal CO_2-emission mitigation systems—A review. *Biotechnol. Adv.* **2011**, *29*, 189–198. [CrossRef]
25. Kim, J.; Yoo, G.; Lee, H.; Lim, J.; Kim, K.; Kim, C.W.; Park, M.S.; Yang, J.-W. Methods of downstream processing for the production of biodiesel from microalgae. *Biotechnol. Adv.* **2013**, *31*, 862–876. [CrossRef]

26. Uduman, N.; Qi, Y.; Danquah, M.K.; Forde, G.M.; Hoadley, A. Dewatering of microalgal cultures: A major bottleneck to algae-based fuels. *J. Renew. Sustain. Energy* **2010**, *2*, 012701. [CrossRef]
27. Besson, A.; Guiraud, P. High-pH-induced flocculation–flotation of the hypersaline microalga *Dunaliella salina*. *Bioresour. Technol.* **2013**, *147*, 464–470. [CrossRef]
28. Milledge, J.J.; Heaven, S. A review of the harvesting of micro-algae for biofuel production. *Rev. Environ. Sci. Bio/Technol.* **2013**, *12*, 165–178. [CrossRef]
29. Wang, Y.; Fan, X.; Gao, G.; Beardall, J.; Inaba, K.; Hall-Spencer, J.M.; Xu, D.; Zhang, X.; Han, W.; McMinn, A. Decreased motility of flagellated microalgae long-term acclimated to CO_2-induced acidified waters. *Nat. Clim. Change* **2020**, *10*, 561–567. [CrossRef]
30. Guillard, R.R. Culture of phytoplankton for feeding marine invertebrates. In *Culture of Marine Invertebrate Animals*; Springer: Berlin/Heidelberg, Germany, 1975; pp. 29–60.
31. Guillard, R.R.; Ryther, J.H. Studies of marine planktonic diatoms: I. *Cyclotella nana* Hustedt, and Detonula confervacea (Cleve) Gran. *Can. J. Microbiol.* **1962**, *8*, 229–239. [CrossRef]
32. Leyendekkers, J. Prediction of the density and viscosity of seawater, its concentrates and other multicomponent solutions using the Tammann-Tait-Gibson (TTG) model. *Desalination* **1979**, *29*, 263–274. [CrossRef]
33. Smayda, T.J. The suspension and sinking of phytoplankton in the sea. *Oceanogr. Mar. Biol. Ann. Rev.* **1970**, *8*, 353–414.
34. Xu, Y.; Milledge, J.J.; Abubakar, A.; Swamy, R.; Bailey, D.; Harvey, P. Effects of centrifugal stress on cell disruption and glycerol leakage from *Dunaliella salina*. *Microalgae Biotechnol.* **2015**, *1*, 20–27. [CrossRef]
35. Ludwig, W. Zur theorie der flimmerbewegung (dynamik, nutzeffekt, energiebilanz). *Z. Für Vgl. Physiol.* **1930**, *13*, 397–504. [CrossRef]
36. Jahn, T.L.; Bovee, E.C. Motile behavior of protozoa. In *Research in Protozoology*; Elsevier: Amsterdam, The Netherlands, 1967; pp. 41–200.
37. Watsuji, T.-o.; Naka, A.; Morita, Y.; Kurahashi, M. Effect of temperature and dissolved oxygen on gravity sedimentation of the unicellular alga *Dunaliella salina*. *Ann. Microbiol.* **2021**, *71*, 1–7.

applied biosciences

MDPI

Communication

Analysis of the Bacterial Community of Metal Scrap Using an Enrichment Culture Approach

Hironaga Akita [1], Yoshiki Shinto [2] and Zen-ichiro Kimura [2,*]

[1] Research Institute for Sustainable Chemistry, National Institute of Advanced Industrial Science and Technology (AIST), 3-11-32 Kagamiyama, Higashi-Hiroshima 739-0046, Hiroshima, Japan
[2] Department of Civil and Environmental Engineering, National Institute of Technology, Kure College, 2-2-11 Aga-minami, Kure 737-8506, Hiroshima, Japan
* Correspondence: z-kimura@kure-nct.ac.jp; Tel.: +81-(0)823-73-8486

Abstract: Microbiologically influenced corrosion (MIC) of metal alloys is promoted by biofilms formed on metal surfaces. In the marine environment, MIC causes serious metal infrastructure problems, which lead to significant economic losses. In this study, we used an enrichment culture approach to examine the bacterial community that grows on metal surface at levels below the detection limit as a preliminary study for developing guidelines to prevent biofilm formation. An enrichment culture approach was employed to analyze the bacterial community on metal surface without biofilms and corrosion. Genomic DNA was extracted from culture sample after incubation in the enrichment culture with a metal piece, and then the V3–V4 variable regions of the bacterial 16S rRNA gene were amplified using the extracted genomic DNA as the template. Subsequently, using a next-generation sequencing approach, the amplified V3–V4 regions were sequenced, and the bacterial community was analyzed using the QIIME 2 microbiome bioinformatics platform. Using this enrichment culture approach, more than 80 bacterial genera were detected with *Sphingomonas* bacteria exhibiting the highest relative abundance (44%). These results demonstrated that this method could be useful for bacterial community analysis for bacteria below detection limits, and will serve as a basis for the development of the guidelines.

Keywords: 16S rRNA gene; bacterial community analysis; microbiologically influenced corrosion; metal piece; enrichment culture approach

Citation: Akita, H.; Shinto, Y.; Kimura, Z.-i. Analysis of the Bacterial Community of Metal Scrap Using an Enrichment Culture Approach. *Appl. Biosci.* **2023**, 2, 23–30. https://doi.org/10.3390/applbiosci2010004

Academic Editor: Robert Henry

Received: 26 October 2022
Revised: 12 January 2023
Accepted: 13 January 2023
Published: 29 January 2023

Correction Statement: This article has been republished with a minor change. The change does not affect the scientific content of the article and further details are available within the backmatter of the website version of this article.

1. Introduction

Microorganisms grow by adapting to environmental conditions. Biofilms are biological membranes enclosed in a matrix composed of polymeric substances produced by multiple species of microorganisms [1]. When microorganisms adhere to the metal surfaces of infrastructure in the marine environment such as bridges, offshore wind-power equipment and ocean-going vessels, biofilms form on the metal surfaces. The polymeric substance matrix of the biofilm protects the microorganisms within the biofilm from stress associated with chemicals and environmental changes [1]. Thus, a favorable environment for the growth of microorganisms is created within the biofilm [1]. When biofilms form over time on metal surfaces of infrastructure, the relative abundance of corrosion-causing microorganisms increases, ultimately leading to microbiologically influenced corrosion (MIC). Marine infrastructure components affected by MIC exhibit progressive metal degradation on the metal surfaces of the infrastructure, which reduces its useful lifetime. It is estimated that MIC is responsible for approximately 20% of metal corrosion affecting marine infrastructure components [2]. For example, in 2013, it was estimated that MIC causes economic losses of $2.5 trillion, which correspond to approximately 3.4% of the gross world product. MIC of the marine infrastructure is therefore a global problem.

High-throughput amplicon sequencing is an effective method for identifying the bacterial community that causes MIC. Sulfate-reducing and sulfur-oxidizing bacteria are the major causative organisms of corrosion for several metal alloys including aluminum alloy [3], carbon steel [4], stainless steel [5] and other engineering steel [6–8]. In some cases, iron-oxidizing, nitrate-reducing and manganese-oxidizing bacteria have been associated with MIC [9]. Those previous characterizations of the bacterial community were carried out on samples with corrosion. On the other hand, a bacterial community is changed significantly during the early stages of biofilm formation to MIC development. In other words, major bacterial species that act on biofilm formation and development as well as MIC are different. Thus, we hypothesized that guidelines to prevent biofilm formation on the metal surfaces of marine infrastructure could be established if the structure of the bacterial community on the metal surfaces could be characterized before biofilm formation. Moreover, the lifetime of marine infrastructure might be extended if the effective utilization of such guidelines could reduce the occurrence of MIC. However, no studies have examined the bacterial community before biofilm formation on metal surfaces of marine infrastructure because the number of bacteria in the community is too low, so it is difficult to prepare the sufficient genomic DNA required for high-throughput amplicon sequencing.

Enrichment culture methods are used to increase the abundance of specific bacterial species within the total bacterial population of an environmental sample, and they are different from environmental enrichment, which is used to elicit normal behavior in captive animals [10–12]. Although the culture condition needs to be set according to the purpose of the research, this method can be used to grow bacteria below the detection limit to a detectable level. Using an enrichment culture approach, previous studies have reported metal corrosion-causing bacteria or bacterial communities growing on steel [13] and stainless steel [14,15]. However, the nutritional conditions used in previous studies differed from the natural environment because organic acids, sugars or yeast extract were added as carbon sources to the enrichment culture media. Those media contained high concentrations of carbon sources compared to the natural environment, and many kinds of environmental bacteria cannot grow in those media due to their growths are inhibited by high carbon concentrations [16]. Thus, the results of previous studies may not accurately reflect the communities in the natural environment, as poor-nutrient media should be used for enrichment culture media to obtain results closer to the actual target environment.

Here we report a preliminary study for developing guidelines to prevent biofilm formation. Using a synthetic medium with low concentrations of carbonate as the single carbon source, an enrichment culture was performed with a metal piece without biofilm or corrosion and genomic DNA was extracted from the culture sample. Subsequently, we amplified the V3–V4 variable regions of the bacterial 16S rRNA gene (amplicon) using the extracted genomic DNA as a template and then analyzed the bacterial community based on high-throughput sequencing of the V3–V4 variable regions. As a result, more than 80 bacterial genera were detected by applying an enrichment culture approach, which demonstrated that this method could be useful in developing the guidelines.

2. Materials and Methods

2.1. Sampling and Enrichment Culture

Several kinds of metal pieces (approximately 2.0 cm^2) consisting of stainless steel SUS304 and free of biofilm or corrosion were collected from a non-corroded area at the Akitsu Port, Hiroshima Prefecture, Japan. The temperature on the sampling day was approximately 20 °C, and there had been no precipitation for 1 week before sample collection. The collected metal pieces were each separated into a sterile tube, immediately placed in a cool box at 4 °C, and transported to our laboratory within 2 h.

To prevent contamination, flasks were autoclaved at 121 °C for 20 min and cooled to room temperature. Subsequently, each of the metal pieces were subjected to an enrichment culture for 24 h at 25 °C in a flask containing a synthetic medium (pH 7.0) consisting of 3.0 g/L NaCl, 1.7 g/L $Na_2HPO_4 \cdot 12H_2O$, 1.0 g/L K_2CO_3, 0.3 g/L KH_2PO_4, 0.1 g/L NH_4Cl, 0.024 g/L $MgSO_4 \cdot 7H_2O$ and 0.0011 g/L $CaCl_2 \cdot 2H_2O$. The growth of bacteria was assessed by monitoring the OD_{600} value in comparison with cell-free control samples using an Eppendorf BioSpectrometer (Eppendorf, Hamburg, Germany).

2.2. Genomic DNA Extraction and PCR Conditions

After genomic DNA was extracted from the culture sample using an illustra™ bacteria genomicPrep Mini Spin kit (GE Healthcare, Chicago, IL, USA), the concentration and purity were measured using a NanoDrop ND-1000 spectrophotometer (Thermo Fisher Scientific, Waltham, MA, USA). The concentration of genomic DNA extracted from the enrichment culture sample with the highest OD_{600} value was 143 ng/μL. Using genomic DNA as the template, the V3–V4 variable regions were amplified by KOD -Plus- Neo (TOYOBO) with the bacterial domain–specific primers 341F (5′-TCGTCGGCAGCGTCAGATGTGTA TAAGAGACAGCCTACGGGNGGCWGCAG-3′) [17] and 805R (5′-GTCTCGTGGGCTC GGAGATGTGTATAAGAGACAGGACTACHVGGGTATCTAATCC-3′) [18]. The PCR protocol for amplification of the V3–V4 variable regions entailed initial denaturation at 94 °C for 2 min, followed by 35 cycles of denaturation at 98 °C for 10 s, annealing at 55 °C for 30 s, and extension at 68 °C for 30 s, with a final extension at 68 °C for 5 min.

2.3. Sequencing Library Preparation, Sequencing, and Bioinformatics Analysis

Using a Wizard SV Gel and PCR Clean-Up System (Promega, Madison, WI, USA), the PCR products (approximately 450 bp) were purified. After libraries for high-throughput amplicon sequencing were prepared using a Nextera XT DNA Library Preparation Kit (Illumina, San Diego, CA, USA), the concentration of the sequencing libraries was determined using a Quanti Fluor™ dsDNA System (Promega). The sequencing of libraries was performed using a MiSeq sequencer (Illumina) with MiSeq Reagent Kit v3 (Illumina). The sequence data have been deposited in the DDBJ database under BioProject number PRJDB13529, BioSample number SAMD00469933 and DRA number DRA014101.

The analysis of the bacterial community was performed using the QIIME 2 microbiome bioinformatics platform [19]. After the quality of sequence data was confirmed using FastQC ver. 0.11.9 [20], the filtered sequence data were yielded. Subsequently, using the DADA2 algorithm [21], amplicon sequence variants (ASVs) were prepared based on the filtered sequence data. Using the Sliva taxonomic database ver. 138 [22,23], operation taxonomic units (OTUs) at 99% sequence identity were grouped based on the ASVs. The neighbor-joining tree based on a dataset consisting of more than 1% of the OTUs was constructed using MEGA 11 software [24].

3. Results and Discussion

When bacteria on metal surfaces are targeted for the analysis, the number of bacteria is too low for preparation of genomic DNA since high-throughput amplicon sequencing cannot be performed. Indeed, we could not extract genomic DNA from a non-cultured sample. Subsequently, to obtain a sufficient amount of bacteria for the preparation of genomic DNA while preventing deviation from the actual target environment, we used filter-sterilized seawater or artificial seawater as the enrichment culture media, but bacterial growth could not be confirmed. To increase the number of bacteria for preparation of genomic DNA, therefore, a synthetic medium containing low concentrations of nutrients was used for the enrichment culture. When this enrichment culture approach using the synthetic medium was performed for each of the metal pieces, the OD_{600} values increased from 0.08–0.14. As a control test, an enrichment culture was also performed without adding metal pieces, but no growth of the bacteria could be confirmed. Moreover, genomic DNA could be extracted from each of the culture samples, and the V3–V4 variable regions could be amplified using the genomic DNA as the template. Thus, genomic DNA extracted from the culture sample with the highest OD_{600} value was used for bacterial community analysis.

OTUs based on the analysis of genomic DNA from the cultured samples indicated the presence of more than 80 bacterial genera with the following genera present at >1% relative abundance: *Sphingomonas* (44%), *Methylobacterium/Methylorubrum* (9.4%), *Mu-*

cilaginibacter (4.9%), *Massilia* (4.8%), *Pseudomonas* (4.1%), *Hymenobacter* (2.9%), *Alkanindiges* (1.3%), and *Chroococcidiopsis* (1.2%) (Figure 1). To confirm the phylogenetic relationship between these bacterial genera, a neighbor-joining tree based on the V3–V4 variable regions was constructed (Figure 2). The neighbor-joining tree included α-Proteobacteria such as *Sphingomonas* sp. and *Methylobacterium* sp., β-Proteobacteria such as *Massilia* sp., γ-Proteobacteria such as *Pseudomonas* sp. and *Alkanindiges* sp., and Sphingobacteria such as *Mucilaginibacter* sp. In particular, *Sphingomonas* sp. 1–10 formed a major cluster with *S. kyungheensis* THG-B283^T and *S. hankookensis* ODN7^T and exhibited a 97.6–99.1% sequence similarity to *S. kyungheensis* THG-B283^T (Table 1). These results suggested that a variety of bacteria existed on the metal surface before biofilm formation, and biofilm formation may be promoted when the bacterial community changes in environmental conditions and the relative abundance of bacteria with biofilm-forming ability increases. In fact, our previous study suggested that the bacterial community changes significantly in biofilms before and after the occurrence of metal corrosion [25]. On the other hand, the proportion of bacteria with a relative abundance of <1% was about 15.4% (Supplementary Table S1), and the decrease in this proportion is considered to be one of the improvements for future study. The proportion could be improved by changing the bacterial domain-specific primer set. According to Bukin et al. [26], the V2–V3 variable regions of the 16S rRNA gene are also useful for the identification of bacterial genera and species in bacterial community analysis. The V4–V5 variable regions of the 16S rRNA gene have a higher resolution for the identification of an archaeal community [27]. Thus, to decrease the proportion of unclassified bacteria, we are going to use those primer sets for bacterial community analysis in our next study.

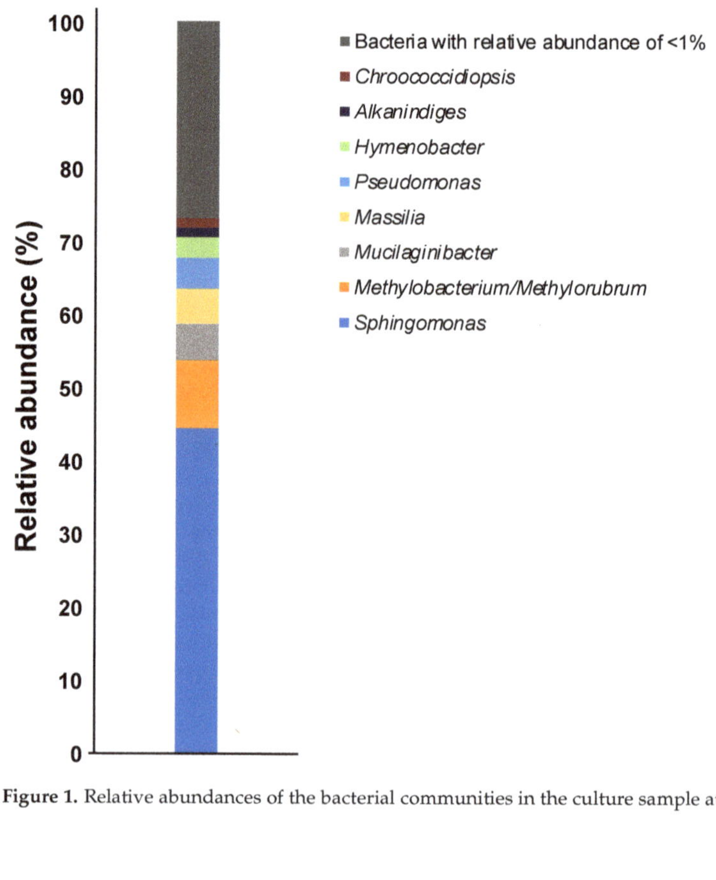

Figure 1. Relative abundances of the bacterial communities in the culture sample at the genus level.

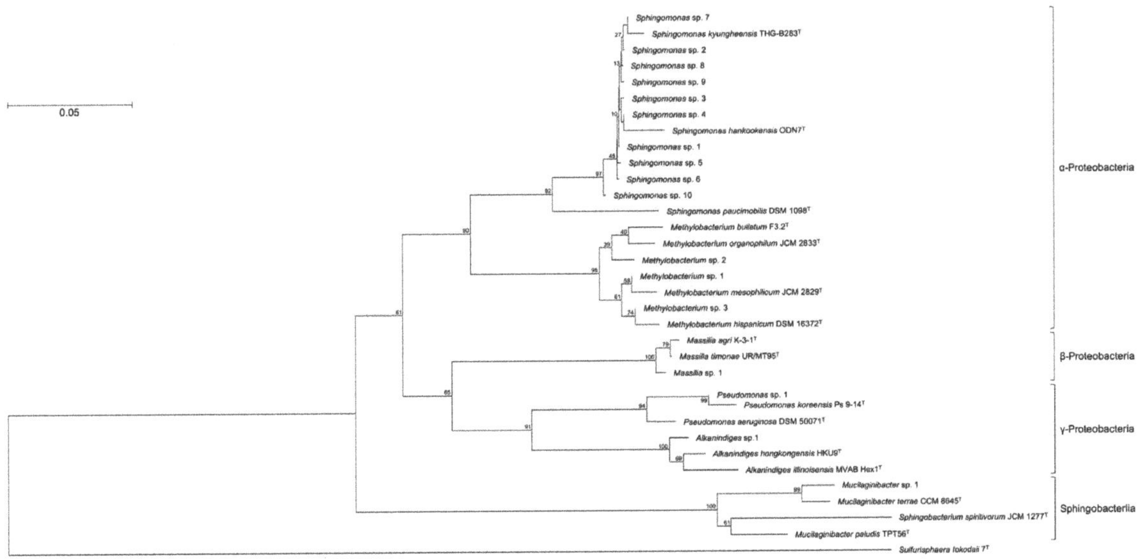

Figure 2. Consensus bootstrap phylogenetic tree constructed from analysis of the 16S rRNA gene sequences showing the relationships between OTUs comprising more than 1% and the related type strains. Each OTU is indicated as its representative type strain. The tree was rooted using *Sulfurisphaera tokodaii* 7^T as the outgroup. The bar indicates a 0.05% nucleotide substitution rate.

Table 1. Bacteria with more than 1% of OTUs observed in the culture sample.

OTU	**Closest Relative**		**Accession No.**	**Identity (%)**	**Relative Abundance (%)**
	Class	**Species**			
Sphingomonas sp. 1	α-Proteobacteria	*Sphingomonas kyungheensis* THG-B283^T	NR_118263	99.1	7.1
Sphingomonas sp. 2	α-Proteobacteria	*Sphingomonas kyungheensis* THG-B283^T	NR_118263	99.1	5.6
Sphingomonas sp. 3	α-Proteobacteria	*Sphingomonas kyungheensis* THG-B283^T	NR_118263	99.1	4.8
Sphingomonas sp. 4	α-Proteobacteria	*Sphingomonas kyungheensis* THG-B283^T	NR_118263	99.3	4.6
Mucilaginibacter sp. 1	Sphingobacteria	*Mucilaginibacter terrae* CCM 8645^T	NR_158094	99.7	4.3
Sphingomonas sp. 5	α-Proteobacteria	*Sphingomonas kyungheensis* THG-B283^T	NR_118263	97.6	4.2
Sphingomonas sp. 6	α-Proteobacteria	*Sphingomonas kyungheensis* THG-B283^T	NR_118263	99.1	4.0
Methylobacterium sp. 1	α-Proteobacteria	*Methylobacterium mesophilicum* JCM 2829^T	NR_115550	99.1	3.7
Pseudomonas sp. 1	γ-Proteobacteria	*Pseudomonas koreensis* Ps 9-14^T	NR_025228	99.1	3.5
Sphingomonas sp. 7	α-Proteobacteria	*Sphingomonas kyungheensis* THG-B283^T	NR_118263	99.3	3.1
Sphingomonas sp. 8	α-Proteobacteria	*Sphingomonas kyungheensis* THG-B283^T	NR_118263	99.3	2.4
Sphingomonas sp. 9	α-Proteobacteria	*Sphingomonas kyungheensis* THG-B283^T	NR_118263	99.1	2.0
Massilia sp. 1	β-Proteobacteria	*Massilia agri* K-3-1^T	NR_157781	98.5	1.9
Methylobacterium sp. 2	α-Proteobacteria	*Methylobacterium bullatum* F3.2^T	NR_116548	98.2	1.4
Sphingomonas sp. 10	α-Proteobacteria	*Sphingomonas hankookensis* ODN7^T	NR_116570	98.6	1.3
Alkanindiges sp. 1	γ-Proteobacteria	*Alkanindiges hongkongensis* HKU9^T	NR_115178	97.4	1.2
Methylobacterium sp. 3	α-Proteobacteria	*Methylobacterium hispanicum* DSM 16372^T	NR_112613	99.1	1.1

Sphingomonas bacteria are one species of the most abundant microorganisms in a variety of environments including freshwater, seawater and soils, due to the capacity of these bacteria to utilize a wide variety of organic compounds [28,29]. When *Sphingomonas* bacteria adhere to metal surfaces, microcolonies form and then biofilms develop over time [30]. Within developed biofilms in which *Sphingomonas* bacteria are the dominant species, an oxygen-free environment is created and the growth of other corrosive bacteria is accelerated, leading to MIC of the metal substrate [31]. In addition to *Sphingomonas* bacteria, other MIC-causing bacteria were observed such as *Massilia* and *Pseudomonas* bacteria (Figure 1). These bacteria reportedly accelerate metal corrosion via oxidation of the metal surfaces to generate energy [32] and accumulate acidic metabolites [33]. These results suggest that *Sphingomonas* bacteria initiate biofilm formation when their growth is not inhibited by environmental stresses or nutrient limitations. The resulting growths of *Massilia* and *Pseudomonas* bacteria inside the biofilm encourage corrosion of the metal scrap surfaces.

In this study, bacterial community analysis was performed on metal pieces without biofilm or corrosion using a synthetic medium with carbonate as the single carbon source as a preliminary study for developing guidelines to prevent biofilm formation. The bacterial community on the metal pieces could not be analyzed without an enrichment culture approach. More than 80 bacterial genera were detected by applying an enrichment culture approach. In addition, many kinds of bacteria in the community were different from the commonly known MIC-causing bacteria, suggesting that the bacteria that act on biofilm formation are different from the MIC-causing bacteria. These results demonstrated that our idea of using low concentrations of carbonate for the enrichment culture media is useful for bacterial community analysis for bacteria below the detection limits. Moreover, the results of this study will serve as a basis for the development of guidelines for the prevention of biofilm formation on metal surfaces of infrastructure in the marine environment. To facilitate the development of the guidelines, we are planning to confirm the correlation between culture conditions and the type of sample that affects the bacterial community, which may be useful in improving the accuracy of the guidelines. These results will be presented in a future report.

4. Conclusions

The bacterial community before biofilm formation on metal surfaces has not been reported, but the analysis is necessary to develop guidelines for the prevention of biofilm formation. In this study, we demonstrated the effectiveness of bacterial community analysis using a low-nutrient media for an enrichment culture approach to identify the bacteria below detection limits. To develop accurate guidelines, it is necessary to clarify how enrichment culture conditions affect the bacterial community.

Supplementary Materials: The following supporting information can be downloaded at: https://www.mdpi.com/article/10.3390/applbiosci2010004/s1, Table S1: Bacterial species with less than 1% of relative abundance observed in the culture sample.

Author Contributions: Conceptualization, H.A.; methodology, H.A. and Y.S.; validation, H.A. and Y.S.; formal analysis, H.A. and Y.S.; investigation, H.A. and Y.S.; resources, H.A.; data curation, H.A. and Y.S.; writing—original draft preparation, H.A.; writing—review and editing, H.A., Y.S. and Z.-i.K.; visualization, H.A.; supervision, H.A. and Z.-i.K.; project administration, H.A.; funding acquisition, H.A. and Z.-i.K. All authors have read and agreed to the published version of the manuscript.

Funding: This work was supported by a grant from The Salt Science Research Foundation (No. 2101).

Institutional Review Board Statement: Not applicable.

Informed Consent Statement: Not applicable.

Data Availability Statement: The 16S rRNA gene sequence of *A. hongkongensis* HKU9^T, *A. illinoisensis* MVAB Hex1^T, *M. agri* K-3-1^T, *M. timonae* UR/MT95^T, *M. bullatum* F3.2^T, *M. hispanicum* DSM 16372^T, *M. mesophilicum* JCM 2829^T, *M. organophilum* JCM 2833^T, *M. paludis* TPT56^T, *M. terrae* CCM 8645^T, *P. aeruginosa* DSM 50071^T, *P. koreensis* Ps 9-14^T, *S. hankookensis* ODN7^T, *S. kyungheensis* THG-B283^T, *S. paucimobilis* DSM 1098^T, *S. spiritivorum* JCM 1277^T, and *Sulfurisphaera tokodaii* 7^T are available in the GenBank/EMBL/DDBJ databases under accession numbers NR_114676, NR_025254, KX672812, U54470, NR_116548, NR_112613, NR_115550, AJ400920, AM490402, NR_158094, NR_026078, NR_025228, NR_116570, NR_118263, X72722, D14026, and NR_028609, respectively.

Acknowledgments: We are grateful to all members of the Bio-conversion Research Group at our Institute [Research Institute for Sustainable Chemistry, National Institute of Advanced Industrial Science and Technology (AIST)] for their technical assistance and valuable discussion.

Conflicts of Interest: The authors declare no conflict of interest.

References

1. Karygianni, L.; Ren, Z.; Koo, H.; Thurnheer, T. Biofilm matrixome: Extracellular components in structured microbial communities. *Trends Microbiol.* **2020**, *28*, 668–681. [CrossRef]
2. Little, B.J.; Lee, J.S. *Microbiologically Influenced Corrosion*; John Wiley and Sons, Inc.: Hoboken, NJ, USA, 2007; pp. 1–279.
3. Dai, X.; Wang, H.; Ju, L.K.; Cheng, G.; Cong, H.; Newby, B.Z. Corrosion of aluminum alloy 2024 caused by *Aspergillus niger*. *Int. Biodeterior. Biodegrad.* **2016**, *115*, 1–10. [CrossRef]
4. Xu, D.; Wen, J.; Fu, W.; Gu, T.; Raad, I.I. D-Amino acids for the enhancement of a binary biocide cocktail consisting of THPS and EDDS against an SRB biofilm. *World J. Microbiol. Biotechnol.* **2012**, *28*, 1641–1646. [CrossRef] [PubMed]
5. Zhang, P.; Xu, D.; Li, Y.; Yang, K.; Gu, T. Electron mediators accelerate the microbiologically influenced corrosion of 304 stainless steel by the *Desulfovibrio vulgaris* biofilm. *Bioelectrochemistry* **2015**, *101*, 14–21. [CrossRef] [PubMed]
6. Xia, J.; Yang, C.; Xu, D.; Sun, D.; Nan, L.; Sun, Z.; Li, Q.; Gu, T.; Yang, K. Laboratory investigation of the microbiologically influenced corrosion (MIC) resistance of a novel Cu-bearing 2205 duplex stainless steel in the presence of an aerobic marine *Pseudomonas aeruginosa* biofilm. *Biofouling* **2015**, *31*, 481–492. [CrossRef]
7. Li, H.; Zhou, E.; Zhang, D.; Xu, D.; Xia, J.; Yang, C.; Feng, H.; Jiang, Z.; Li, X.; Gu, T.; et al. Microbiologically influenced corrosion of 2707 hyper-duplex stainless steel by marine *Pseudomonas aeruginosa* biofilm. *Sci. Rep.* **2016**, *6*, 20190. [CrossRef]
8. Li, H.; Yang, C.; Zhou, E.; Yang, C.; Feng, H.; Jiang, Z.; Xu, D.; Gu, T.; Yang, K. Microbiologically influenced corrosion behavior of S32654 super austenitic stainless steel in the presence of marine *Pseudomonas aeruginosa* biofilm. *J. Mater. Sci. Technol.* **2017**, *33*, 1596–1603. [CrossRef]
9. Ma, Y.; Zhang, Y.; Zhang, R.; Guan, F.; Hou, B.; Duan, J. Microbiologically influenced corrosion of marine steels within the interaction between steel and biofilms: A brief view. *Appl. Microbiol. Biotechnol.* **2020**, *104*, 515–525. [CrossRef]
10. Näslund, J.; Johnsson, J.I. Environmental enrichment for fish in captive environments: Effects of physical structures and substrates. *Fish Fish.* **2016**, *17*, 1–30. [CrossRef]
11. Arechavala-Lopez, P.; Cabrera-Álvarez, M.J.; Maia, C.M.; Saraiva, J.L. Environmental enrichment in fish aquaculture: A review of fundamental and practical aspects. *Rev. Aquacult.* **2022**, *14*, 704–728. [CrossRef]
12. Zhang, Z.; Gao, L.; Zhang, X. Environmental enrichment increases aquatic animal welfare: A systematic review and meta-analysis. *Rev. Aquacult.* **2022**, *14*, 1120–1135. [CrossRef]
13. Guangfeng, X.; Xiaodong, Z.; Shuai, W.; Jie, Y.; Jie, S.; Zhongyi, A.; Yan, L.; Xinlei, Q. Synergistic effect between sulfate-reducing bacteria and *Pseudomonas aeruginosa* on corrosion behavior of Q235 steel. *Int. J. Electrochem. Sci.* **2020**, *15*, 361–370.
14. Tran, T.T.T.; Kannoorpatti, K.; Padovan, A.; Thennadil, S. A study of bacteria adhesion and microbial corrosion on different stainless steels in environment containing *Desulfovibrio vulgaris*. *R. Soc. Open Sci.* **2021**, *8*, 201577. [CrossRef]
15. Prasanna, J.; Rosalie, C.; Sun, S.; Martin, S.; Florentin, C.; Sudesh, W.L.; Dominique, T.; Daniel, B.J.; Diane, M.; Scott, R.A.; et al. Onset of microbial influenced corrosion (MIC) in stainless steel exposed to mixed species biofilms from equatorial seawater. *J. Electrochem. Soc.* **2017**, *164*, C532–C538.
16. Bloomfield, S.F.; Stewart, G.S.A.B.; Dodd, C.E.R.; Booth, I.R.; Power, E.G.M. The viable but non-culturable phenomenon explained? *Microbiology* **1998**, *144*, 1–3. [CrossRef]
17. Muyzer, G.; de Waal, E.C.; Uitterlinden, A.G. Profiling of complex microbial populations by denaturing gel electrophoresis analysis of polymerase chain reaction amplified genes coding for 16S rRNA. *Appl. Environ. Microbiol.* **1993**, *59*, 695–700. [CrossRef]
18. Herlemann, D.P.; Labrenz, M.; Jürgens, K.; Bertilsson, S.; Waniek, J.J.; Andersson, A.F. Transitions in bacterial communities along the 2000 km salinity gradient of the Baltic Sea. *ISME J.* **2011**, *5*, 1571–1679. [CrossRef]
19. Bolyen, E.; Rideout, J.R.; Dillon, M.R.; Bokulich, N.A.; Abnet, C.C.; Al-Ghalith, G.A.; Alexander, H.; Alm, E.J.; Arumugam, M.; Asnicar, F.; et al. Reproducible, interactive, scalable and extensible microbiome data science using QIIME 2. *Nat. Biotechnol.* **2019**, *37*, 852–857. [CrossRef]
20. Andrews, S. FastQC: A Quality Control Tool for High Throughput Sequence Data. 2010. Available online: http://www.bioinformatics.babraham.ac.uk/projects/fastqc (accessed on 15 December 2021).
21. Callahan, B.J.; McMurdie, P.J.; Rosen, M.J.; Han, A.W.; Johnson, A.J.; Holmes, S.P. DADA2: High-resolution sample inference from Illumina amplicon data. *Nat. Methods* **2016**, *13*, 581–583. [CrossRef]
22. Quast, C.; Pruesse, E.; Yilmaz, P.; Gerken, J.; Schweer, T.; Yarza, P.; Peplies, J.; Glöckner, F.O. The SILVA ribosomal RNA gene database project: Improved data processing and web-based tools. *Nucleic Acids Res.* **2013**, *41*, D590–D596. [CrossRef]
23. Bokulich, N.A.; Kaehler, B.D.; Rideout, J.R.; Dillon, M.; Bolyen, E.; Knight, R.; Huttley, G.A.; Caporaso, J.G. Optimizing taxonomic classification of marker-gene amplicon sequences with QIIME 2's q2-feature-classifier plugin. *Microbiome* **2018**, *6*, 90. [CrossRef] [PubMed]
24. Tamura, K.; Stecher, G.; Kumar, S. MEGA11: Molecular evolutionary genetics analysis version 11. *Mol. Biol. Evol.* **2021**, *38*, 3022–3027. [CrossRef] [PubMed]

25. Akita, H.; Shinto, Y.; Kimura, Z.-I. Bacterial Community Analysis of Biofilm Formed on Metal Joint. *Appl. Biosci.* **2022**, *1*, 221–228. [CrossRef]
26. Bukin, Y.S.; Galachyants, Y.P.; Morozov, I.V.; Bukin, S.V.; Zakharenko, A.S.; Zemskaya, T.I. The effect of 16S rRNA region choice on bacterial community metabarcoding results. *Sci. Data* **2019**, *6*, 190007. [CrossRef]
27. Fadeev, E.; Cardozo-Mino, M.G.; Rapp, J.Z.; Bienhold, C.; Salter, I.; Salman-Carvalho, V.; Molari, M.; Tegetmeyer, H.E.; Buttigieg, P.L.; Boetius, A. Comparison of two 16S rRNA primers (V3–V4 and V4–V5) for studies of arctic microbial communities. *Front. Microbiol.* **2021**, *12*, 637526. [CrossRef]
28. White, D.C.; Sutton, S.D.; Ringelberg, D.B. The genus *Sphingomonas*: Physiology and ecology. *Curr. Opin. Biotechnol.* **1996**, *7*, 301–306. [CrossRef]
29. Waigi, M.G.; Sun, K.; Gao, Y. Sphingomonads in microbe-assisted phytoremediation: Tackling soil pollution. *Trends Biotechnol.* **2017**, *35*, 883–899. [CrossRef]
30. Venugopalan, V.P.; Kuehn, M.; Hausner, M.; Springael, D.; Wilderer, P.A.; Wuertz, S. Architecture of a nascent *Sphingomonas* sp. biofilm under varied hydrodynamic conditions. *Appl. Environ. Microbiol.* **2005**, *71*, 2677–2686. [CrossRef]
31. Beale, D.J.; Morrison, P.D.; Key, C.; Palombo, E.A. Metabolic profiling of biofilm bacteria known to cause microbial influenced corrosion. *Water Sci. Technol.* **2014**, *69*, 1–8. [CrossRef]
32. Meiying, L.; Min, D. A review: Microbiologically influenced corrosion and the effect of cathodic polarization on typical bacteria. *Rev. Environ. Sci. Biotechnol.* **2018**, *17*, 431–446.
33. Kaewyaia, J.; Noophana, P.; Wantawina, C.; Munakata-Marrb, J. Microbiologically influenced corrosion and current mitigation strategies: A state of the art review. *Int. Biodeterior. Biodegrad.* **2019**, *137*, 42–58.

applied biosciences

MDPI

Article

Biochar Extracts Can Modulate the Toxicity of Persistent Free Radicals in the Nematode *Caenorhabditis elegans*

Xuchao Zhang [1,2], Nadine Saul [3,4], Thora Lieke [3,5,6], Yi Chen [1], Min Wu [1,*], Bo Pan [1] and Christian E. W. Steinberg [1,3,*]

1 Yunnan Provincial Key Lab of Soil Carbon Sequestration and Pollution Control, Faculty of Environmental Science and Engineering, Kunming University of Science and Technology, Kunming 650500, China
2 Faculty of Life Sciences, Ecology Group, Humboldt Universität zu Berlin, Philippstraße 13, 10115 Berlin, Germany
3 Faculty of Life Sciences, Freshwater and Stress Ecology Group, Humboldt Universität zu Berlin, 12437 Berlin, Germany
4 Faculty of Life Sciences, Molecular Genetics Group, Humboldt Universität zu Berlin, 10115 Berlin, Germany
5 Institute of Aquaculture and Protection of Waters, South Bohemian Research Center of Aquaculture and Biodiversity of Hydrocenoses, Faculty of Fisheries and Protection of Waters, University of South Bohemia in Ceske Budejovice, Na Sádkách 1780, 370 05 České Budějovice, Czech Republic
6 Institute of Parasitology, Biology Centre of the Czech Academy of Sciences, 370 05 Ceske Budejovice, Czech Republic
* Correspondence: minwup@kust.edu.cn (M.W.); christian_ew_steinberg@web.de (C.E.W.S.); Tel.: +86-871-6510-2829 (M.W.); +49-30-614-2746 (C.E.W.S.)

Abstract: As an effective soil amendment, biochars require a comprehensive ecological evaluation before they can be widely used in agriculture because endogenous contaminants, such as environmentally persistent free radicals (EPFRs), certainly pose an ecological risk to soil invertebrates. In this study, *Caenorhabditis elegans* (*C. elegans*) was used as a model organism to investigate the neurotoxicity of two rice straw biochars pyrolyzed at 500 and 700 °C. After 24 h exposure to unwashed biochar, washed biochar, and leaching fluids (supernatants), the neurobehavioral parameters of *C. elegans* were determined in a liquid toxicity test. The results showed that the washed 700 °C biochar particles significantly impaired locomotion and prolonged the defecation interval at a biochar concentration of 4 g·well^{-1}, while the unwashed biochar and supernatants caused no apparent impairment. Supporting this, electron paramagnetic resonance (EPR) results showed that the intensity of EPFRs in unwashed 700 °C biochar was stronger than that of the corresponding washed particles. This indicates that, in the liquid test, the EPR signal alone is not indicative of particle toxicity. The accessibility and activity of the EPFRs should be considered. Dissolved organic matter (DOM) was observed in the leaching fluids. The neurotoxic activity of the washed biochar was alleviated after the re-addition of leaching fluids to the washed biochar, suggesting that the dissolved organic materials modulate the reactivity of the EPFRs in the liquid phase. This study suggests that the leaching process may increase the risk of biochar when used in the field environment.

Keywords: environmental persistent free radicals; *Caenorhabditis elegans*; neurotoxicity; biochar

Citation: Zhang, X.; Saul, N.; Lieke, T.; Chen, Y.; Wu, M.; Pan, B.; Steinberg, C.E.W. Biochar Extracts Can Modulate the Toxicity of Persistent Free Radicals in the Nematode *Caenorhabditis elegans*. *Appl. Biosci.* **2023**, 2, 71–83. https://doi.org/10.3390/applbiosci2010007

Academic Editor: Robert Henry

Received: 30 November 2022
Revised: 10 February 2023
Accepted: 24 February 2023
Published: 28 February 2023

1. Introduction

Biochar is an environmentally friendly carbonaceous material to improve soil quality [1,2]. Biochar is expected to mitigate global warming through carbon sequestration [3]; enhance crop production by improving soil quality, including fertility and water balance [4,5]; and sequester contaminants due to its absorbability [6]. However, biochar can also be critical to organisms due to endogenous contaminants [7]. Although conventional endogenous contaminants in biomass-derived biochars, such as polycyclic aromatic hydrocarbons (PAHs) and metals, have not posed significant toxicity risks to soil organisms due to their relatively low concentrations [8–10]. An emerging contaminant, environmentally persistent free radicals (EPFRs), has been detected in biochar and is of increasing concern due to its long-term existence and potential risks in environmental compartments such as soil, atmospheric particulate matter, and coal dust [11,12]. In general, EPFRs have typically long been overlooked as elicitors of adverse effects in individuals or populations, probably because traditional analytical techniques did not detect EPFRs and there was a lack of understanding of their fate in the environment [13]. In previous studies, EPFRs found in biochar were able to damage seed membranes and inhibit seedling growth and urease activity in soil [8,14], and were even neurotoxic to an invertebrate model organism, *Caenorhabditis elegans* (*C. elegans*) [15]. This implies that the potential adverse effects of exposure to biochar-containing EPFRs should be thoroughly evaluated before large-scale application.

Dissolved organic matter (DOM) is a ubiquitous component in terrestrial and aquatic environments. It plays a critical role in the fate, transport, and toxicity of environmental pollutants [2,16–18]. DOM has been recognized as a signaling compound in the environment due to its ability to interact with molecular signaling structures in invertebrates and fish [19–22]. Studies found that the nematode *C. elegans* prefers to live in DOM-rich environments, both in nature and under laboratory conditions, and that DOM extends their lifespan and increases their reproductive capacity [20,21]. Biochar, as a source of DOM, can release significant amounts of DOM into the soil under rainfall and irrigation conditions [23–26]. DOM has a positive effect on soil quality by providing essential nutrients and promoting the growth and diversity of microorganisms. Still, it could also have adverse effects, such as increased toxicity of environmental pollutants [15,27]. However, the effect of DOM and its impact on organisms has not yet been fully explored or understood. It is important to have a better understanding of the impact of DOM leaching from biochar on the environment and its biota, as this will provide a more comprehensive evaluation of using biochar in soil applications.

To further investigate the potential toxicity of EPFRs on organisms and the role of DOM, experiments were conducted using *C. elegans* and biochars with varying intensities of EPFR and DOM content. *C. elegans* has proven to be an effective and powerful tool in evaluating the risks of environmental pollutants, such as nanomaterials and xenobiotics [15,27]. First, neurobehavior, including autonomic behavior (movement length, body bend, defecation interval) and sensory behavior (touch response and chemotaxis), was determined to evaluate the neurotoxic potential of the biochar [15,27]. The results were compared to the effects of EPFRs to determine the cause of stress. Finally, DOM leached from the biochar was analyzed to verify its impact on the nematodes. We hypothesize that DOM reduces the direct neurotoxicity of biochar-derived EPFRs to *C. elegans*.

2. Material and Method

2.1. Strains and Maintenance

Wild-type *C. elegans* nematode N2 (var. Bristol) and its feed *Escherichia coli* strain OP50 were originally obtained from the *Caenorhabditis* Genetics Center (University of Minnesota, Minneapolis, MN, USA). After culturing OP50 in 200 mL LB medium at 37 °C for 12 h, the optical density (OD) was detected at 600 nm spectrophotometrically (Techne Specgene, Model FSPECGE, Stone, UK). The OD_{600} value used in this study was 2.6 ± 0.3. Before exposure tests, all *C. elegans* were kept in nematode growth plates (NGM) at 20 °C and seeded with OP50 bacteria as a diet.

2.2. Preparation and Characterization of Biochars

Rice straw was collected from the Wujiaying Residential District, Chenggong, Kunming, China (24.8°N; 102.8°E) as raw biomass for biochar production. The biochar was produced as described in our previous study [15].

The samples were dried, ground, and passed through a 60-mesh sieve. They were then pyrolyzed at 500 and 700 °C for 2 h under N_2 atmosphere in a muffle furnace. The resulting biochars were ground and sieved through a 300-mesh sieve. According to the pyrolysis temperature, the unwashed biochars were labeled B500 and B700. The washed particles were prepared by adding 40 mg of unwashed biochar to 30,000 μL of sterilized double-distilled water (sddH_2O), then dried and labeled as P500 and P700. To obtain the supernatant, 40 mg of unwashed biochar was added to 15,000 μL of sddH_2O and shaken for 24 h at 20 °C without light; the leachate was passed through a 0.45 μm filter and labeled as S500 and S700. All supernatants were stored at 4 °C.

Hydroxyl radicals (•OH) were scavenged with 5,5-dimethyl-1-pyrroline N-oxide (DMPO, Aladdin). Briefly, for unwashed and washed biochar, 4 mg of solid powder was mixed with 300 μL of sddH_2O, vortexed for 1 min, and the solid powder was allowed to settle. Then, 100 μL of 0.3 M fresh DMPO was mixed with 100 μL supernatant of the solid samples; for liquid samples, 100 μL of the sample was mixed with 100 μL of fresh DMPO. Finally, 50 μL of each sample was used to determine the free radical content by absorption in a micropipette (1.5 mm in outer diameter, 0.5 mm in wall thickness, and 125 mm in length). Electron paramagnetic resonance (EPR) signals were recorded using an EPR spectrometer (X-Band A300-6/1, Bruker, Billerica, MA, USA) [8]. All EPR detection experiments were performed at room temperature.

The DOM concentration in the supernatants was determined with a fluorescence excitation emission matrix (EEM). Spectra were recorded with an excitation wavelength range of 200–400 nm and an emission wavelength range of 250–550 nm using an F-7000 FL spectrophotometer (HITACHI, Hitachi, Ibaraki, Japan) at a scan speed of 12,000 $nm \cdot min^{-1}$ and a voltage of 550 V. All supernatants were freshly prepared before testing, as described above.

2.3. Exposure Condition

C. elegans individuals were exposed to three treatments: unwashed biochar, washed biochar, and supernatant, respectively. The unwashed and washed biochars were mixed directly with the exposure medium (K medium: M9 buffer and OP50 in a 3:3:1 ratio, see Text S1 for more details). In each well of a sterile 24-well tissue culture plate (Costar, Corning, New York, NY, USA), 300 μL of exposure medium was added, including 0, 0.5, 1, 2, or 4 mg of unwashed and washed biochar according to our previous experiment [15] and the germination assay of biochar at the ratio of 0, 0.5, 1, 2, 4 g per Petri dish [8]. For unwashed biochar of 500 °C (B500), it equals about 0, 550, 1100, 2200, 4400 mg $C \cdot L^{-1}$; for unwashed biochar of 700 °C (B700), it equals about 0, 400, 800, 1600, 3200 mg $C \cdot L^{-1}$. All the carbon data are based on the elemental analyses (Table S2).

For the supernatant tests, 150 μL was added to 150 μL of exposure medium (to obtain the same solid–liquid ratio of 40 mg: 30,000 μL was used to obtain the washed biochar). *C. elegans* worms were synchronized and cultured to L4 stage at 20 °C on NGM plates before being transferred to the 24-well exposure plates. All test plates were cultured at 20 °C for 24 h, and *C. elegans* were transferred to new NGM plates before neurotoxic assays.

2.4. Neurotoxic Assays

Neurobehavioral experiments, including autonomic and sensory behaviors, were performed as previously described [15,27]. For the autonomic behavioral tests, a single worm was transferred to a new NGM plate covered with OP50, and the crawling path was observed for 20 s under a microscope (Nikon SMZ 1500, Tokyo, Japan). Body bends were counted as changes in the direction of movement in the sinusoidal motion of the worm, and the relative length of movement was measured using the NIS-Elements D software (Nikon Eclipse E100, Tokyo, Japan) and normalized to the average body size of each group. The defecation intervals were calculated as the time between two contractions of the posterior body wall under the microscope. For the mechanical sensory behavior, the anterior tip of the worm was touched 10 times with a fine hair, and each stimulus response was recorded as a reversal movement. The chemotaxis assay is based on the worm's attraction to sodium chloride (NaCl), as its memory is associated with feeding. Ninety-six mm chemotaxis assay plates were prepared without NaCl and stored at 4 °C (Text S2). Twenty-four hours before an assay, an agar plug (approximately 5 mm × 5 mm × 3 mm) containing 100 mM NaCl was placed on the NaCl spot, with the NaCl spot and the control spot equidistant from the starting spot (3 cm), and their distance was 4 cm. The agar plug was removed just before the assay, then 1 μL 0.5 M sodium azide was added to the NaCl spot and the control spot. Thirty worms were transferred to the starting spot and incubated at 20 °C for 1 h. Then, the number of nematodes in the circular area (r = 1.5 cm) of the NaCl spot and the control spot was counted as N_n and N_c, respectively. The chemotaxis index was calculated as the difference between the number of N_n and N_c divided by the total number. The test order for each day was body bending, relative movement length, touch response, and defecation; the chemotaxis test was constantly tested on a different day and was not performed with the four endpoints mentioned above. For the body bending, defecation, and mechanical sensory assays, at least 10 worms were randomly selected from each exposure concentration. All experiments were repeated at least three times.

2.5. The Autonomic Behavior Assay of Selected Supernatant

Supernatants S500 and S700 were selected to test their potential effects after a short exposure time (12 h, test 1 in Figure 1). Autonomous behavior was performed as described in Section 2.4. Next, S500 or S700 was added to the M9 buffer containing OP50 (see Section 2.3), and body curvature and relative movement length were determined after 12 h of exposure. In test 2 (Figure 1), *C. elegans* were first treated with S500 or S700 for 12 h, then P500 and P700 were added to the appropriate portion of the supernatant (S+P500 and S+P700). Additional controls were exposed to P500 and P700 only (C+P500 and C+P700). The concentration of P500 and P700 was 4 mg·well^{-1}, and the corresponding supernatants were used.

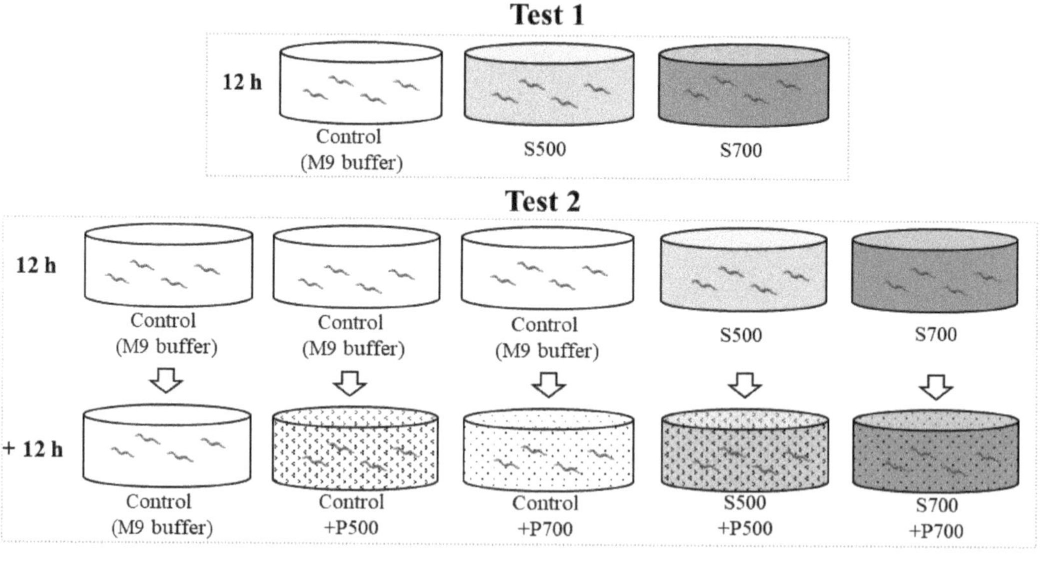

Figure 1. Diagram of the autonomic assay with supernatant (S500 and S700) and washed biochar particles (P500 and P700). Test 1: Exposure to the supernatant for 12 h. Test 2: Exposure to the supernatant for 12 h with the subsequent addition of washed particles.

2.6. Statistical Analysis

The statistical significance of the changes was determined using GraphPad Prism 7 (GraphPad Software, San Diego, CA, USA) and the one-way ANOVA test (Holm–Sidak's multiple comparison test). Probability levels of 0.05 (*) and 0.01 (**) were considered statistically significant. All neurobehavioral data are expressed as mean ± SEM (standard error of the mean). Neurobehavioral figures were generated using Graphpad Prism 7. For the EPR signal and fluorescence of DOM, the figures were generated using Origin 2018 (OriginLab Corporation, Northampton, MA, USA).

3. Results and Discussion

3.1. Only Washed Particles from 700 °C Pyrolysis Caused Neurobehavioral Changes

To investigate the neurotoxic potential of different biochars on *C. elegans*, two different pyrolysis temperatures (500 and 700 °C) were chosen for biochar preparation. *C. elegans* were exposed to unwashed biochars, washed biochar particles, and supernatants. After 24 h exposure to unwashed biochars and supernatants, none of the neurological parameters tested was significantly altered by any biochar concentration (Figure S1 in Supplementary). For washed biochar particles, body curvatures decreased by 15.1% ($p < 0.01$) and 21.0% ($p < 0.01$) after treatment with 1 mg·well^{-1} and 4 mg·well^{-1} of P700, respectively (Figure 2B). After exposure to 2 and 4 mg·well^{-1} of P700, the defecation interval increased by 13.3% ($p < 0.01$) and 20.9% ($p < 0.01$), respectively (Figure 2C). Relative movement length, touch response, and chemotaxis did not change for any particle exposures tested (Figure 2A,D,E).

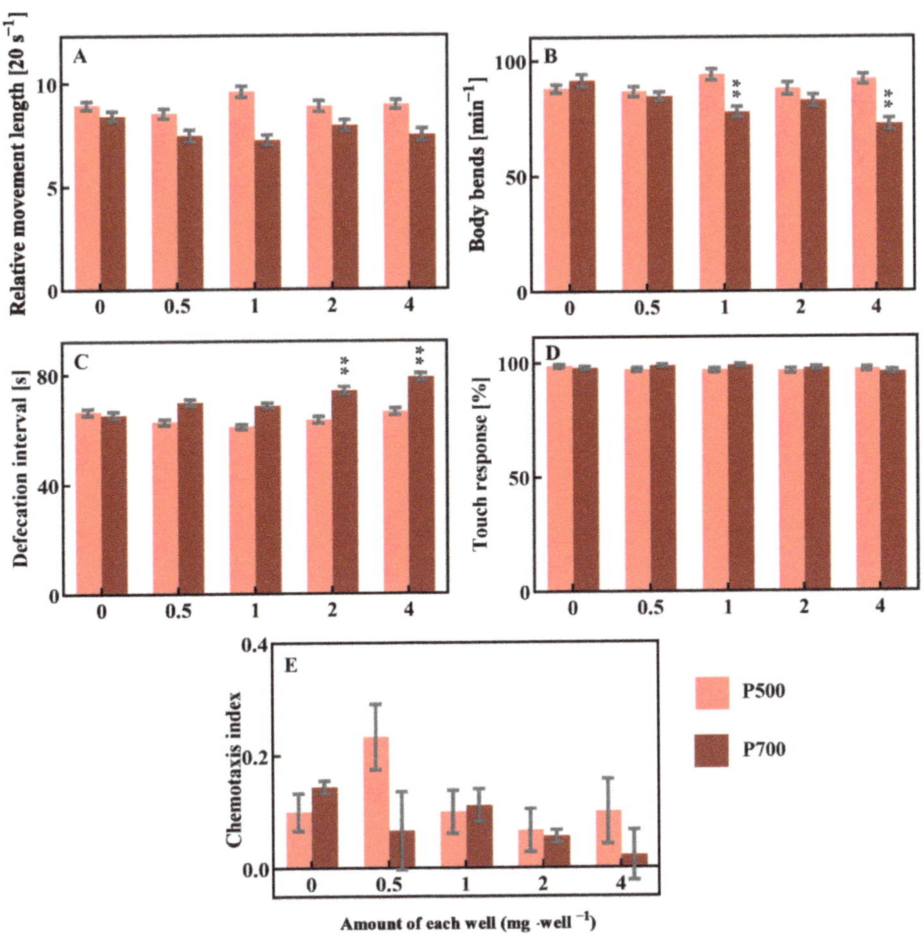

Figure 2. (**A–E**) Neurobehavior of *Caenorhabditis elegans* after 24 h exposure to washed biochar particles of 500 °C and 700 °C (P500 and P700). ** means $p < 0.01$, which indicates a significant difference.

In the literature, the avoidance behavior of earthworms to biochar could be alleviated in the liquid phase [28], and the phytotoxicity of corncob biochar could also be reduced by washing out the water-soluble toxic compounds [29]. However, the question arises: how does biochar change during the washing process? Thus, to find the underlying cause of the phenotypical observations, we compare the different characteristics before and after washing the biochar (details are shown in Figure S2, Tables S1 and S2).

The consequences of the reduced rate of body bends could be an impaired ability to reach food and avoid stressful or risky places. The prolonged defecation interval was considered a possibility of neuronal damage [30]. In this study, unwashed biochar had no adverse effects on nematodes, while the washed particle fraction at 700 °C caused significant inhibition of the autonomic behavior. These results contradict our initial assumption that EPFRs could trigger the production of free oxygen radicals (•OH) in solution and, therefore, exhibit higher toxicity in solution than on the NGM plate. Compared to our previous study using solid agar surface plates for exposure, in which 500 °C rice biochar with high-intensity EPFRs decreased locomotion behavior and prolonged the defecation interval of *C. elegans* [15], the present study found that the two unwashed biochars did not significantly affect the behavioral parameters tested. According to the literature, earthworm avoidance behavior towards biochar was mitigated in the liquid phase [28], and the phytotoxicity of biochar was also reduced by washing [29]. However, the question remains on how biochar changes during the washing process. To find the reason for the phenotypic observations, we compare the different traits before and after washing the biochar (details are shown in Figure S2, Tables S1 and S2).

3.2. EPFRs Reactivity Might Play a Crucial Role in Neurotoxicity

Many contaminants, such as PAHs and metals, remain or are formed during the pyrolysis of biochar [11]. Our previous work with plate cultures showed that PAHs and metals were not the main cause of the observed neurotoxicity due to their low content in biochar [15]. EPFRs have been found in biochars and have been shown to have the potential to inhibit grain seedling growth and modulate *C. elegans* behavior [8,14,15]. In this word, EPFRs were also determined in all samples before and after washing. The *g*-factor and EPR signal intensity parameters can be used to identify the type of free radicals and the concentration of EPR-active species [8,31]. As shown in Figure 3, both the unwashed biochar and the washed biochar particles exhibited significantly higher intensities than the blank control. The supernatants did not show any EPFR signals. With increasing pyrolysis temperature, the EPR intensity of the unwashed biochar decreased from 4.8×10^5 spins·mg^{-1} (500 °C) to 5.9×10^4 spins·mg^{-1} (700 °C), and the same phenomenon was observed for the washed particles (Figure 3A,C). The toxicity experiments showed that only P700 decreased body bends and prolonged defecation intervals (Figure 2B,C). At the same time, P500 and B500, with the strongest EPR intensities, did not affect the neurobehavior of *C. elegans*. Moreover, Yang, et al. [32] showed that the degradation of 700 °C biochar with low EPR intensity was even better than that of 500 °C biochar with relatively high intensity of EPR in the first 100 h, indicating that the 700 °C biochar is likely to be more reactive. Thus, a stronger EPR signal does not necessarily indicate higher toxicity; rather, the reactivity of EPFRs should be considered.

The EPFRs could stimulate the formation of •OH in an aqueous solution, causing a deleterious effect [8,33,34]. In this work, •OH was trapped using DMPO, and the typical spectrum of the DMPO-OH adduct is a quartet shape with a relative intensity ratio of 1:2:2:1 [8]. As shown in Figure 3B,D, both unwashed (B500 and B700) and washed biochar (P500 and P700) exhibit strong DMPO-OH signals. This indicates that both 500 °C and 700 °C biochar can induce the production of •OH. Researchers shows that •OH is the main cause of organic chemical degradation and biochar toxicity [8,35], indicating that •OH might be the reason for the toxicity. However, only P700 showed a significant inhibitory effect (Figure 2), suggesting that •OH may not be responsible for this observation.

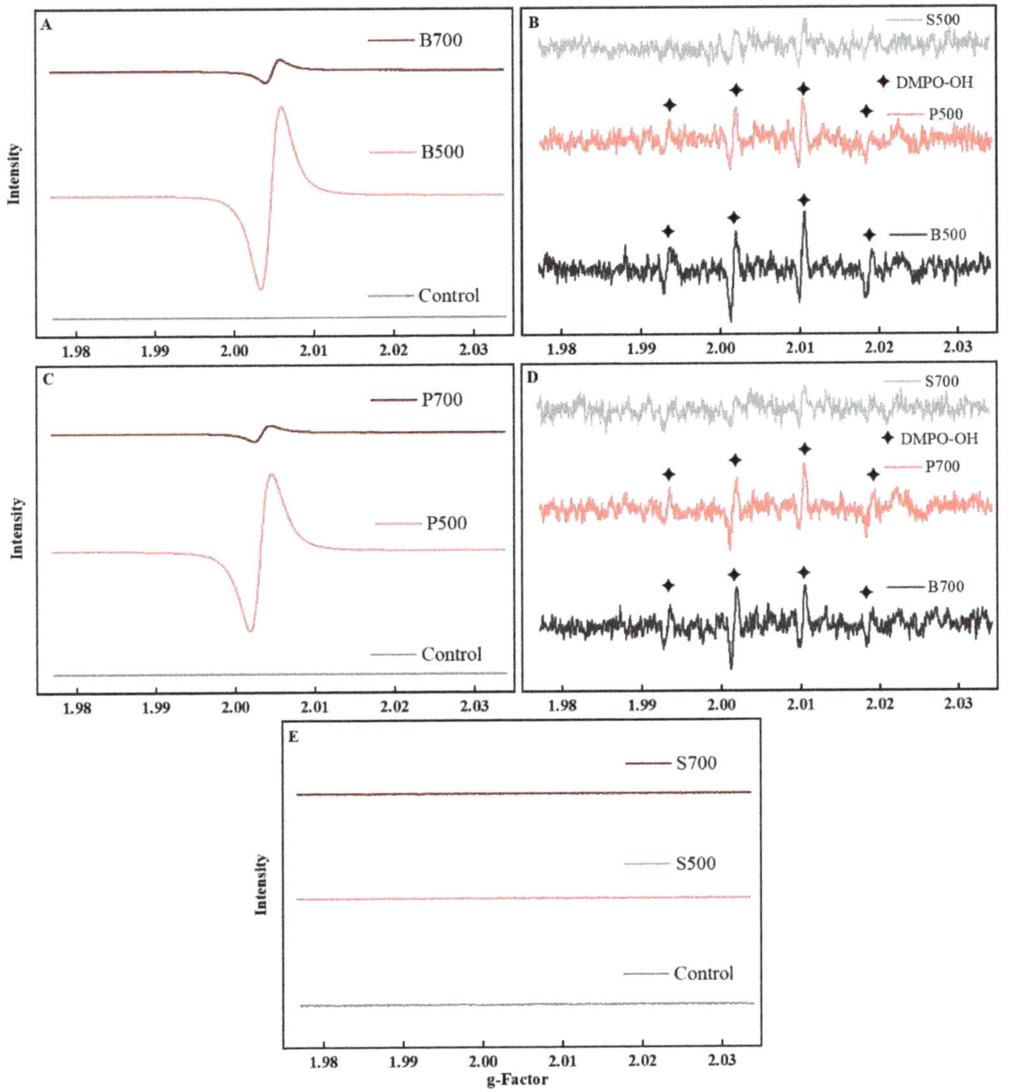

Figure 3. The EPR spectra of biochar at 500 and 700 °C. The resulting intensity of unwashed biochar (**A**), washed particle (**C**), and supernatant (**E**) was divided by their mass (mg^{-1}). EPR signals of DMPO-OH (**B**,**D**) are marked as four-point stars (✦).

3.3. DOM Can Modulate the Toxicity of EPFRs

Dissolved organic material plays an essential role in the formation and stability of EPFRs and can affect EPFR activity [13,36,37]. In this work, DOM was observed in all supernatants leached from biochar using the EEM spectrometer, and the peaks at the excitation/emission wavelength pair (Ex/Em) of S500 and S700 were 270 nm/395 nm and 250 nm/380 nm, respectively (Figure 4). Peaks at these two Ex/Em pairs are commonly associated with the presence of humic acid-like substances [38,39]. The relative fluorescence intensity of S500 was stronger than that of S700, showing that the amount of DOM leached from B500 was higher than that from B700.

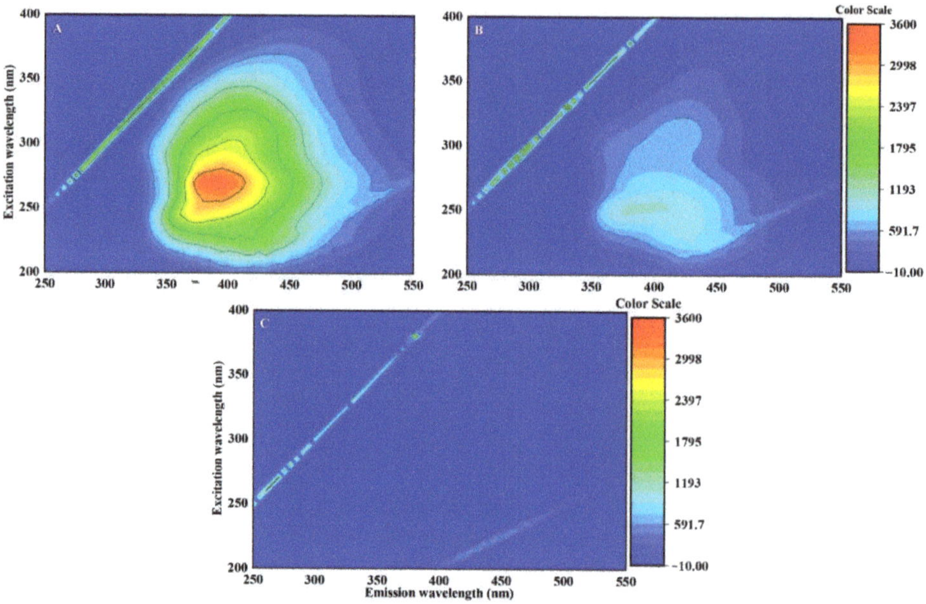

Figure 4. Fluorescence excitation-emission matrices of S500 (**A**), S700 (**B**), and H_2O (**C**).

Applying solid-state ^{13}C NMR, Liu, et al. [40] showed that biochar DOM contains abundant small aromatics with fused rings, aliphatic C, and carboxyl-C, and that water-extractable DOM can reach up to 0.5–40 mg C g^{-1} biochar. This coarse characterization is similar to that of humic substances. Pyrolyzing manure, protein-like and humic acid-like substances were the major components of DOM. The proteinaceous matter first decreased and then increased with an increase in the pyrolysis temperature, while an opposite trend was observed for the humic acid-like matter [41]. By sequential extraction, Liu, et al. [42] confirmed the humic-like character of DOM extractable from biochars. They showed that water-extractable DOM components are mostly composed of low aromatic and highly hydrophilic fulvic acid-like substances, which are the most labile and diversified components. This statement was confirmed by recent studies [43–45], which also detected fulvic acid-like compounds. In contrast, Tomczyk, et al. [46] found that extractable DOM from biochars produced from potato and raspberry stems was predominantly constituted of substances of large molecular weight and high aromaticity. This illustrates that there is (still) no unifying model that describes under which conditions, and of what quality, leachable organic carbon is produced by the pyrolysis of biowaste. However, some effects of humic and humic-like substances on soil organisms can be predicted, which are relevant for the present study since mainly fulvic acids interact with organisms [45,47].

Humic substances primarily reduce the bioavailability of organic chemicals, metals, and nanomaterials to *C. elegans*, and the extent of the reduction depends on the quantity and quality of the humic substances [21,46,48,49]. In addition, Lieke, et al. [50] found that fulvic acid protects fish gills from oxidative stress, probably by directly scavenging reactive oxygen species (ROS). Based on the above studies, DOM in unwashed biochar and supernatant can be expected to modulate the fate and reactivity of EPFRs. However, humic substances in higher internalized concentrations can block exporter proteins [51] and disproportionately increase bioconcentrations of persistent organic pollutants [52]. This could be detrimental if these persistent organic pollutants can also be toxic via the release of EPFRs.

To test whether DOM can modulate the toxicity of EPFRs, supernatants and particles of 500 °C and 700 °C biochar were selected to determine the effect of DOM. In test 1, the nematodes were directly exposed to the supernatant for 12 h. The aim is to check whether the nematode adapts to the DOM environment after 24 h exposure since no impairments were observed (unshown data). The relative movement length and body bends of nematodes exposed to S700 increased significantly compared to the control (Figure 5A,C). In addition, there was a significant difference between S500 and S700, and exposure to S700 improved locomotion. In test 2, nematodes were first exposed to supernatants for 12 h, then corresponding particles were added for another 12 h. After exposure to C+P500 and C+P700, the rate of body bends decreased compared to the control, but there was no significant difference between P500 and P700 (Figure 5B). The results demonstrate that the washed particles are harmful to the nematodes. After 12 h exposure to the supernatant and 12 h exposure to the particles, the body bends of the nematodes exposed to S+P500 were significantly different from C+P500 but not significantly different from the control. The supernatant S500 significantly reduced the damage caused by EPFRs in the particles. A comparison of DOM levels and EPFR intensities in biochar, particles, and supernatants indicates that DOM can modulate the toxicity of EPFRs.

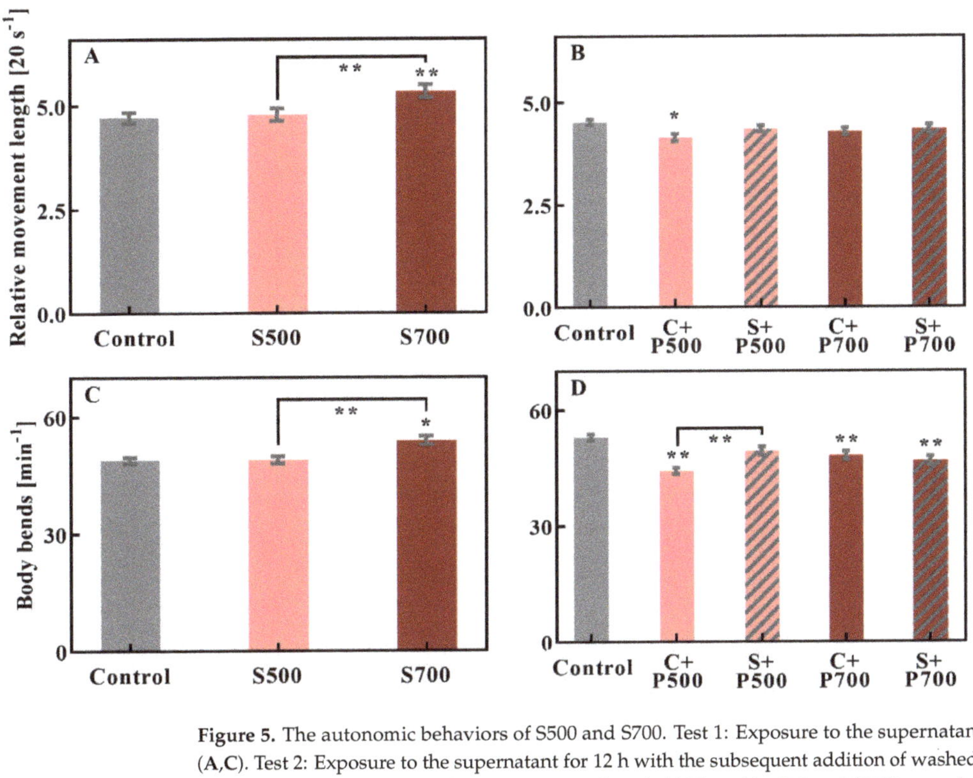

Figure 5. The autonomic behaviors of S500 and S700. Test 1: Exposure to the supernatant for 12 h (**A**,**C**). Test 2: Exposure to the supernatant for 12 h with the subsequent addition of washed particles (**B**,**D**). After exposure, the relative movement length (**A**,**B**) and body bend (**C**,**D**) were determined. * $p < 0.05$, ** $p < 0.01$.

It is evident that DOM has modulatory effects on environmental substances and their effects on organisms; however, the direction of modulation is not uniform. In a model analysis, Steinberg and Brüggemann [53] showed that humic substances can react in seemingly opposite ways, depending on their structure and chemical characterization. In fact, the effect of organic matter is a trade-off between promotion and inhibition due to its diverse nature, and this contrasting effect appeared to be intrinsic. Moreover, dissolved humic substances could facilitate fish life in rather acidic environments and prolong the lifespan of *C. elegans* [17]. Therefore, it is not unlikely that DOM, which is leachable from biochar, can protect nematodes from damage by EPFRs. However, a promoting effect of S700 was observed in test 1, while no significant reduction in toxicity was observed in test 2. The reason may be the different nature and reactivity of EPFRs in the solid particles and the leached DOM between 500 °C and 700 °C, which should be clarified in future studies

4. Conclusions

The present study aimed to determine the reactivity of EPFRs in different biochars with or without the presence of different DOM. Increased neurotoxicity was only present in EPFRs with relatively high reactivity (P700). Although the potential risks of EPFRs in biochar have been a growing concern, our findings suggest that the presence of organic matter in the environment can significantly influence the toxicity of EPFRs, potentially overestimating the danger they pose. The DOM released from biochar can alleviate the harm from biochar particles, such as through (1) decreasing the probability of direct contact between biochar and *C. elegans*, and (2) partially quenching EPFRs by some active components of DOM. This may be the reason why EPFRs have been overlooked and understudied. To better understand how environmental factors affect the reactivity of EPFRs, future research should focus on using different types of biochar and developing methods to quantify reactivity. In addition, this study was only considering short-term exposure; long-term exposure may lead to chronic toxicity or affect the next generations, which also should be of more concern in further studies. This could contribute to developing effective strategies for reducing the risks associated with EPFRs and promoting the safe use of biochar in various applications.

Supplementary Materials: The following supporting information can be downloaded at: https://www.mdpi.com/article/10.3390/applbiosci2010007/s1, Figure S1. Neurobehavior of *C. elegans* after 24 h exposure to unwashed biochars (A, D, G, J, and M), washed biochar particles (B, E, H, K, and N, these data were copied from Figure 2A–E for a better comparison), and supernatants (C, F, I, L, and O) at 20 °C. Relative move length (A, B, and C), body bend (D, E, and F), defecation interval (G, H, and I), touch response (J, K, and L), and chemotaxis index (M, N, and O) were determined. * $p < 0.05$, ** $p < 0.01$; Figure S2. Fourier transform infrared (FTIR) spectra of unwashed biochar and washed biochar. Samples were prepared with potassium bromide (mass ratio = 1:1500) and detected by ThermoFisher spectrometer (NICOLET iS50 FT-IR, USA). The dark line means unwashed biochar, the grey line means washed biochar particles.; Table S1. Total of carbon concentration in the supernatant; Table S2. The elemental composition in unwashed biochar and washed biochar.

Author Contributions: Conceptualization, X.Z. and M.W.; Data curation, B.P.; Formal analysis, X.Z., N.S., T.L., Y.C., M.W., B.P. and C.E.W.S.; Funding acquisition, M.W. and B.P.; Investigation, X.Z. and Y.C.; Methodology, X.Z., N.S., T.L., B.P. and C.E.W.S.; Project administration, M.W. and B.P.; Resources, X.Z., N.S., T.L. and C.E.W.S.; Supervision, M.W. and B.P.; Validation, X.Z.; Visualization, M.W. and B.P.; Writing—original draft, X.Z.; Writing—review and editing, X.Z., N.S., T.L., Y.C., M.W., B.P. and C.E.W.S. All authors have read and agreed to the published version of the manuscript.

Funding: This research was supported by the National Natural Scientific Foundation of China (41977334, and 42067055), the Yunnan Province Basic Research Project (202001AS070015), and the joint fund between NSFC-NCN (41961134002).

Institutional Review Board Statement: Not applicable.

Informed Consent Statement: Not applicable.

Data Availability Statement: Not applicable.

Conflicts of Interest: The authors declare no competing interests.

References

1. El-Naggar, A.; Lee, S.S.; Rinklebe, J.; Farooq, M.; Song, H.; Sarmah, A.K.; Zimmerman, A.R.; Ahmad, M.; Shaheen, S.M.; Ok, Y.S. Biochar application to low fertility soils: A review of current status, and future prospects. *Geoderma* **2019**, *337*, 536–554. [CrossRef]
2. Yu, S.; Liu, J.; Yin, Y.; Shen, M. Interactions between engineered nanoparticles and dissolved organic matter: A review on mechanisms and environmental effects. *J. Environ. Sci. (China)* **2018**, *63*, 198–217. [CrossRef]
3. Spokas, K.A.; Cantrell, K.B.; Novak, J.M.; Archer, D.W.; Ippolito, J.A.; Collins, H.P.; Boateng, A.A.; Lima, I.M.; Lamb, M.C.; McAloon, A.J.; et al. Biochar: A synthesis of its agronomic impact beyond carbon sequestration. *J. Environ. Qual.* **2012**, *41*, 973–989. [CrossRef] [PubMed]
4. Beesley, L.; Moreno-Jimenez, E.; Gomez-Eyles, J.L.; Harris, E.; Robinson, B.; Sizmur, T. A review of biochars' potential role in the remediation, revegetation and restoration of contaminated soils. *Environ. Pollut.* **2011**, *159*, 3269–3282. [CrossRef] [PubMed]
5. Mohan, D.; Sarswat, A.; Ok, Y.S.; Pittman, C.U., Jr. Organic and inorganic contaminants removal from water with biochar, a renewable, low cost and sustainable adsorbent–a critical review. *Bioresour. Technol.* **2014**, *160*, 191–202. [CrossRef]
6. Hussain, M.; Farooq, M.; Nawaz, A.; Al-Sadi, A.M.; Solaiman, Z.M.; Alghamdi, S.S.; Ammara, U.; Ok, Y.S.; Siddique, K.H.M. Biochar for crop production: Potential benefits and risks. *J. Soils Sediments* **2017**, *17*, 685–716. [CrossRef]
7. Hale, S.E.; Lehmann, J.; Rutherford, D.; Zimmerman, A.R.; Bachmann, R.T.; Shitumbanuma, V.; O'Toole, A.; Sundqvist, K.L.; Arp, H.P.; Cornelissen, G. Quantifying the total and bioavailable polycyclic aromatic hydrocarbons and dioxins in biochars. *Environ. Sci. Technol.* **2012**, *46*, 2830–2838. [CrossRef]
8. Liao, S.; Pan, B.; Li, H.; Zhang, D.; Xing, B. Detecting free radicals in biochars and determining their ability to inhibit the germination and growth of corn, wheat and rice seedlings. *Environ. Sci. Technol.* **2014**, *48*, 8581–8587. [CrossRef]
9. Malev, O.; Contin, M.; Licen, S.; Barbieri, P.; De Nobili, M. Bioaccumulation of polycyclic aromatic hydrocarbons and survival of earthworms (*Eisenia andrei*) exposed to biochar amended soils. *Environ. Sci. Pollut. Res.* **2016**, *23*, 3491–3502. [CrossRef]
10. Wang, P.; Pan, B.; Li, H.; Huang, Y.; Dong, X.; Ai, F.; Liu, L.; Wu, M.; Xing, B. The overlooked occurrence of environmentally persistent free radicals in an area with low-rank coal burning, Xuanwei, China. *Environ. Sci. Technol.* **2018**, *52*, 1054–1061. [CrossRef]
11. Lian, F.; Xing, B. Black carbon (biochar) in water/soil environments: Molecular structure, sorption, stability, and potential risk. *Environ. Sci. Technol.* **2017**, *51*, 13517–13532. [CrossRef] [PubMed]
12. Zhang, C.; Zeng, G.; Huang, D.; Lai, C.; Chen, M.; Cheng, M.; Tang, W.; Tang, L.; Dong, H.; Huang, B.; et al. Biochar for environmental management: Mitigating greenhouse gas emissions, contaminant treatment, and potential negative impacts. *Chem. Eng. J.* **2019**, *373*, 902–922. [CrossRef]
13. Pan, B.; Li, H.; Lang, D.; Xing, B. Environmentally persistent free radicals: Occurrence, formation mechanisms and implications. *Environ. Pollut.* **2019**, *248*, 320–331. [CrossRef] [PubMed]
14. Liu, Y.; Dai, Q.; Jin, X.; Dong, X.; Peng, J.; Wu, M.; Liang, N.; Pan, B.; Xing, B. Negative impacts of biochars on urease activity: High pH, heavy metals, polycyclic aromatic hydrocarbons, or free radicals? *Environ. Sci. Technol.* **2018**, *52*, 12740–12747. [CrossRef] [PubMed]
15. Lieke, T.; Zhang, X.; Steinberg, C.E.W.; Pan, B. Overlooked risks of biochars: Persistent free radicals trigger neurotoxicity in *Caenorhabditis elegans*. *Environ. Sci. Technol.* **2018**, *52*, 7981–7987. [CrossRef]
16. Steinberg, C.E.W.; Sturm, A.; Kelbel, J.; Lee, S.K.; Hertkorn, N.; Freitag, D.; Kettrup, A.A. Changes of Acute Toxicity of Organic Chemicals to Daphnia magnain the Presence of Dissolved Humic Material (DHM). *Acta Hydrochim. Hydrobiol.* **1992**, *20*, 326–332. [CrossRef]
17. Steinberg, C.E.; Saul, N.; Pietsch, K.; Meinelt, T.; Rienau, S.; Menzel, R. Dissolved humic substances facilitate fish life in extreme aquatic environments and have the potential to extend the lifespan of Caenorhabditis elegans. *Ann. Environ. Sci.* **2007**, *1*, 81–90.
18. Yang, X.; Jiang, C.; Hsu-Kim, H.; Badireddy, A.R.; Dykstra, M.; Wiesner, M.; Hinton, D.E.; Meyer, J.N. Silver nanoparticle behavior, uptake, and toxicity in Caenorhabditis elegans: Effects of natural organic matter. *Environ. Sci. Technol.* **2014**, *48*, 3486–3495. [CrossRef]
19. Lutz, I.; Jie, Z.; Opitz, R.; Kloas, W.; Ying, X.; Menzel, R.; Steinberg, C.E. Environmental signals: Synthetic humic substances act as xeno-estrogen and affect the thyroid system of *Xenopus laevis*. *Chemosphere* **2005**, *61*, 1183–1188. [CrossRef]
20. Menzel, R.; Rodel, M.; Kulas, J.; Steinberg, C.E. CYP35: Xenobiotically induced gene expression in the nematode Caenorhabditis elegans. *Arch. Biochem. Biophys.* **2005**, *438*, 93–102. [CrossRef]
21. Menzel, R.; Sturzenbaum, S.; Barenwaldt, A.; Kulas, J.; Steinberg, C.E. Humic material induces behavioral and global transcriptional responses in the nematode Caenorhabditis elegans. *Environ. Sci. Technol.* **2005**, *39*, 8324–8332. [CrossRef] [PubMed]
22. Menzel, R.; Menzel, S.; Tiedt, S.; Kubsch, G.; Stosser, R.; Bahrs, H.; Putschew, A.; Saul, N.; Steinberg, C.E. Enrichment of humic material with hydroxybenzene moieties intensifies its physiological effects on the nematode Caenorhabditis elegans. *Environ. Sci. Technol.* **2011**, *45*, 8707–8715. [CrossRef] [PubMed]
23. Duan, W.; Oleszczuk, P.; Pan, B.; Xing, B. Environmental behavior of engineered biochars and their aging processes in soil. *Biochar* **2019**, *1*, 339–351. [CrossRef]
24. Rajapaksha, A.U.; Ok, Y.S.; El-Naggar, A.; Kim, H.; Song, F.; Kang, S.; Tsang, Y.F. Dissolved organic matter characterization of biochars produced from different feedstock materials. *J. Environ. Manag.* **2019**, *233*, 393–399. [CrossRef] [PubMed]

25. Rombola, A.G.; Torri, C.; Vassura, I.; Venturini, E.; Reggiani, R.; Fabbri, D. Effect of biochar amendment on organic matter and dissolved organic matter composition of agricultural soils from a two-year field experiment. *Sci. Total Environ.* **2022**, *812*, 151422. [CrossRef] [PubMed]
26. Sun, Y.; Xiong, X.; He, M.; Xu, Z.; Hou, D.; Zhang, W.; Ok, Y.S.; Rinklebe, J.; Wang, L.; Tsang, D.C.W. Roles of biochar-derived dissolved organic matter in soil amendment and environmental remediation: A critical review. *Chem. Eng. J.* **2021**, *424*, 130387. [CrossRef]
27. Ju, J.; Saul, N.; Kochan, C.; Putschew, A.; Pu, Y.; Yin, L.; Steinberg, C.E. Cyanobacterial xenobiotics as evaluated by a *Caenorhabditis elegans* neurotoxicity screening test. *Int. J. Environ. Res. Public Health* **2014**, *11*, 4589–4606. [CrossRef]
28. Li, D.; Hockaday, W.C.; Masiello, C.A.; Alvarez, P.J.J. Earthworm avoidance of biochar can be mitigated by wetting. *Soil Biol. Biochem.* **2011**, *43*, 1732–1737. [CrossRef]
29. Intani, K.; Latif, S.; Islam, M.; Müller, J. Phytotoxicity of Corncob Biochar before and after Heat Treatment and Washing. *Sustainability* **2018**, *11*, 30. [CrossRef]
30. Lieke, T.; Steinberg, C.E.; Ju, J.; Saul, N. Natural Marine and Synthetic Xenobiotics Get on Nematode's Nerves: Neuro-Stimulating and Neurotoxic Findings in Caenorhabditis elegans. *Mar. Drugs* **2015**, *13*, 2785–2812. [CrossRef]
31. Dellinger, B.; Lomnicki, S.; Khachatryan, L.; Maskos, Z.; Hall, R.W.; Adounkpe, J.; McFerrin, C.; Truong, H. Formation and stabilization of persistent free radicals. *Proc. Combust. Inst.* **2007**, *31*, 521–528. [CrossRef]
32. Yang, J.; Pan, B.; Li, H.; Liao, S.; Zhang, D.; Wu, M.; Xing, B. Degradation of *p*-nitrophenol on biochars: Role of persistent free radicals. *Environ. Sci. Technol.* **2016**, *50*, 694–700. [CrossRef]
33. Khachatryan, L.; Dellinger, B. Environmentally persistent free radicals (EPFRs)-2. Are free hydroxyl radicals generated in aqueous solutions? *Environ. Sci. Technol.* **2011**, *45*, 9232–9239. [CrossRef] [PubMed]
34. Khachatryan, L.; Vejerano, E.; Lomnicki, S.; Dellinger, B. Environmentally persistent free radicals (EPFRs). 1. Generation of reactive oxygen species in aqueous solutions. *Environ. Sci. Technol.* **2011**, *45*, 8559–8566. [CrossRef] [PubMed]
35. Wen, G.; Wang, S.J.; Ma, J.; Huang, T.L.; Liu, Z.Q.; Zhao, L.; Xu, J.L. Oxidative degradation of organic pollutants in aqueous solution using zero valent copper under aerobic atmosphere condition. *J. Hazard. Mater.* **2014**, *275*, 193–199. [CrossRef] [PubMed]
36. Paul, A.; Hackbarth, S.; Vogt, R.D.; Roder, B.; Burnison, B.K.; Steinberg, C.E. Photogeneration of singlet oxygen by humic substances: Comparison of humic substances of aquatic and terrestrial origin. *Photochem. Photobiol. Sci.* **2004**, *3*, 273–280. [CrossRef] [PubMed]
37. Huang, D.; Luo, H.; Zhang, C.; Zeng, G.; Lai, C.; Cheng, M.; Wang, R.; Deng, R.; Xue, W.; Gong, X.; et al. Nonnegligible role of biomass types and its compositions on the formation of persistent free radicals in biochar: Insight into the influences on Fenton-like process. *Chem. Eng. J.* **2019**, *361*, 353–363. [CrossRef]
38. Chen, W.; Westerhoff, P.; Leenheer, J.A.; Booksh, K. Fluorescence excitation-emission matrix regional integration to quantify spectra for dissolved organic matter. *Environ. Sci. Technol.* **2003**, *37*, 5701–5710. [CrossRef]
39. Helms, J.R.; Stubbins, A.; Perdue, E.M.; Green, N.W.; Chen, H.; Mopper, K. Photochemical bleaching of oceanic dissolved organic matter and its effect on absorption spectral slope and fluorescence. *Mar. Chem.* **2013**, *155*, 81–91. [CrossRef]
40. Liu, C.-H.; Chu, W.; Li, H.; Boyd, S.A.; Teppen, B.J.; Mao, J.; Lehmann, J.; Zhang, W. Quantification and characterization of dissolved organic carbon from biochars. *Geoderma* **2019**, *335*, 161–169. [CrossRef]
41. Gui, X.; Liu, C.; Li, F.; Wang, J. Effect of pyrolysis temperature on the composition of DOM in manure-derived biochar. *Ecotoxicol. Environ. Saf.* **2020**, *197*, 110597. [CrossRef] [PubMed]
42. Liu, H.; Zhao, B.; Zhang, X.; Li, L.; Zhao, Y.; Li, Y.; Duan, K. Investigating Biochar-Derived Dissolved Organic Carbon (DOC) Components Extracted Using a Sequential Extraction Protocol. *Materials* **2022**, *15*. [CrossRef] [PubMed]
43. Yu, S.; Zhang, H.; Ni, J.; Xiang, Y.; Wei, R.; Qian, W.; Chen, W. Spectral characteristics coupled with self-organizing maps analysis on different molecular size-fractionated water-soluble organic carbon from biochar. *Sci. Total Environ.* **2023**, *857*, 159424. [CrossRef]
44. Zhang, H.; Qian, W.; Wu, L.; Yu, S.; Wei, R.; Chen, W.; Ni, J. Spectral characteristics of dissolved organic carbon (DOC) derived from biomass pyrolysis: Biochar-derived DOC versus smoke-derived DOC, and their differences from natural DOC. *Chemosphere* **2022**, *302*, 134869. [CrossRef] [PubMed]
45. Zhang, P.; Huang, P.; Xu, X.; Sun, H.; Jiang, B.; Liao, Y. Spectroscopic and molecular characterization of biochar-derived dissolved organic matter and the associations with soil microbial responses. *Sci. Total Environ.* **2020**, *708*, 134619. [CrossRef] [PubMed]
46. Tomczyk, A.; Kubaczynski, A.; Szewczuk-Karpisz, K. Assessment of agricultural waste biochars for remediation of degraded water-soil environment: Dissolved organic carbon release and immobilization of impurities in one- or two-adsorbate systems. *Waste Manag.* **2023**, *155*, 87–98. [CrossRef]
47. Reemtsma, T.; These, A.; Linscheid, M.; Leenheer, J.; Spitzy, A. Molecular and structural characterization of dissolved organic matter from the deep ocean by FTICR-MS, including hydrophilic nitrogenous organic molecules. *Environ. Sci. Technol.* **2008**, *42*, 1430–1437. [CrossRef]
48. Collin, B.; Tsyusko, O.V.; Starnes, D.L.; Unrine, J.M. Effect of natural organic matter on dissolution and toxicity of sulfidized silver nanoparticles to Caenorhabditis elegans. *Environ. Sci. Nano* **2016**, *3*, 728–736. [CrossRef]
49. Haitzer, M.; Abbt-Braun, G.; Traunspurger, W.; Steinberg, C.E.W. Effects of humic substances on the bioconcentration of polycyclic aromatic hydrocarbons: Correlations with spectroscopic and chemical properties of humic substances. *Environ. Toxicol. Chem.* **1999**, *18*, 2782–2788. [CrossRef]

50. Lieke, T.; Steinberg, C.E.W.; Pan, B.; Perminova, I.V.; Meinelt, T.; Knopf, K.; Kloas, W. Phenol-rich fulvic acid as a water additive enhances growth, reduces stress, and stimulates the immune system of fish in aquaculture. *Sci. Rep.* **2021**, *11*, 174. [CrossRef]
51. Timofeyev, M.A.; Shatilina, Z.M.; Bedulina, D.S.; Menzel, R.; Steinberg, C.E. Natural organic matter (NOM) has the potential to modify the multixenobiotic resistance (MXR) activity in freshwater amphipods *Eulimnogammarus cyaneus* and *E. verrucosus*. *Comp. Biochem. Physiol. B Biochem. Mol. Biol.* **2007**, *146*, 496–503. [CrossRef] [PubMed]
52. Haitzer, M.; Hoss, S.; Traunspurger, W.; Steinberg, C. Effects of dissolved organic matter (DOM) on the bioconcentration of organic chemicals in aquatic organisms–a review. *Chemosphere* **1998**, *37*, 1335–1362. [CrossRef] [PubMed]
53. Steinberg, C.E.W.; Brüggemann, R. Ambiguous Ecological Control by Dissolved Humic Matter (DHM) and Natural Organic Matter (NOM): Trade-offs between Specific and Non-specific EffectsWe dedicate this paper to Prof. Dr. Fritz H. Frimmel on the occasion of his 60th birthday anniversary. *Acta Hydrochim. Et Hydrobiol.* **2001**, *29*, 399. [CrossRef]

applied biosciences

MDPI

Review

Nature-Inspired Cellulose-Based Active Materials: From 2D to 4D

Marta I. Magalhães [1] and Ana P. C. Almeida [1,2,3,*]

[1] Laboratório Associado Para a Química Verde/Requimte, Department of Chemistry, NOVA School of Science and Technology, 2829-516 Caparica, Portugal
[2] CENIMAT | i3N, Department of Materials Science, School of Science and Technology, NOVA University Lisbon, 2829-516 Caparica, Portugal
[3] iBET—Instituto de Biologia Experimental e Tecnológica, Apartado 12, 2780-901 Oeiras, Portugal
* Correspondence: ana.almeida@fct.unl.pt

Abstract: Multifunctional materials and devices with captivating properties can be assembled from cellulose and cellulose-based composite materials combining functionality with structural performance. Cellulose is one of the most abundant renewable materials with captivating properties, such as mechanical robustness, biocompatibility, and biodegradability. Cellulose is a low-cost and abundant biodegradable resource, CO_2 neutral, with a wide variety of fibers available all over the world. Over thousands of years, nature has perfected cellulose-based materials according to their needs, such as function vs. structure. Mimicking molecular structures at the nano-, micro-, and macroscales existing in nature is a great strategy to produce synthetic cellulose-based active materials. A concise background of cellulose and its structural organization, as well as the nomenclature of cellulose nanomaterials, are first addressed. Key examples of nature-designed materials with unique characteristics, such as "eternal" coloration and water-induced movement are presented. The production of biomimetic fiber and 2D fiber-based cellulosic materials that have attracted significant attention within the scientific community are represented. Nature-inspired materials with a focus on functionality and response to an external stimulus are reported. Some examples of 3D-printed cellulosic materials bioinspired, reported recently in the literature, are addressed. Finally, printed cellulosic materials that morph from a 1D strand or 2D surface into a 3D shape, in response to an external stimulus, are reported. The purpose of this review is to discuss the most recent developments in the field of "nature-inspired" cellulose-based active materials regarding design, manufacturing, and inspirational sources that feature existing tendencies.

Keywords: cellulose; biomimetic; bioinspired; fibers; films; membranes; 4D printing

Citation: Magalhães, M.I.; Almeida, A.P.C. Nature-Inspired Cellulose-Based Active Materials: From 2D to 4D. *Appl. Biosci.* **2023**, *2*, 94–114. https://doi.org/10.3390/applbiosci2010009

Academic Editor: Robert Henry

Received: 20 January 2023
Revised: 8 March 2023
Accepted: 9 March 2023
Published: 15 March 2023

1. Introduction

Anselme Payen isolated cellulose, a water-insoluble substance, from green plants in 1838. Payen observed isomerism with starch and determined the empirical formula, $C_6H_{10}O_5$ by elemental analysis [1,2]. Some species of bacteria, such as the nonpathogenic bacteria *Komagataeibacter xylinus*, former Acetobacter, and Gluconacetobacter, can also produce cellulose [3–5]. In addition, cellulose can also be produced by some algae, oomycetes, the amoeboid protozoa *Dictyostelium discoideum*, and by a group of special marine animals, such as tunicates [6–8]. Clarification of the polymeric structure of cellulose can be traced back to 1920 with the revolutionary work of Hermann Staudinger [9]. Cellulose is composed of a *β-D*-glucopyranose unit linked by (1→4) glycosidic bonds, as can be observed in Figure 1. Cellulose is a main-chain polymer that results from the reaction of glucose molecules with the condensation of water. Cellulose is a linear-chain polymer that presents a large number of hydroxy groups (three per anhydroglucose unit (AGU)). In order to adjust the preferred bond angles of acetal oxygen bridges, every second AGU ring is rotated 180° in the plane. The repetitive unit of cellulose is cellobiose [10–12].

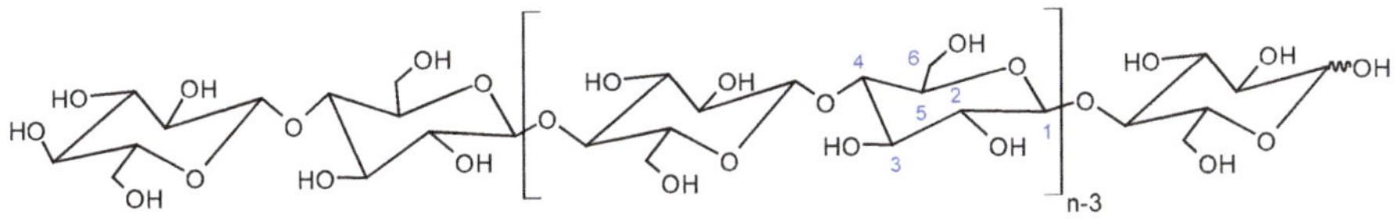

Figure 1. Molecular structure of cellulose (n = degree of polymerization). Cellobiose is the repetitive unit of cellulose, and the formation of hydrogen bonds is due to the existence of –OH.

The supramolecular structure of cellulose is responsible for its insolubility in water and in a large number of organic solvents. The supramolecular structure can be defined by the degree of crystallinity, crystallographic parameters, crystallite dimensions and presence of defects, structural indices of amorphous domains, and dimensions of fibrillar formations [13]. The physicochemical properties of cellulose are established by a particular hierarchical order in supramolecular structure and organization. The molecular structure conveys cellulose with its distinguishing properties, such as chirality, hydrophilicity, degradability, and extensive chemical variability, introduced by the high donor reactivity of the –OH groups [12]. One of the explanations for the stability of cellulose is that it exists generally in the form of crystals that have broad van der Waals attractive forces, as well as hydrogen bonds [14–16].

Interest in biobased materials, such as cellulosic materials, is continuously increasing in response to growing environmental awareness. Natural cellulose fibers that exhibit a wide range of sizes are excellent candidates to pave the way for the development of biobased high-performance materials with low environmental impact.

Cellulose Micro-/Nano- and Macroscale

Several studies have pursued isolation, characterization, and new applications for cellulose. Innovative approaches for their production scales from top-down methods, including enzymatic/chemical/physical methodologies using wood and forest/agricultural residues, to the bottom-up production of cellulose microfibrils by bacteria, have been presented. The cellulosic materials accomplished using these methodologies are commonly referred to as nanocelluloses [17]. The American Paper & Pulp Association (TAPPI WI 3021, Peachtree Corners, GA, USA) established a classification for nanocellulose based on the nanocellulose dimensions. The nomenclature, abbreviation, and dimensions defined for each group are described in Figure 2.

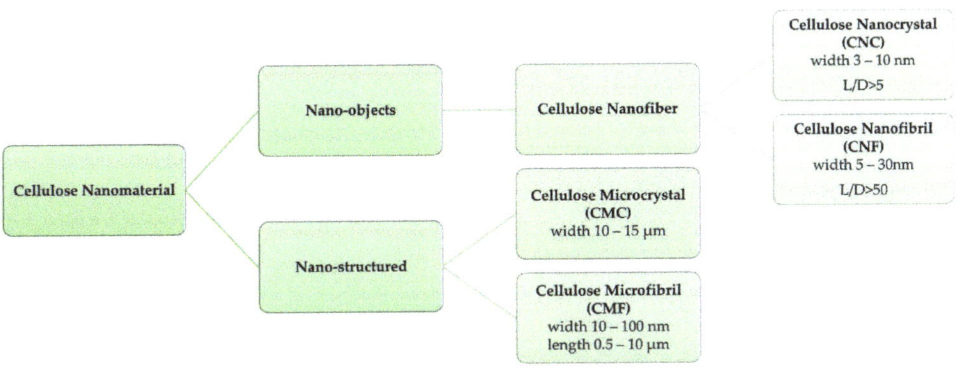

Figure 2. Standard terms established by The American Paper & Pulp Association (TAPPI WI 3021). Adapted from Ref. [18].

Cellulose microfibrils (CMF) can be obtained from suspensions of wood-based cellulose fibers using extensive mechanical methods, such as high-pressure homogenizers. This mechanical treatment delaminates the fibers and microfibrils with widths ranging from

10 to 100 μm and lengths ranging from 0.5 to 10 μm [18–20]. Cellulose nanofibrils (CNFs) are stretched aggregates of elementary nanofibrils that alternate with amorphous domains. The most common method to produce CNFs is through the mechanical delamination of softwood pulp in high-pressure homogenizers without any pretreatment or after chemical or enzymatic pretreatment. Pretreatments such as 2,2,6,6-tetramethylpiperidine-1-oxyl radical (TEMPO)-mediated oxidation, carboxymethylation, and mildly acidic or enzymatic cellulose hydrolysis, among others have been used to facilitate the mechanical disintegration process [21].

Cellulose nanocrystals (CNCs), also designated as nanowhiskers, consist of short, stiff, rod-like cellulose crystals [17,22–24]. According to The American Paper & Pulp Association (TAPPI WI 3021), the CNCs present widths in the range of 3–10 nm and aspect ratios larger than 5 and less than 50. They are obtained by the removal of amorphous sections of a purified cellulose source through acid hydrolysis [25], often followed by ultrasonic treatment. Disordered or amorphous regions of cellulose are favorably hydrolyzed, while crystalline regions that have higher resistance to acidic attack and cellulose rod-like nanocrystals are produced, as illustrated in the scheme presented in Figure 3.

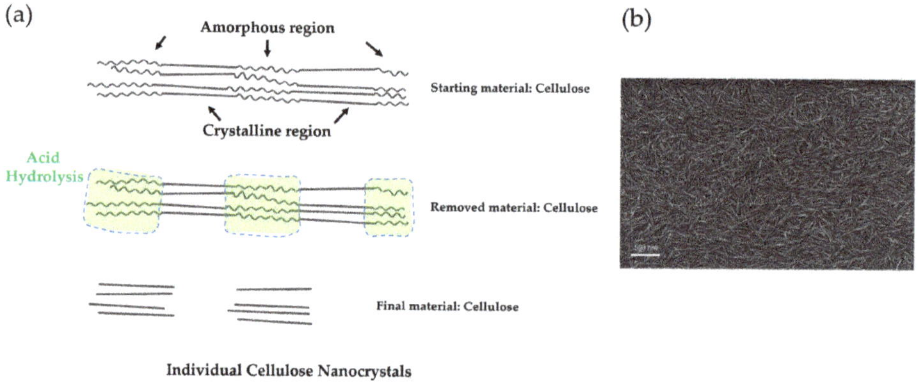

Figure 3. Cellulose nanomaterials: (**a**) schematic representation of the process to produce cellulose nanocrystals by means of acid hydrolysis, in which the amorphous regions are hydrolyzed and removed by the acid while the crystalline fractions are maintained. Adapted from Ref. [26]; (**b**) scanning electron microscopy image shows the top surface of a nanocrystalline cellulose freestanding film obtained by solvent casting of a CNC water suspension. Adapted with permission from Ref. [27]. Copyright 2020 Springer Nature B.V.

One of the oldest and most conventional procedures used to obtain CNCs consists of subjecting pure cellulosic material to strong acid hydrolysis under rigorously controlled conditions of temperature, agitation, and time [28]. For example, if the hydrolysis is performed with hydrogen chloride, HCl, weakly negatively charged particles are obtained, but if sulfuric acid, H_2SO_4, is used, the particles are more negatively charged, and approximately one-tenth of the glucose units can be functionalized with sulfate ester groups [17,24]. In both cases, the CNC suspension must be successively diluted with water and rinsed with consecutive centrifugations, to control the value of the pH of the suspension. Dialysis using distilled water is then accomplished to remove any free acid molecules from the dispersion. Additional steps such as filtration, differential centrifugation, or ultracentrifugation can also be used [17]. The dimensions of the crystals depend on the duration of the hydrolysis, and a longer reaction time produces shorter crystals [29]. The dimensions of the crystals produced are also dependent on the source and degree of crystallinity of the origin of cellulose [17]. The preparation of nanocellulose from cellulose requires two main stages. The first stage is dedicated to the pretreatment (chemical, physical, physicochemical, biological, or combined) of feedstocks to obtain pure cellulose, whereas the second is dedicated to the transformation of cellulose to nanocellulose [28].

Bacterial nanocellulose (BNC) is produced by some species of bacteria, such as *Acetobacter*, *Acanthamoeba*, and *Achromobacter* spp. The molar mass, the molar mass distribution, and the supramolecular structure of the produced BNC can be controlled by selecting the substrates, cultivation conditions, various additives, and finally the bacterial strain. Quite a few strains of *K. xylinus* produce extracellular cellulose, a biofilm of varying thickness that is responsible for maintaining a high oxygenation of the colonies near the surface is formed, which also acts as a protective barrier against dehydration, natural enemies, and radiation [5]. In contrast to C and CNC isolated from cellulose sources, BNC is formed by the bacteria as a polymer and nanomaterial by biotechnological assembly processes, which involve a large number of genes coding individual enzymes and regulatory proteins [30], from low-molecular-weight carbon sources, such as *D*-glucose. BNC comprises a nonwoven nanofiber network, with fiber diameters spanning from 20 to 100 nm, and presents a remarkable water content of 99%. A high degree of polymerization and high crystallinity with values of 60–90% are also important features to mention in BNC [12,17].

In recent years, cellulose nanomaterials have been proven to be one of the most promising green materials in modern times. Their attractive characteristics, such as abundance, mechanical properties, renewability, biocompatibility, and high aspect ratio, have gained growing interest. Table 1 summarizes some examples of recent and emerging uses of cellulose nanomaterials in the field of biomedical engineering and materials science, which present high potential to be used in the near future by industry.

Table 1. Bioapplications using cellulose nanomaterials.

Cellulose Nanomaterial	Field of Application	Ref
Cellulose Nanocrystal (CNC)	Tissue regeneration	[31–33]
	Scaffolds	[34–38]
	Drug delivery systems	[39–43]
	Food industry	[44–48]
Cellulose Nanofibril (CNF)	Wound healing	[49–51]
	3D cell culture	[52–54]
	Drug delivery systems	[55–58]
	Food industry	[59,60]
Bacterial Nanocellulose (BNC)	Tissue regeneration	[61–63]
	Scaffolds	[64,65]
	Food packing	[66–68]
	Wound dressing	[69–71]
	Drug delivery systems	[72–75]

2. Nature-Designed Materials

Millions of years of natural selection have resulted in the design of cellulose-based structures that combine structure–function and are an inspiration for the production of synthetic materials. In this section, we describe some key examples of structures found in nature with unique characteristics, such as support, coloration, and movement. Plants present one interesting feature: the outer layer known as the cuticle, made of lipids embedded on a typically hydrophobic polymeric matrix [76], which can confer several properties to the plant. One of the most recognized examples is the self-cleaning mechanism present in the common lotus leaf. The micro- and nanostructured surface of the lotus leaf provides a highly effective antiadhesive effect, allowing the plant to protect itself against contamination [76]. Plants can also comprise, at different scales, anisotropic cellulose structures that are responsible for color and/or movement. These interesting features of plants inspired and continue to inspire scientists and engineers to pursue new biomimetic and actuating materials/devices that can change, for example, color and/or shape [77]. The majority of the colors that plants exhibit are due to pigments and dye molecules that are ephemeral and disappear with time [78,79]. Another approach used in the plant kingdom to exhibit lively and attractive colors is accomplished by the interaction of light with ordered structures at the micron and nanoscale [80], defined as structural color. Iridescent colors, similar to the interference colors observed in soap bubbles or thin soap films [81,82], are colors that differ with the observing angle or lighting geometry; these features can also be observed in plants and result from ordered structures existing in the plants [83].

Whitney et al. [84] identified and explained that the iridescence observed in the *Hibiscus trionum* flower is attributed to a regular nanoscale pattern (striations or wrinkles) sculpted into the cuticle that covers the surface of the petal. These patterns present a biological purpose, since it was found that it could interact with its pollinators, predominantly bumblebees, through iridescent signals due to diffraction gratings [84]. Other authors determined that the microstructures embedded in the surface of the petals can act as a tactile signal used by the bees in the course of pollination to discriminate between different textures [85]. Structural colors do not disappear, and, in fruits, this permanent bright coloration increases the probability of dispersal [86]. A well-known example of structural coloration is the fruit from *Pollia condensata* preserved by Clark in the herbarium of the Royal Botanic Gardens, Kew, UK [86]. The fruit collected in 1974, continues to display its strong blue coloration. Vignolini et al. investigated its photonic response and reported that a multilayer helical structure composed of cellulose microfibrils is associated with the structural color presented by the fruit [86]. L *Margaritaria nobilis* fruit presents a structural iridescent green-blue color when fresh and a pearlescent color when dry. The color variation is reversible and is associated with the presence/absence of water in the fruit structure [87]. This performance is also assigned to a helical structure in the cell wall of the fruit. In the absence of water, the fruit dries, the seeds shrink, and a layer of air between the seeds and the endocarp is formed, blocking light absorption and decreasing the contrast. In the presence of water, the layer of air disappears due to fruit expansion, and the seeds interact with the endocarp, boosting the appearance of the blue-green color. Structural color has been more extensively studied in the animal kingdom, since there are several examples such as the cuticles of beetles, the wings of butterflies, and the feathers of numerous birds [88]. Animals shift their color for different reasons and by varied mechanisms. A well-known example is chameleons, which can suddenly modify their color to adjust to a new environment, to communicate, and to adjust temperature [89].

Micro- and nanofibers produced by plants are fundamental elements that assist in vital functions, for example, reproduction or weight support. Climbing plants are a notable example of organisms that use filaments as support; these structures are designated as tendrils. Darwin reported in 1865 that tendrils can create 2D structures (spirals) if they do not find a support point or 3D structures (helices) if they do [90]. Micro-nano-helical structures can also be found in water and the nutrient transport system and can be isolated from the leaves of different plants. For example, in *Agapanthus africanus* and *Ornithogalum thyrsoides*, the filaments are tightly coiled and present the same cellulosic skeleton but with different mechanical properties and surface morphology. These differences were revealed by the texture observed in nematic liquid crystal droplets pierced by these microfilaments [91]. Some advantages described for plants to exhibit helical structures are the enhancement of the mechanical properties and also the ability to present iridescent colors. In 2010, Murugesan et al. [92] developed a model based on the Landau–de Gennes theory, in which the helical arrangement of the plant cell wall presents improved mechanical resistance performance against the propagation of fractures by diverting the direction of the spread of cracks. The enhancement of the mechanical properties in structures composed of fibers that present helicoidal alignment was also reported by Tan et al. in 2017 [93].

Plant cell walls with crystalline cellulose microfibrils engrained in an amorphous matrix of polysaccharides, aromatic compounds, and structural proteins are the most common plant tissues that present hygroscopic behavior. A broad range of complex movements in plants can be triggered by numerous water-controlled actuators that combine different layers of cellulose [94,95]. The movements observed are due to anisotropic cellulose-based micro- and nanostructures that are specifically assembled to change shape when the environmental conditions change [77]. Plant movements rely on differential water content within the plant cell tissues, i.e., the water is capable of flowing differently through the tissue, inflating the cells on one side and diminishing the cells on the other, therefore generating movement that can also exist in nonliving sections of plants, such as seeds and awns [96,97]. The steering forces for these mechanisms are the formation or breaking of hydrogen bonds and the entropic forces linked with the dilution of the molecular or macromolecular components. These swelling processes are easily mutable by controlling the humidity variations of the environment or changing the chemical potential of the swelling agent, since no intramolecular bonds are broken [97].

The seed-dispersal method of the true rose of Jericho, *Anastatica hierochuntica*, a desert plant, is correlated with the water-absorption-driven motion of the entire dead skeleton. The dead branches, which consist of dead cellulosic tissue, present a curling motion in the presence of water, allowing the fruit valves to open and release seeds. The closing process is associated with the rapid drying of the xylem vessels existent on the upper side of the stem [98,99]. The spike moss *Selaginella lepidophyla* also presents a water-driven motion, due to a different response to water by the inner and outer stems of the spike moss (Figure 4a). This type of movement is vital for plant survival, since it avoids photoinhibitory and thermal damage being a mechanism that overcomes stress due to sun radiation, high temperatures, and water scarcity. The water-driven movement of the stems can occur for several cycles without structural impairment. Water loss within the internal capillary spaces of parallel cellulose fibers in the cell walls, and the consequent shrinking of the cell wall is the driving force responsible for stem curling [100]. Pine cones can also fold their scales when moist air is present, and seed dispersion cannot occur, whereas in the absence of water, the scales open, and the seeds are released to be carried by the wind [101]. The asymmetrically oriented design of the cellulose fibrils in pine-cone scales is responsible for the conversion of local swelling/shrinking to a large-scale bending movement [102–104]. The presence of hygroscopic motion in coalified pine cones from the Eemian Interglacial period ($\approx$126,000–113,000 years ago) and from the Middle Miocene period ($\approx$16.5–11.5 million years ago) was reported by Poppinga et al. in 2017 [105].

Moist-driven movement is also abundant in plants that present spore dispersion. In the absence of water, the spores dry and deform, the sporangia (capsule in which the reproductive spores are produced and stored) opens, and the spores are released [106–108]. An interesting example is *Equisetum* spores, which present four flexible ribbon-like elaters (flexible ribbon-like limbs, designated elaters, that upon dehydration, unfold, and impel the spore). The elaters fold back in moist air; this mechanism allows the spores to move randomly [109]. Seedpods also present macroscopic deformations due to the nanoscale anisotropic expansion of their tissues. *Leucaena*, *Jacaranda*, and *Erythrina* pods open up over gradual bending and twisting in response to moisture absence (Figure 4b) [110]. The two initially flat pod valves curl into helical strips of opposite handedness. The pods present a layered structure composed of microfibrils oriented perpendicular to each other and at 45° to the pod axis. When the content of water in the tissues that constitute the pods decreases, the two layers shrink along perpendicular axes and shape distortion occurs [111,112]. A group of plants that also presents hygroscopic movement is the *Geraniaceae* family [113]. These plants are distinguished by the presence of a beak-like fruit. The moisture-driven helical movement of the *Erodium* (Figure 4c), a representative of this family, is due to the presence of the tilted helical configuration of the microfibrils in which the axis of the helix is at an angle to the long axis of the cell [113,114]. Consequently, changes in water content result in the coiling and uncoiling of the awn and self-burial of the seed [115–118].

Figure 4. Macroscopic moisture-driven deformation in plants: (**a**) Spike moss *Selaginella lepidophylla* hydrated state (**left**) and desiccate state (**right**). Adapted from Ref. [119]; (**b**) (**i**) *Leucaena*, (**ii**) *Jacaranda* and (**iii**) *Erythrina* seed pods; (**c**) *Erodium* seed and awn from the *Geraniaceae* family. Adapted from Ref. [113].

The term "nature inspired" is linked with the desire to understand, synthesize, and emulate a natural motif or phenomenon, to better understand nature. The scope of this review is to clarify and consolidate the knowledge and highlight emerging opportunities in cellulose-based active materials. The examples presented in the following sections, by no means exhaustive, were chosen among the innumerous works already reported, to exemplify the rich diversity of works on the production and development of fascinating multifunctional cellulose-based materials that merge functionality with structural performance using nature as inspiration.

3. Cellulose Biomimetic Fiber for Developing Functional Materials

Nature is full of extraordinary examples, some of them already mentioned previously in this review, of active materials with remarkable properties and performances. In this topic, the production of biomimetic fiber and fiber-based cellulosic materials that have attracted significant attention within the scientific community will be presented. The main strategy of biomimetic materials is to observe and understand the function of unique structures present in nature, and then use that knowledge to obtain similar structures and functionality. Jiang et al. [120] prepared cellulose membranes that can present a variety of predetermined deformations in the presence of moisture. Thin cellulose membranes from balsa wood fibers through a series of procedures that include chemical, mechanical, and heat treatment were prepared. The high humidity response presented by the thin membrane is due to a self-maintained moisture gradient induced by an asymmetric design of the membrane surfaces, strengthened by the hygroscopic swelling of the cellulose matrix. The authors designed and built a hydro-driven robot hand using these cellulose membranes. The hydro-driven robot hand could controllably grab and release objects 40 times heavier than its weight (Figure 5a). The high capability of load uptake is attributed to the fairly rigid cellulose elements in the membranes. These results are presented as a good option to achieve new smart structures and devices, such as flexible robots and self-unfolding structures produced from an environmentally friendly and recyclable source [120].

The scientific community has shown increased interest in the manufacture of electrospinning fibers. Using this process, it is possible to produce membranes and three-dimensional constructions with a wide range of polymers of different sizes and shapes and small pores. The fibers have diameters in the submicron range, and the surface area is remarkably high. Cao et al. [121] reported that it is possible to produce fiber-based nanoporous membranes that mimic spider webs using jute cellulose nanocrystals (CNCs) in their composition. For this purpose, different polyacrylonitrile (PAN), crosslinked polyvinyl alcohol (PVA), and silica membranes were produced using electrospinning. The electrospun membranes (Figure 5b) were later immersed in a solution of dodecyl trimethyl ammonium bromide (DTAB) and jute CNCs. The membranes obtained were produced using a low-energy consumption method and presented pores in the submicron range, very useful for ultrafiltration [121]. Later, also inspired by spider webs, Wang et al. [122] manufactured a fiber-based membrane composed of 48% functionalized sulfhydryl cellulose (SC) and polyacrylonitrile (PAN). The reusable membrane produced via electrospinning presented a great water–oil separation capability. To increase the concentration of cellulose nanocrystals that could be used, functionalized sulfhydryl cellulose nanocrystals were also added to the membrane composition. This strategy made it possible to manufacture ultrathin and porous membranes with better mechanical properties and with a separation efficiency of 99.9% [122].

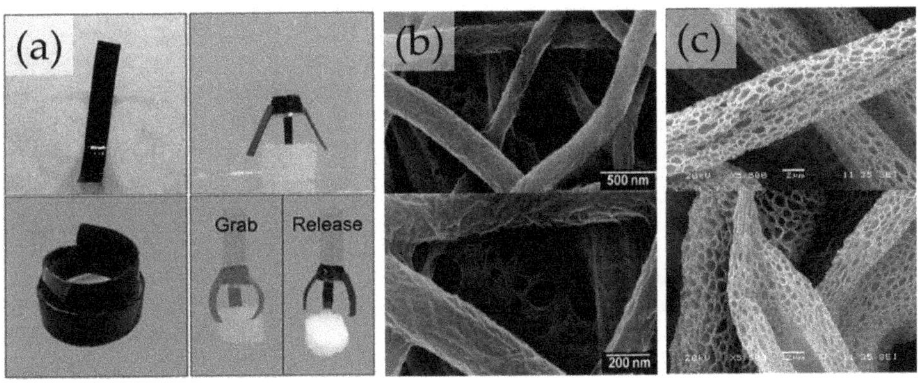

Figure 5. Biomimetic cellulosic fiber and fiber-based materials: (**a**) moisture-driven robot hand produced with balsa wood fibers. Adapted from Ref. [120]; (**b**) field-emission scanning electron micrograph of porous jute cellulose nanocrystals nanofibrous membranes. Adapted with permission from Ref. [121]. Copyright 2012 Elsevier; (**c**) scanning electron microscopy micrograph of porous cellulose acetate electrospun fibers. Adapted from Ref. [123].

Shigezawa et al. [124] were inspired by the water collection mechanism used by *Berkheya purpurea*. This plant, found in arid areas, has hydrophobic hairs (60–80 μm thick) and fine fibers (3–8 μm) to collect fresh water from the air. The authors mimicked this system using electrospinning to produce cellulose acetate fibers on a nylon mesh as support to capture water from fog. They studied different CA electrospinning times (10 s, 1 min, 5 min, 10 min) and concluded that the fibers electrospun for 1 min were the ones that had the best performance to capture water [124].

Electrospinning membranes have also begun to be used in tissue engineering with the aim of mimicking the extracellular matrix (ECM) to restore, maintain, or improve tissue function. Cellulose acetate membranes, produced using electrospinning, capable of mimicking this three-dimensional structure were reported by Han et al. [125]. These membranes will help in cell differentiation, tissue, and organ growth and also facilitate the cellular response [126]. Over the years, electrospun fibers of cellulose acetate combined with other polymers such as gelatin [127], polycaprolactone [128], polyvinyl alcohol [129], zein [130], and polyurethane [131] have been used in the field of wound-healing applications. An interesting example reported by Beikzadeh et al. [132] was the development of electrospinning membranes of cellulose acetate with encapsulated essential oil of lemon myrtle (LMEO), a natural antibacterial agent. It was proven that these membranes eradicated *Escherichia coli* and *Staphylococcus aureus* bacteria. In addition, the cellulose acetate membrane s released LMEO for a prolonged time, allowing the fibers to maintain their antimicrobial activity [132]. Laboy-López et al. [123] reported a bioactive cellulose acetate electrospun mat as a scaffold to replicate bone structure and promote tissue regeneration. A porous CA fiber mat (Figure 5c) was developed and chemically modified to bioactivate its surface in order to promote new tissue formation [123].

A smart moisture-wicking fabric produced with biomimetic micro- and nanofibrous Murray cellulosic membranes, with antigravity directional water transport and quick-dry performance for uninterrupted sweat release, was described by Wang et al. [133]. Moisture wicking is correlated with the capillarity force, which is responsible for driving the sweat through tiny capillary pores within the fabric. The authors used as inspiration the directional water transport systems present in nature, such as the hierarchical multibranching porous structure responsible for the transpiration in vascular plants, where the antigravity water movement from the soil to the plant stems and leaves happens by a passive wicking effect. The hierarchically porous membranes are produced using a bottom-up layer-by-layer deposition of three-layered fibrous membranes with multibranching macro-, micro-, and sub-micro-sized pores that comply with Murray's law. Murray's law states that the volumetric flow rate is proportional to the cube of the radius in a cylindrical channel optimized to require the minimum work to drive and maintain the fluid [134]. Murray membranes were prepared based on three criteria: antigravity directional water transport under differential capillary forces, ultrafast water transport and evaporation through hierarchical multibranching porous matrix, and extremely dry inner layer due to the surface energy gradient. Taking this into consideration the authors produced first an electrospun cellulose acetate fibrous membrane which is deposited on a polylactic acid nonwoven substrate. After that, the bilayer is dip-coated with microfibrilated cellulose, resulting in a three-layered leaf-vein-like nanofibrous membrane [133].

Guan et al. [135] reported a bacterial-cellulose-inspired lotus-fiber-like spiral-structure hydrogel, which they named biomimetic hydrogel fiber (BHF). BHF is described as a promising hydrogel fiber in biomedicine applications. Its exceptional stretchability and energy dissipation characteristics make it an ideal candidate for surgical sutures. BHF with lotus-fiber-mimetic spiral structure is built taking advantage of the bacterial cellulose hydrogel 3D cellulose nanofiber network [135].

4. Biomimetic Cellulosic Active Films

The development of nature-inspired materials with a focus on functionality and response to an external stimulus has been reported for several years. A one-step method to produce monolayer cellulose-based Janus-like membranes with reversible behavior of the micro-/nanostructures and solvent-responsive properties was reported by Liu and coworkers [136]. The structural difference among the top and bottom surfaces, due to the anisotropic self-assembly of cellulose nanocrystals (Figure 6a), enables the solvent-responsive curling of the film in different solvents with a rapidly reversible bending motion for many cycles [136]. Gevorkian et al. [137] reported a planar single-layer cellulose-based hydrogel with anisotropic structures. Shear-induced-orientation cellulose nanocrystals resulted in anisotropic mechanical and swelling properties of the cellulose hydrogel. Multiple complex 3D shapes from the same hydrogel were prepared according to the degree of its structural anisotropy [137]. Inspired by the plant movements triggered by the water content within nano- and mesoscale cellulose fibrillar structures, Wang and coworkers [138] developed a thin film of cellulose nanofibrils (CNFs) that presents humidity-controlled reversible actuation. To mimic the water-dependent curling and blooming of the glory flower, the authors prepared a four-petal CNF film "flower" that gradually folded the petals and that bloomed and recovered its original shape as a consequence of the presence or absence of moisture, respectively. The curling and blooming mechanism is due to the formation of a dynamic bilayer-like structure in the thin film that comprises the difference between the water-swollen surface and the other surface [138].

Freestanding, flexible, and porous moisture-driven actuators inspired by *M. pudica* and pine cone were reported by Zhu et al. [139]. The active films consisted of poly (vinyl alcohol-*co*-ethylene) nanofibers and cellulose nanocrystals. The films prepared were applied to switches, soft robots, and humidity regulation. The locomotion of the actuator produced could be adjusted by cutting the film in different directions, and is a reflection of the orientation of the CNC existing in the film. The film could bend to an angle of 180° and recover its shape in less than 1 s for more than 100 cycles when the environment moisture decreased [139]. A flexible and flat photonic CNC with poly(ethylene glycol) (PEG) or glycerol composite films that present structural colors ranging from blue to red was reported by Duan et al. [140]. The CNC-PEG composite film presents reversible structural color change as the structure swells and dehydrates. The CNC-glycerol film presented diversified color changes in response to an external stimulus, depending on the content of the film additive. These films are described as low-cost materials suitable for colorimetric biosensors, optically active ingredients, inks, and decorative coatings [140].

Wu et al. developed a hyper-reflective, photonic cellulose nanocrystal-based nanocomposite film that mimics the shell structure of the *Chrysina* genus of beetles. The authors produced an asymmetric sandwich-like film by assembling hydrophilic cellulose nanocrystal (CNC) and poly(ethylene glycol) diacrylate (PEGDA) layers with a uniaxial orientation of the polyamide-6 layers. Humidity-triggered movement and variation of the reflected color is observed as a result of the asymmetric swelling of the hydrophilic components in the film, CNC and PEGDA. These interesting characteristics make this material suitable for color-changing sensors and actuators with a response in function of the environment [141].

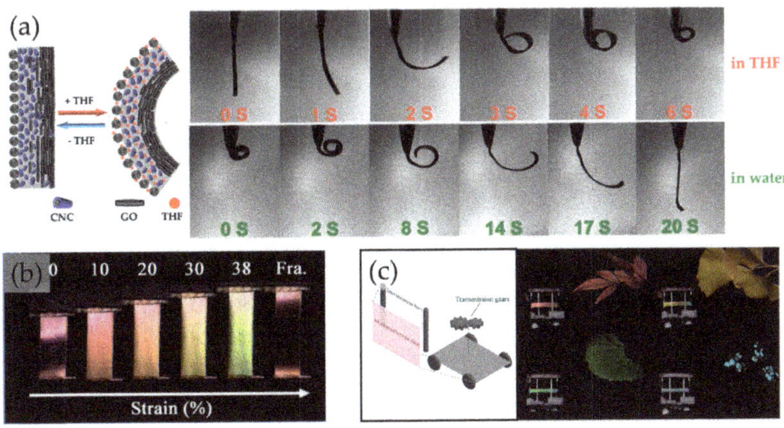

Figure 6. Cellulose-based active films: (**a**) cellulose-based Janus-like membranes composed of acylated cellulose nanocrystals (CNCs) and graphene oxide (GO) presenting solvent-responsive bending movement after exposed to tetrahydrofuran (THF) and water. Adapted from Ref. [136]; (**b**) optical photographs of the CNC/PEGDA film under deformation. Adapted with permission from Ref. [142]. Copyright 2020 American Chemical Society; (**c**) diagram of stealthy robot and photographs of the camouflage process of the CNC elastic film. Adapted with permission from Ref. [143]. Copyright 2020 American Chemical Society.

Motivated by the elastic skin of chameleons, Zhang et al. [142] developed a freestanding highly flexible film with structural color produced with cellulose nanocrystals in a poly(ethylene glycol)-based network. The biomimetic structural film showed reversible structural color change when stretched, visible to the naked eye (Figure 6b). The prepared films also presented water- and compression-responsive properties in controlled environmental conditions. This bioinspired and biobased photonic film is described as suitable for chromogenic sensing, encryption, and anticounterfeit applications [142]. More recently, Zang et al. [143] also reported a ternary coassembly and post-UV-irradiation polymerization strategy to build a flexible and elastic CNC composite film that mimics chameleon skin. The smart skin produced presented the ability to adapt to the surrounding environment for camouflage simply by adjusting the stretching of the film (Figure 6c). The film presented vivid structural coloration and stretching-induced reversible color change from red to blue with relatively low stretching strain [143].

A biomimetic hydrogel based on CNCs inspired by human natural skin was produced by Lin et al. [144]. The simple method to develop a stretchable electronic device is by incorporating conductive polymers into a hydrogel matrix. The conductive polymers cannot tolerate large deformations; therefore, the authors added Ag nanoparticles into the hydrogel matrix composed of a polyvinyl alcohol (PVA) CNC system. To immobilize and stabilize the Ag nanoparticles on CNCs, the CNCs were initially coated with tannic acid (TA). The composite hydrogel presented enhanced conductive capability and stretchability, exceptional antibacterial properties, and repeatable self-healing capability. The hydrogels were manufactured into a flexible/self-healable capacitive sensor, and that was capable of sensing both large and subtle body motions, such as bending of joints, face expression, and breath, important features to replicate the tactual sensation of human skin [144].

Inspired by stomata, the structure existing in plant leaves responsible for the control of the rate of gas exchange, Hou et al. [145] prepared a bilayer intelligent hydrogel film. The authors presented a one-step casting strategy to produce an N-isopropylacrylamide/clay gel reinforced with a bacterial cellulose network. The anisotropic swelling promoted by the bilayer structure provided a rapid temperature-induced tunable shape modification with reversible behavior. As result, a bionic leaf stoma was built where the vapor could pass and fill the space above to achieve moisture circulation, as well as temperature and air balance. Moreover, taking into consideration the rapid shape transformation of the film, a circuit was built, and the film was used as a smart switch, where it was possible to accurately control the bending direction for turning the circuit on and off in a hygrothermal environment. This work presents a simple strategy to produce biomimetic actuators with applications in soft robotics, such as environmental regulation of temperature and moisture [145].

5. Printed 3D Structures

Traditional fabrication technologies struggle to accurately reproduce or imitate the complex structures existing in nature, which have specific mechanical, hydrodynamic, optical, and electrical properties. Three-dimensional (3D_ printing is reported as a suitable technology to fabricate structures with complex and arbitrary geometry [146,147]. The classification standards of the American Society for Testing and Materials (ASTM) define seven categories in 3D printing technology: powder bed fusion molding (PBF), material extrusion molding (ME), binder injection molding (BJ), photopolymerization curing, material inject forming (MJ), direct energy deposition (DED), and sheet lamination (SL) [148,149]. The standard term for 3D printing technologies and a description of each generalized term is represented in Table 2.

The most commonly used 3D bioprinting technologies are extrusion, injection, laser-assisted, fusion deposition, and direct ink writing [148]. In this topic, we will address some bioinspired examples of 3D-printed cellulosic materials recently reported in the literature.

In 2019, Kam et al. [150] reported a method called direct cryo writing (DCW) that combines freeze casting and 3D printing in a single step. The authors used an aqueous mixture of cellulose nanocrystals and xyloglucan to produce an aerogel with an internal structure mimicking plant cell walls. Directly printing onto a cold platform enables the direct post-processing of the resulting 3D-printed aerogel with aligned structures via lyophilization. DCW allows obtaining structures with a low solid content coupled with controlled porosity and tunable architecture, making this method suitable for the production of custom-designed biological scaffolds [150].

Table 2. Standard terms for 3D printing technologies. Please refer to the ISO/ASTM standard for a complete description of each generalized term [149], adapted with permission under a Creative Commons Attribution 4.0 International License [148].

Standard Term	Commercial Term	Description
PBF (Powder bed fusion molding)	SLS—Selective Laser Sintering SLM—Selective Laser Melting DMP—Direct Metal Printing DMLS—Direct Metal Laser Sintering EBM—Electron Beam Melting MJF—Multi Jet Fusion	Powder media is deposited on a build platform and subsequently bonded together through a heating process.
ME (Material extrusion molding)	FDM—Fused Deposition Modeling FFF—Fused Filament Fabrication	Material is dispensed, usually through a heated nozzle, onto a build platform.
BJ (Binder injection molding)	CJP—ProJet Color Jet Printing	Liquid agents are selectively dropped onto powder media. Subsequent infiltration or heating may be required.
Photopolymerization curing	SLA—Stereolithography apparatus DLP—Direct Light Processing CLIP—Continuous liquid interface production	Liquid photopolymer is selectively exposed to a light source facilitating layer-by-layer curing.
MJ (Material inject forming)	NPJ—Nanoparticle Jetting DOD—Drop-on-Demand PolyJet MJP—PolyJet Multijet Printing	A print head dispenses droplets of media, usually a photopolymer, onto a build platform where each layer is solidified or cured.
DED (Direct energy deposition)	LENS—Laser Engineered Net Shape EBAM—Electron Beam Additive Manufacture	Focused application of energy and material selectively melted and fused on a build platform or part.
SL (Sheet lamination)	LOM—Laminated Object Manufacturing	Discrete layers of material are fused or glued together to form a 3D object.

Three-dimensional (3D) bioprinting of a hierarchical nano- to macrofibrillary artificial human tissue with high resolution and integrity was reported by Mendes et al. [151]. The authors developed a nanocomposite bioink of platelet lysate hydrogel reinforced by cellulose nanocrystals, presented as suitable to print dynamic and personalized 3D living constructs that mimic the structure and composition of native tissues [151].

A lightweight (~90 mg/cm^3) and super-strong (16.6 MPa compressive Young's modulus) all-cellulose honeycomb structure, mimicking a paper wasp nest, was fabricated using direct-ink-writing 3D printing technology. The produced structure demonstrates superb elasticity and flexibility when wet, whereas the dried structure is a rigid and strong material capable of supporting 15,800 times its own weight. The authors describe this material as being applicable in biomedical, environmental, and structural engineering areas [152].

Two-layer cellulose-based scaffolds were 3D printed to mimic the osteochondral structure (bone structure), as reported by Guo et al. [153]. The scaffold consisted of a top pristine cellulose and a bottom cellulose/bioactive glass hydrogel. The printed scaffold exhibited superior performance for repairing the osteochondral defect by promoting osteoinduction and the formation of new bone inside the implanted scaffold. These results indicate that this cellulose-based printing system is suitable for application in tough tissue engineering applications [153].

Liu et al. reported a 3D-printed cellulose acetate nanoporous network that mimics water absorption in plant roots. The authors describe that this artificial root model produced using biomaterial-derived ink has immense potential for applications in plant science and bioengineering [154].

The possibility of 3D printing materials with nano/microscale orders was reported recently by Esmaeili et al. [155]. Three-dimensional (3D) complex geometries inspired by Bouligand structures, i.e., chiral arrangement of fibrous materials responsible for structural color in some beetles and berries, were produced using cellulose nanocrystal-based inks. The chiral assembly during the printing process is related to the shear forces inside the 3D printer's nozzle [155]. This biomimetic approach expands 3D printing beyond what has been reported so far.

6. 4D-Printed Cellulosic Responsive Materials

Four-dimensional (4D) printing is a one-step process, where time is considered the fourth dimension, in which the properties of the printed materials allow morphing a 1D strand or 2D surface into a 3D shape, or converting a 3D shape into another 3D shape, in response to an external stimulus such as the presence of water [156]. This technique was first introduced in 2013 by Skylar Tibbits at a TED conference [157]. As mentioned above, 4D printing can produce dynamic structures with adjustable shapes, properties, or functionality [158,159]. The 4D printing market is predicted to reach USD 510 million in the 2022–2030 period [160]. Gladman et al. [161] printed a series of functional folding-flower architectures using a biomimetic composite hydrogel with nanofibrillated cellulose. The printed material successfully reproduces the complex architecture of the orchid *Dendrobium helix*. The materials used present encoded anisotropy that promotes the appearance of an intricate shape when the printed structure is immersed in water [161]. Throughout the printing process, when the ink flows, fibril arrangement is induced, and anisotropic stiffness and longitudinal swelling are introduced in the structure. The printed programmable bilayer architecture enables petals to be closed or twisted upon swelling according to the orientation of the printed layers [161].

Siqueira et al. reported a monomer-based ink with cellulose CNCs dispersed in a mixture of 2-hydroxyethyl methacrylate (HEMA) monomer, polyether urethane acrylate, and a UV-light-cured photoinitiator, suitable for 3D printing. The structures printed with the inks showed shear-induced alignment of the CNC and enhanced stiffness along the printing direction. The authors state that this type of formulation is a suitable option for the biomimetic 4D printing of programmable reinforcer materials that can be reactive to external stimuli [162]. Natural fiber biocomposites are reported as a new class of smart materials, as they can be used in 4D printing to build unique shape-changing materials and structures. The fibers used in these biocomposites can be derived from hemp, flax, coconut, or wood [163–165].

Correia et al. described the printing of 4D pine scale and flap structures capable of multiphase movement [166]. This printed hydromorphic structure was structurally programmed and presented multiphase motion similar to that observed during the desiccation of natural pine cones (Figure 7a). The authors used 3D printing with a fused filament of wood polymer composite (fibrous filler from wood-derived fibers combined with co-polyester polymer matrix) and acrylonitrile butadiene styrene (ABS). The prepared composite mimics the properties of the swellable layer of the pine-cone scale, and the ABS acts as the stiffer and much less swellable layer, the resistance layer. The 4D-printed scales successfully exhibited a two-phase movement capable of recreating the natural scale behavior. The procedure used allowed defining local hygroscopic anisotropies and local nonhygroscopic movements that paved the way for the development of new advanced and passively actuating systems, with highly tailored shape-changing capabilities [166]. Mulakkal et al. developed a cellulose hydrogel composite ink to print a complex structure with predetermined layout rules to respond differently to hydration/dehydration. They used carboxymethylcellulose hydrocolloid with cellulose pulp fibers, also adding montmorillonite clay to the mixture to enhance the storage stability of the ink formulations and to facilitate the extrusion process during printing [167].

Wang et al. developed a starch, cellulose, and protein-based 2D film that changed to a 3D shape in the presence of moisture. The use of 4D printing in the food industry is of great interest since it can help to customize the product and develop unique characteristics that change in response to external stimuli such as flavors, textures, and aroma [168–170]. More recently, Lai et al. were capable of printing out a series of simple and complex morphing structures, using single ink, composed of a mixture of alginate and methylcellulose. By controlling the 2D architecture, the authors were able to obtain prescribed 3D morphologies after immersion in a calcium chloride solution (Figure 7b). The excellent printability combined with anisotropic swelling and consequent shape morphing of this cellulose-based hydrogel makes it adequate to be used in tissue engineering, biomedical devices, and soft robotic fields [171].

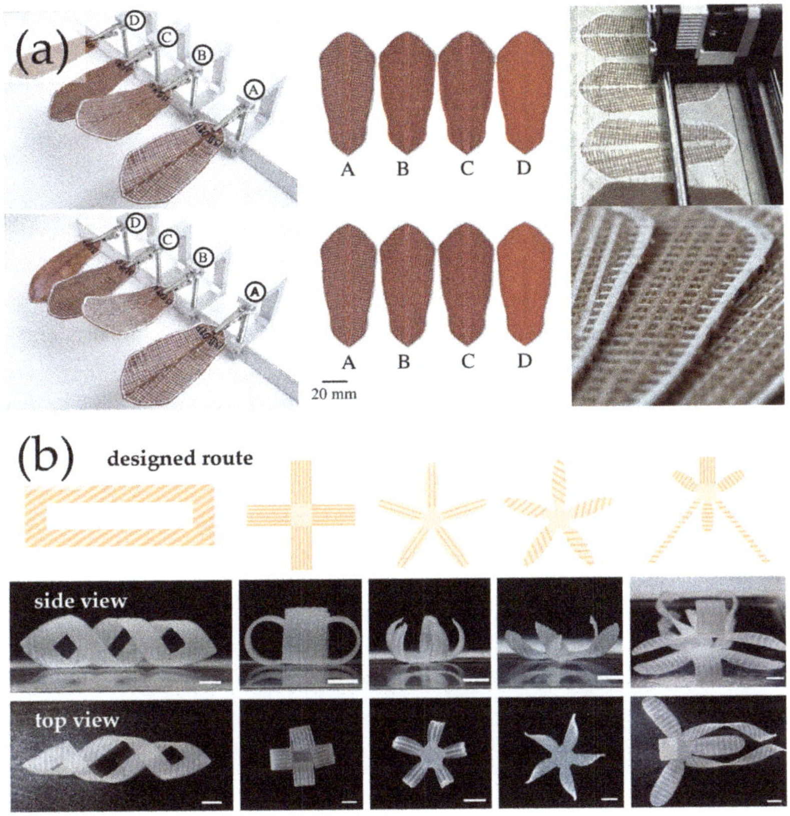

Figure 7. Four-dimensional (4D) printing inspired by nature structures of cellulosic materials: (**a**) 4D pine-scale-like structures capable of multiphase movement (**left**), according to several different predetermined patterns (**center**) using a 3D printer and a multimaterial process (**right**). Adapted with permission from Ref. [166]. Copyright 2020 The Royal Society; (**b**) designated 4D printed pattern alginate and methylcellulose hydrogel and their corresponding 3D complex structures from side and top view immersed in 0.1 M calcium chloride. Adapted from Ref. [171].

7. Conclusions and Future Directions

Stimuli-responsive nature-inspired cellulose-based materials have paved the way for a wide variety of applications, namely in actuators, sensors, soft robotics, biomedicine, tissue engineering, and smart fabrics. Cellulose and cellulose derivatives are easily available and low-cost polymers that have been widely used in commercial applications. Here, we have shown that cellulose and cellulose-based materials are excellent candidates to mimic the structures and functions existing in some plants and animals. The development of stable and innovative active materials/devices requires: (1) a full understanding of the underlying mechanisms after being exposed to external stimuli; (2) tunable fabrication of the smart cellulose-based materials; and (3) evaluation and continuous improvement of the active materials/devices performance regarding the desired sensitivity, specificity, and stability.

Cellulose has been used for centuries and will continue to be one of the most abundant natural polymeric sources for structural and functional materials. There is, therefore, no doubt that the sprouting of the next generation of cellulose-based interactive materials inspired by nature will be in the markets within the next few years or decades.

Author Contributions: Conceptualization A.P.C.A.; writing—original draft preparation M.I.M. and A.P.C.A.; writing—review and editing A.P.C.A. All authors have read and agreed to the published version of the manuscript.

Funding: This work was supported by the Associate Laboratory for Green Chemistry—LAQV, which is financed by national funds from FCT/MCTES (UIDB/50006/2020 and UIDP/50006/2020) and financed by national funds from FCT—Fundação para a Ciência e a Tecnologia, I.P., in the scope of the projects LA/P/0037/2020, UIDP/50025/2020, and UIDB/50025/2020 of the Associate Laboratory Institute of Nanostructures, Nanomodelling and Nanofabrication—i3N. Ana Almeida is grateful for the financial support from Fundação para a Ciência e a Tecnologia (FCT), Portugal, through project 2022.01619.PTDC.

Institutional Review Board Statement: Not applicable.

Informed Consent Statement: Not applicable.

Data Availability Statement: Not applicable.

Conflicts of Interest: The authors declare no conflict of interest.

References

1. Payen, A. Mémoire sur la composition du tissu propre des plantes et du ligneux. *J. Comptes Rendus* **1838**, *7*, 1052–1056.
2. Fisher, C.H. Anselm Payen Pioneer in Natural Polymers and Industrial Chemistry. In *Pioneers in Polymer Science*; Seymour, R.B., Ed.; Springer: Dordrecht, The Netherlands, 1989; pp. 47–61. [CrossRef]
3. Brown, A.J. XIX.—The chemical action of pure cultivations of bacterium aceti. *J. Chem. Soc. Trans.* **1886**, *49*, 172–187. [CrossRef]
4. Ohad, I.; Danon, D.; Hestrin, S. Synthesis of Cellulose by Acetobacter Xylinum: V. Ultrastructure of Polymer. *J. Cell Biol.* **1962**, *12*, 31–46. [CrossRef] [PubMed]
5. Portela, R.; Leal, C.R.; Almeida, P.L.; Sobral, R.G. Bacterial cellulose: A versatile biopolymer for wound dressing applications. *Microb. Biotechnol.* **2019**, *12*, 586–610. [CrossRef] [PubMed]
6. Hol, H.R.; Jehli, J. The presence of cellulose microfibrils in the proteinaceous slime track of Dictyostelium discoideum. *Arch. Für Mikrobiol.* **1973**, *92*, 179–187. [CrossRef]
7. De Souza Lima, M.M.; Borsali, R. Static and Dynamic Light Scattering from Polyelectrolyte Microcrystal Cellulose†. *Langmuir* **2002**, *18*, 992–996. [CrossRef]
8. Dunlop, M.J.; Acharya, B.; Bissessur, R. Isolation of nanocrystalline cellulose from tunicates. *J. Environ. Chem. Eng.* **2018**, *6*, 4408–4412. [CrossRef]
9. Staudinger, H. Über Polymerisation. *Ber. Dtsch. Chem. Ges.* **1920**, *53*, 1073–1085. [CrossRef]
10. O'Sullivan, A.C. Cellulose: The structure slowly unravels. *Cellulose* **1997**, *4*, 173–207. [CrossRef]
11. Kamide, K. Characterization of Molecular Structure of Cellulose Derivatives. In *Cellulose and Cellulose Derivatives*; Kamide, K., Ed.; Elsevier: Amsterdam, The Netherlands, 2005; pp. 25–188. [CrossRef]
12. Klemm, D.; Heublein, B.; Fink, H.-P.; Bohn, A. Cellulose: Fascinating Biopolymer and Sustainable Raw Material. *Angew. Chem. Int. Ed.* **2005**, *44*, 3358–3393. [CrossRef]
13. Ioyelovich, M.Y. Supermolecular structure of native and isolated cellulose. *Polym. Sci. USSR* **1991**, *33*, 1670–1676. [CrossRef]
14. Bergenstråhle, M.; Wohlert, J.; Himmel, M.E.; Brady, J.W. Simulation studies of the insolubility of cellulose. *Carbohydr. Res.* **2010**, *345*, 2060–2066. [CrossRef]
15. French, A.D.; Concha, M.; Dowd, M.K.; Stevens, E.D. Electron (charge) density studies of cellulose models. *Cellulose* **2014**, *21*, 1051–1063. [CrossRef]
16. French, A.D.; Pérez, S.; Bulone, V.; Rosenau, T.; Gray, D. Cellulose. In *Encyclopedia of Polymer Science and Technology*; John Wiley & Sons: Hoboken, NJ, USA, 2018; pp. 1–69. [CrossRef]
17. Klemm, D.; Kramer, F.; Moritz, S.; Lindström, T.; Ankerfors, M.; Gray, D.; Dorris, A. Nanocelluloses: A New Family of Nature-Based Materials. *Angew. Chem. Int. Ed.* **2011**, *50*, 5438–5466. [CrossRef]
18. Mariano, M.; El Kissi, N.; Dufresne, A. Cellulose nanocrystals and related nanocomposites: Review of some properties and challenges. *J. Polym. Sci. Pol. Phys.* **2014**, *52*, 791–806. [CrossRef]
19. Turbak, A.F.; Snyder, F.W.; Sandberg, K.R. Microfibrillated cellulose, a new cellulose product: Properties, uses, and commercial potential. *J. Appl. Polym. Sci. Appl. Polym. Symp.* **1983**, *37*, 815–827.
20. Pääkkö, M.; Ankerfors, M.; Kosonen, H.; Nykänen, A.; Ahola, S.; Österberg, M.; Ruokolainen, J.; Laine, J.; Larsson, P.T.; Ikkala, O.; et al. Enzymatic Hydrolysis Combined with Mechanical Shearing and High-Pressure Homogenization for Nanoscale Cellulose Fibrils and Strong Gels. *Biomacromolecules* **2007**, *8*, 1934–1941. [CrossRef]

21. Ganapathy, V.; Muthukumaran, G.; Sudhagar, P.E.; Rashedi, A.; Norrrahim, M.N.F.; Ilyas, R.A.; Goh, K.L.; Jawaid, M.; Naveen, J. Mechanical properties of cellulose-based multiscale composites: A review. *Polym. Compos.* **2023**, *44*, 734–756. [CrossRef]
22. Ciolacu, D.E.; Darie, R.N. Nanocomposites Based on Cellulose, Hemicelluloses, and Lignin. In *Nanomaterials and Nanocomposites*; John Wiley & Sons: Hoboken, NJ, USA, 2016; pp. 391–424. [CrossRef]
23. Rojas, O.J. *Cellulose Chemistry and Properties: Fibers, Nanocelluloses and Advanced Materials*; Springer International Publishing: Berlin/Heidelberg, Germany, 2016.
24. Visakh, P.M. Introduction for Nanomaterials and Nanocomposites: State of Art, New Challenges, and Opportunities. In *Nanomaterials and Nanocomposites*; Wiley: Hoboken, NJ, USA, 2016; pp. 1–20. [CrossRef]
25. Bangar, S.P.; Harussani, M.M.; Ilyas, R.A.; Ashogbon, A.O.; Singh, A.; Trif, M.; Jafari, S.M. Surface modifications of cellulose nanocrystals: Processes, properties, and applications. *Food Hydrocoll.* **2022**, *130*, 107689. [CrossRef]
26. Mishra, R.K.; Sabu, A.; Tiwari, S.K. Materials chemistry and the futurist eco-friendly applications of nanocellulose: Status and prospect. *J. Saudi Chem. Soc.* **2018**, *22*, 949–978. [CrossRef]
27. Almeida, A.P.C.; Oliveira, J.; Fernandes, S.N.; Godinho, M.H.; Canejo, J.P. All-cellulose composite membranes for oil microdroplet collection. *Cellulose* **2020**, *27*, 4665–4677. [CrossRef]
28. Trache, D.; Tarchoun, A.F.; Derradji, M.; Hamidon, T.S.; Masruchin, N.; Brosse, N.; Hussin, M.H. Nanocellulose: From Fundamentals to Advanced Applications. *Front. Chem.* **2020**, *8*, 392. [CrossRef] [PubMed]
29. Beck-Candanedo, S.; Roman, M.; Gray, D.G. Effect of Reaction Conditions on the Properties and Behavior of Wood Cellulose Nanocrystal Suspensions. *Biomacromolecules* **2005**, *6*, 1048–1054. [CrossRef] [PubMed]
30. Barja, F. Bacterial nanocellulose production and biomedical applications. *J. Biomed. Res.* **2021**, *35*, 310–317. [CrossRef]
31. Maturavongsadit, P.; Paravyan, G.; Shrivastava, R.; Benhabbour, S.R. Thermo-/pH-responsive chitosan-cellulose nanocrystals based hydrogel with tunable mechanical properties for tissue regeneration applications. *Materialia* **2020**, *12*, 100681. [CrossRef]
32. Kamel, R.; El-Wakil, N.A.; Abdelkhalek, A.A.; Elkasabgy, N.A. Topical cellulose nanocrystals-stabilized nanoemulgel loaded with ciprofloxacin HCl with enhanced antibacterial activity and tissue regenerative properties. *J. Drug Deliv. Sci. Technol.* **2021**, *64*, 102553. [CrossRef]
33. Yang, W.; Zheng, Y.; Chen, J.; Zhu, Q.; Feng, L.; Lan, Y.; Zhu, P.; Tang, S.; Guo, R. Preparation and characterization of the collagen/cellulose nanocrystals/USPIO scaffolds loaded kartogenin for cartilage regeneration. *Mater. Sci. Eng. C* **2019**, *99*, 1362–1373. [CrossRef]
34. Patel, D.K.; Dutta, S.D.; Ganguly, K.; Lim, K.-T. Multifunctional bioactive chitosan/cellulose nanocrystal scaffolds eradicate bacterial growth and sustain drug delivery. *Int. J. Biol. Macromol.* **2021**, *170*, 178–188. [CrossRef]
35. Dutta, S.D.; Hexiu, J.; Patel, D.K.; Ganguly, K.; Lim, K.-T. 3D-printed bioactive and biodegradable hydrogel scaffolds of alginate/gelatin/cellulose nanocrystals for tissue engineering. *Int. J. Biol. Macromol.* **2021**, *167*, 644–658. [CrossRef]
36. Patel, D.K.; Dutta, S.D.; Hexiu, J.; Ganguly, K.; Lim, K.-T. Bioactive electrospun nanocomposite scaffolds of poly(lactic acid)/cellulose nanocrystals for bone tissue engineering. *Int. J. Biol. Macromol.* **2020**, *162*, 1429–1441. [CrossRef]
37. Shaheen, T.I.; Montaser, A.S.; Li, S. Effect of cellulose nanocrystals on scaffolds comprising chitosan, alginate and hydroxyapatite for bone tissue engineering. *Int. J. Biol. Macromol.* **2019**, *121*, 814–821. [CrossRef]
38. Luo, W.; Cheng, L.; Yuan, C.; Wu, Z.; Yuan, G.; Hou, M.; Chen, J.Y.; Luo, C.; Li, W. Preparation, characterization and evaluation of cellulose nanocrystal/poly(lactic acid) in situ nanocomposite scaffolds for tissue engineering. *Int. J. Biol. Macromol.* **2019**, *134*, 469–479. [CrossRef]
39. Yusefi, M.; Soon, M.L.-K.; Teow, S.-Y.; Monchouguy, E.I.; Neerooa, B.N.H.M.; Izadiyan, Z.; Jahangirian, H.; Rafiee-Moghaddam, R.; Webster, T.J.; Shameli, K. Fabrication of cellulose nanocrystals as potential anticancer drug delivery systems for colorectal cancer treatment. *Int. J. Biol. Macromol.* **2022**, *199*, 372–385. [CrossRef]
40. Long, W.; Ouyang, H.; Zhou, C.; Wan, W.; Yu, S.; Qian, K.; Liu, M.; Zhang, X.; Feng, Y.; Wei, Y. Simultaneous surface functionalization and drug loading: A novel method for fabrication of cellulose nanocrystals-based pH responsive drug delivery system. *Int. J. Biol. Macromol.* **2021**, *182*, 2066–2075. [CrossRef]
41. Kumari, P.; Seth, R.; Meena, A.; Sharma, D. Enzymatic synthesis of cellulose nanocrystals from lemongrass and its application in improving anti-cancer drug release, uptake and efficacy. *Ind. Crops Prod.* **2023**, *192*, 115933. [CrossRef]
42. Kumar, R.; Chauhan, S. Cellulose nanocrystals based delivery vehicles for anticancer agent curcumin. *Int. J. Biol. Macromol.* **2022**, *221*, 842–864. [CrossRef]
43. Gabriel, T.; Belete, A.; Hause, G.; Neubert, R.H.H.; Gebre-Mariam, T. Nanocellulose-based nanogels for sustained drug delivery: Preparation, characterization and in vitro evaluation. *J. Drug Deliv. Sci. Technol.* **2022**, *75*, 103665. [CrossRef]
44. Patiño Vidal, C.; Velásquez, E.; Galotto, M.J.; López de Dicastillo, C. Development of an antibacterial coaxial bionanocomposite based on electrospun core/shell fibers loaded with ethyl lauroyl arginate and cellulose nanocrystals for active food packaging. *Food Packag. Shelf Life* **2022**, *31*, 100802. [CrossRef]
45. Wu, Y.; Wu, W.; Farag, M.A.; Shao, P. Functionalized cellulose nanocrystal embedded into citrus pectin coating improves its barrier, antioxidant properties and potential application in food. *Food Chem.* **2023**, *401*, 134079. [CrossRef]
46. Halloub, A.; Raji, M.; Essabir, H.; Chakchak, H.; Boussen, R.; Bensalah, M.-o.; Bouhfid, R.; Qaiss, A.e.k. Intelligent food packaging film containing lignin and cellulose nanocrystals for shelf life extension of food. *Carbohydr. Polym.* **2022**, *296*, 119972. [CrossRef]
47. Lei, Y.; Yao, Q.; Jin, Z.; Wang, Y.-C. Intelligent films based on pectin, sodium alginate, cellulose nanocrystals, and anthocyanins for monitoring food freshness. *Food Chem.* **2023**, *404*, 134528. [CrossRef] [PubMed]

48. Mu, R.; Hong, X.; Ni, Y.; Li, Y.; Pang, J.; Wang, Q.; Xiao, J.; Zheng, Y. Recent trends and applications of cellulose nanocrystals in food industry. *Trends Food Sci. Technol.* **2019**, *93*, 136–144. [CrossRef]
49. Dong, D.; Chen, R.; Jia, J.; Zhao, C.; Chen, Z.; Lu, Q.; Sun, Y.; Huang, W.; Wang, C.; Li, Y.; et al. Tailoring and application of a multi-responsive cellulose nanofibre-based 3D nanonetwork wound dressing. *Carbohydr. Polym.* **2023**, *305*, 120542. [CrossRef] [PubMed]
50. Shi, X.; Chen, Z.; He, Y.; Lu, Q.; Chen, R.; Zhao, C.; Dong, D.; Sun, Y.; He, H. Dual light-responsive cellulose nanofibril-based in situ hydrogel for drug-resistant bacteria infected wound healing. *Carbohydr. Polym.* **2022**, *297*, 120042. [CrossRef] [PubMed]
51. Zhang, Q.; Zhu, J.; Jin, S.; Zheng, Y.; Gao, W.; Wu, D.; Yu, J.; Dai, Z. Cellulose-nanofibril-reinforced hydrogels with pH sensitivity and mechanical stability for wound healing. *Mater. Lett.* **2022**, *323*, 132596. [CrossRef]
52. Liu, J.; Cheng, F.; Grénman, H.; Spoljaric, S.; Seppälä, J.; Eriksson, J.E.; Willför, S.; Xu, C. Development of nanocellulose scaffolds with tunable structures to support 3D cell culture. *Carbohydr. Polym.* **2016**, *148*, 259–271. [CrossRef]
53. Li, Q.; Hatakeyama, M.; Kitaoka, T. Bioadaptive Porous 3D Scaffolds Comprising Cellulose and Chitosan Nanofibers Constructed by Pickering Emulsion Templating. *Adv. Funct. Mater.* **2022**, *32*, 2200249. [CrossRef]
54. Kim, H.J.; Oh, D.X.; Choy, S.; Nguyen, H.-L.; Cha, H.J.; Hwang, D.S. 3D cellulose nanofiber scaffold with homogeneous cell population and long-term proliferation. *Cellulose* **2018**, *25*, 7299–7314. [CrossRef]
55. Orasugh, J.T.; Saha, N.R.; Rana, D.; Sarkar, G.; Mollick, M.M.R.; Chattoapadhyay, A.; Mitra, B.C.; Mondal, D.; Ghosh, S.K.; Chattopadhyay, D. Jute cellulose nano-fibrils/hydroxypropylmethylcellulose nanocomposite: A novel material with potential for application in packaging and transdermal drug delivery system. *Ind. Crops Prod.* **2018**, *112*, 633–643. [CrossRef]
56. Zhao, J.; Lu, C.; He, X.; Zhang, X.; Zhang, W.; Zhang, X. Polyethylenimine-Grafted Cellulose Nanofibril Aerogels as Versatile Vehicles for Drug Delivery. *ACS Appl. Mater. Interfaces* **2015**, *7*, 2607–2615. [CrossRef]
57. Löbmann, K.; Svagan, A.J. Cellulose nanofibers as excipient for the delivery of poorly soluble drugs. *Int. J. Pharm.* **2017**, *533*, 285–297. [CrossRef]
58. Bhandari, J.; Mishra, H.; Mishra, P.K.; Wimmer, R.; Ahmad, F.J.; Talegaonkar, S. Cellulose nanofiber aerogel as a promising biomaterial for customized oral drug delivery. *Int. J. Nanomed.* **2017**, *12*, 2021–2031. [CrossRef]
59. Yu, Z.; Wang, W.; Kong, F.; Lin, M.; Mustapha, A. Cellulose nanofibril/silver nanoparticle composite as an active food packaging system and its toxicity to human colon cells. *Int. J. Biol. Macromol.* **2019**, *129*, 887–894. [CrossRef]
60. Yu, Z.; Dhital, R.; Wang, W.; Sun, L.; Zeng, W.; Mustapha, A.; Lin, M. Development of multifunctional nanocomposites containing cellulose nanofibrils and soy proteins as food packaging materials. *Food Packag. Shelf Life* **2019**, *21*, 100366. [CrossRef]
61. Kowalska-Ludwicka, K.; Cala, J.; Grobelski, B.; Sygut, D.; Jesionek-Kupnicka, D.; Kolodziejczyk, M.; Bielecki, S.; Pasieka, Z. Special paper—New method Modified bacterial cellulose tubes for regeneration of damaged peripheral nerves. *Arch. Med. Sci.* **2013**, *9*, 527–534. [CrossRef]
62. Pértile, R.; Moreira, S.; Andrade, F.; Domingues, L.; Gama, M. Bacterial cellulose modified using recombinant proteins to improve neuronal and mesenchymal cell adhesion. *Biotechnol. Prog.* **2012**, *28*, 526–532. [CrossRef]
63. Hwang, D.W.; Park, J.B.; Sung, D.; Park, S.; Min, K.-A.; Kim, K.W.; Choi, Y.; Kim, H.Y.; Lee, E.; Kim, H.S.; et al. 3D graphene-cellulose nanofiber hybrid scaffolds for cortical reconstruction in brain injuries. *2D Mater.* **2019**, *6*, 045043. [CrossRef]
64. Fey, C.; Betz, J.; Rosenbaum, C.; Kralisch, D.; Vielreicher, M.; Friedrich, O.; Metzger, M.; Zdzieblo, D. Bacterial nanocellulose as novel carrier for intestinal epithelial cells in drug delivery studies. *Mater. Sci. Eng. C* **2020**, *109*, 110613. [CrossRef]
65. Jaberifard, F.; Ghorbani, M.; Arsalani, N.; Mostafavi, H. A novel insoluble film based on crosslinked-starch with gelatin containing ZnO-loaded halloysite nanotube and bacterial nanocellulose for wound healing applications. *Appl. Clay Sci.* **2022**, *230*, 106667. [CrossRef]
66. Li, Q.; Gao, R.; Wang, L.; Xu, M.; Yuan, Y.; Ma, L.; Wan, Z.; Yang, X. Nanocomposites of Bacterial Cellulose Nanofibrils and Zein Nanoparticles for Food Packaging. *ACS Appl. Nano Mater.* **2020**, *3*, 2899–2910. [CrossRef]
67. Cazón, P.; Vázquez, M. Bacterial cellulose as a biodegradable food packaging material: A review. *Food Hydrocoll.* **2021**, *113*, 106530. [CrossRef]
68. Kamal, T.; Ul-Islam, M.; Fatima, A.; Ullah, M.W.; Manan, S. Cost-Effective Synthesis of Bacterial Cellulose and Its Applications in the Food and Environmental Sectors. *Gels* **2022**, *8*, 552. [CrossRef] [PubMed]
69. Almeida, A.P.C.; Saraiva, J.N.; Cavaco, G.; Portela, R.P.; Leal, C.R.; Sobral, R.G.; Almeida, P.L. Crosslinked bacterial cellulose hydrogels for biomedical applications. *Eur. Polym. J.* **2022**, *177*, 111438. [CrossRef]
70. Zhang, S.; Yang, Z.; Hao, J.; Ding, F.; Li, Z.; Ren, X. Hollow nanosphere-doped bacterial cellulose and polypropylene wound dressings: Biomimetic nanocatalyst mediated antibacterial therapy. *Chem. Eng. J.* **2022**, *432*, 134309. [CrossRef]
71. Resch, A.; Staud, C.; Radtke, C. Nanocellulose-based wound dressing for conservative wound management in children with second-degree burns. *Int. Wound J.* **2021**, *18*, 478–486. [CrossRef]
72. Jantarat, C.; Muenraya, P.; Srivaro, S.; Nawakitrangsan, A.; Promsornpason, K. Comparison of drug release behavior of bacterial cellulose loaded with ibuprofen and propranolol hydrochloride. *RSC Adv.* **2021**, *11*, 37354–37365. [CrossRef]
73. Beekmann, U.; Schmölz, L.; Lorkowski, S.; Werz, O.; Thamm, J.; Fischer, D.; Kralisch, D. Process control and scale-up of modified bacterial cellulose production for tailor-made anti-inflammatory drug delivery systems. *Carbohydr. Polym.* **2020**, *236*, 116062. [CrossRef]

74. Meneguin, A.B.; da Silva Barud, H.; Sábio, R.M.; de Sousa, P.Z.; Manieri, K.F.; de Freitas, L.A.P.; Pacheco, G.; Alonso, J.D.; Chorilli, M. Spray-dried bacterial cellulose nanofibers: A new generation of pharmaceutical excipient intended for intestinal drug delivery. *Carbohydr. Polym.* **2020**, *249*, 116838. [CrossRef]
75. Park, D.; Kim, J.W.; Shin, K.; Kim, J.W. Bacterial cellulose nanofibrils-reinforced composite hydrogels for mechanical compression-responsive on-demand drug release. *Carbohydr. Polym.* **2021**, *272*, 118459. [CrossRef]
76. Barthlott, W.; Neinhuis, C. Purity of the sacred lotus, or escape from contamination in biological surfaces. *Planta* **1997**, *202*, 1–8. [CrossRef]
77. Oliver, K.; Seddon, A.; Trask, R.S. Morphing in nature and beyond: A review of natural and synthetic shape-changing materials and mechanisms. *J. Mater. Sci.* **2016**, *51*, 10663–10689. [CrossRef]
78. Ball, P. Botanists' blues. *Nature* **2007**, *449*, 982. [CrossRef]
79. Lee, D. *Nature's Palette: The Science of Plant Color*; University of Chicago Press: Chicago, IL, USA, 2010.
80. Vukusic, P.; Sambles, J.R. Photonic structures in biology. *Nature* **2003**, *424*, 852–855. [CrossRef]
81. Doucet, S.M.; Meadows, M.G. Iridescence: A functional perspective. *J. R. Soc. Interface* **2009**, *6* (Suppl. S2), S115–S132. [CrossRef]
82. Gruson, H.; Andraud, C.; Marcillac, W.D.d.; Berthier, S.; Elias, M.; Gomez, D. Quantitative characterization of iridescent colours in biological studies: A novel method using optical theory. *Interface Focus* **2019**, *9*, 20180049. [CrossRef]
83. Morris, R.B. Iridescence from diffraction structures in the wing scales of Callophrys rubi, the Green Hairstreak. *J. Entomol. Ser. A Gen. Entomol.* **1975**, *49*, 149–154. [CrossRef]
84. Whitney, H.M.; Kolle, M.; Andrew, P.; Chittka, L.; Steiner, U.; Glover, B.J. Floral iridescence, produced by diffractive optics, acts as a cue for animal pollinators. *Science* **2009**, *323*, 130–133. [CrossRef]
85. Kevan, P.G.; Lane, M.A. Flower petal microtexture is a tactile cue for bees. *Proc. Natl. Acad. Sci. USA* **1985**, *82*, 4750–4752. [CrossRef]
86. Vignolini, S.; Rudall, P.J.; Rowland, A.V.; Reed, A.; Moyroud, E.; Faden, R.B.; Baumberg, J.J.; Glover, B.J.; Steiner, U. Pointillist structural color in Pollia fruit. *Proc. Natl. Acad. Sci. USA* **2012**, *109*, 15712–15715. [CrossRef]
87. Vignolini, S.; Gregory, T.; Kolle, M.; Lethbridge, A.; Moyroud, E.; Steiner, U.; Glover, B.J.; Vukusic, P.; Rudall, P.J. Structural colour from helicoidal cell-wall architecture in fruits of *Margaritaria nobilis*. *J. R. Soc. Interface* **2016**, *13*, 20160645. [CrossRef]
88. Almeida, A.P.C.; Canejo, J.P.; Fernandes, S.N.; Echeverria, C.; Almeida, P.L.; Godinho, M.H. Cellulose-Based Biomimetics and Their Applications. *Adv. Mater.* **2018**, *30*, 1703655. [CrossRef]
89. Ligon, R.A.; McGraw, K.J. Chameleons communicate with complex colour changes during contests: Different body regions convey different information. *Biol. Lett.* **2013**, *9*, 20130892. [CrossRef] [PubMed]
90. Darwin, C. *The Movements and Habits of Climbing Plants*; John Murray: London, UK, 1875.
91. Almeida, A.P.; Canejo, J.; Mur, U.; Čopar, S.; Almeida, P.L.; Žumer, S.; Godinho, M.H. Spotting plants' microfilament morphologies and nanostructures. *Proc. Natl. Acad. Sci. USA* **2019**, *116*, 13188–13193. [CrossRef] [PubMed]
92. Murugesan, Y.K.; Rey, A.D. Modeling Textural Processes during Self-Assembly of Plant-Based Chiral-Nematic Liquid Crystals. *Polymers* **2010**, *2*, 766–785. [CrossRef]
93. Tan, T.; Ribbans, B. A bioinspired study on the compressive resistance of helicoidal fibre structures. *Proc. R. Soc. A Math. Phys. Eng. Sci.* **2017**, *473*, 20170538. [CrossRef]
94. Fratzl, P.; Elbaum, R.; Burgert, I. Cellulose fibrils direct plant organ movements. *Faraday Discuss.* **2008**, *139*, 275–282. [CrossRef]
95. Elbaum, R.; Abraham, Y. Insights into the microstructures of hygroscopic movement in plant seed dispersal. *Plant Sci.* **2014**, *223*, 124–133. [CrossRef]
96. Lacey, E.P.; Kaufman, P.B.; Dayanandan, P. The Anatomical Basis for Hygroscopic Movement in Primary Rays of Daucus carota Ssp. carota (Apiaceae). *Bot. Gaz.* **1983**, *144*, 371–375. [CrossRef]
97. Bertinetti, L.; Fischer, F.D.; Fratzl, P. Physicochemical Basis for Water-Actuated Movement and Stress Generation in Nonliving Plant Tissues. *Phys. Rev. Lett.* **2013**, *111*, 238001. [CrossRef]
98. Friedman, J.; Gunderman, N.; Ellis, M. Water response of the hygrochastic skeletons of the true rose of Jericho (*Anastatica hierochuntica* L.). *Oecologia* **1978**, *32*, 289–301. [CrossRef]
99. Hegazy, A.K.; Barakat, H.N.; Kabiel, H.F. Anatomical significance of the hygrochastic movement in Anastatica hierochuntica. *Ann. Bot.* **2006**, *97*, 47–55. [CrossRef]
100. Lebkuecher, J.G.; Eickmeier, W.G. Reduced photoinhibition with stem curling in the resurrection plant Selaginella lepidophylla. *Oecologia* **1991**, *88*, 597–604. [CrossRef]
101. Le Duigou, A.; Castro, M. Evaluation of force generation mechanisms in natural, passive hydraulic actuators. *Sci. Rep.* **2016**, *6*, 18105. [CrossRef]
102. Dawson, C.; Vincent, J.F.V.; Rocca, A.-M. How pine cones open. *Nature* **1997**, *390*, 668. [CrossRef]
103. Song, K.; Yeom, E.; Seo, S.-J.; Kim, K.; Kim, H.; Lim, J.-H.; Joon Lee, S. Journey of water in pine cones. *Sci. Rep.* **2015**, *5*, 9963. [CrossRef]
104. Song, K.; Chang, S.-S.; Lee, S.J. How the pine seeds attach to/detach from the pine cone scale? *Front. Life Sci.* **2017**, *10*, 38–47. [CrossRef]
105. Poppinga, S.; Nestle, N.; Šandor, A.; Reible, B.; Masselter, T.; Bruchmann, B.; Speck, T. Hygroscopic motions of fossil conifer cones. *Sci. Rep.* **2017**, *7*, 40302. [CrossRef]
106. Newcombe, F.C.J.B.G. Spore-dissemination of Equisetum. *Bot. Gaz.* **1888**, *13*, 173–178. [CrossRef]

107. Katifori, E.; Alben, S.; Cerda, E.; Nelson, D.R.; Dumais, J. Foldable structures and the natural design of pollen grains. *Proc. Natl. Acad. Sci. USA* **2010**, *107*, 7635–7639. [CrossRef]
108. Noblin, X.; Rojas, N.O.; Westbrook, J.; Llorens, C.; Argentina, M.; Dumais, J. The Fern Sporangium: A Unique Catapult. *Science* **2012**, *335*, 1322. [CrossRef]
109. Marmottant, P.; Ponomarenko, A.; Bienaimé, D. The walk and jump of Equisetum spores. *Proc. R. Soc. B Biol. Sci.* **2013**, *280*, 20131465. [CrossRef] [PubMed]
110. Bar-On, B.; Sui, X.; Livanov, K.; Achrai, B.; Kalfon-Cohen, E.; Wiesel, E.; Wagner, H.D. Structural origins of morphing in plant tissues. *Appl. Phys. Lett.* **2014**, *105*, 033703. [CrossRef]
111. Ghafouri, R.; Bruinsma, R. Helicoid to Spiral Ribbon Transition. *Phys. Rev. Lett.* **2005**, *94*, 138101. [CrossRef] [PubMed]
112. Armon, S.; Efrati, E.; Kupferman, R.; Sharon, E. Geometry and mechanics in the opening of chiral seed pods. *Science* **2011**, *333*, 1726–1730. [CrossRef] [PubMed]
113. Abraham, Y.; Elbaum, R. Hygroscopic movements in Geraniaceae: The structural variations that are responsible for coiling or bending. *New Phytol.* **2013**, *199*, 584–594. [CrossRef]
114. Almeida, A.P.C.; Querciagrossa, L.; Silva, P.E.S.; Gonçalves, F.; Canejo, J.P.; Almeida, P.L.; Godinho, M.H.; Zannoni, C. Reversible water driven chirality inversion in cellulose-based helices isolated from Erodium awns. *Soft Matter* **2019**, *15*, 2838–2847. [CrossRef]
115. Jung, W.; Kim, W.; Kim, H.Y. Self-burial mechanics of hygroscopically responsive awns. *Integr. Comp. Biol.* **2014**, *54*, 1034–1042. [CrossRef]
116. Zhao, C.; Liu, Q.; Ren, L.; Song, Z.; Wang, J. A 3D micromechanical study of hygroscopic coiling deformation in Pelargonium seed: From material and mechanics perspective. *J. Mater. Sci.* **2017**, *52*, 415–430. [CrossRef]
117. Stamp, N.E. Self-Burial Behaviour of Erodium Cicutarium Seeds. *J. Ecol.* **1984**, *72*, 611–620. [CrossRef]
118. Elbaum, R.; Gorb, S.; Fratzl, P. Structures in the cell wall that enable hygroscopic movement of wheat awns. *J. Struct. Biol.* **2008**, *164*, 101–107. [CrossRef]
119. Rafsanjani, A.; Brulé, V.; Western, T.L.; Pasini, D. Hydro-Responsive Curling of the Resurrection Plant Selaginella lepidophylla. *Sci. Rep.* **2015**, *5*, 8064. [CrossRef]
120. Jiang, X.; Tian, B.; Xuan, X.; Zhou, W.; Zhou, J.; Chen, Y.; Lu, Y.; Zhang, Z. Cellulose membranes as moisture-driven actuators with predetermined deformations and high load uptake. *Int. J. Smart Nano Mater.* **2021**, *12*, 146–156. [CrossRef]
121. Cao, X.; Wang, X.; Ding, B.; Yu, J.; Sun, G. Novel spider-web-like nanoporous networks based on jute cellulose nanowhiskers. *Carbohydr. Polym.* **2013**, *92*, 2041–2047. [CrossRef]
122. Wang, Q.; Wang, D.; Cheng, W.; Huang, J.; Cao, M.; Niu, Z.; Zhao, Y.; Yue, Y.; Han, G. Spider-web-inspired membrane reinforced with sulfhydryl-functionalized cellulose nanocrystals for oil/water separation. *Carbohydr. Polym.* **2022**, *282*, 119049. [CrossRef]
123. Laboy-López, S.; Méndez Fernández, P.O.; Padilla-Zayas, J.G.; Nicolau, E. Bioactive Cellulose Acetate Electrospun Mats as Scaffolds for Bone Tissue Regeneration. *Int. J. Biomater.* **2022**, *2022*, 3255039. [CrossRef]
124. Shigezawa, N.; Ito, F.; Murakami, Y.; Yamanaka, S.; Morikawa, H. Development of combination textile of thin and thick fiber for fog collection bioinspired by Burkheya purpurea. *J. Text. Inst.* **2016**, *107*, 1014–1021. [CrossRef]
125. Han, D.; Gouma, P.-I. Electrospun bioscaffolds that mimic the topology of extracellular matrix. *Nanomed. Nanotechnol. Biol. Med.* **2006**, *2*, 37–41. [CrossRef]
126. Gruppuso, M.; Turco, G.; Marsich, E.; Porrelli, D. Polymeric wound dressings, an insight into polysaccharide-based electrospun membranes. *Appl. Mater. Today* **2021**, *24*, 101148. [CrossRef]
127. Vatankhah, E.; Prabhakaran, M.P.; Jin, G.; Mobarakeh, L.G.; Ramakrishna, S. Development of nanofibrous cellulose acetate/gelatin skin substitutes for variety wound treatment applications. *J. Biomater. Appl.* **2014**, *28*, 909–921. [CrossRef]
128. Liao, N.; Unnithan, A.R.; Joshi, M.K.; Tiwari, A.P.; Hong, S.T.; Park, C.-H.; Kim, C.S. Electrospun bioactive poly (ε-caprolactone)–cellulose acetate–dextran antibacterial composite mats for wound dressing applications. *Colloids Surf. A Physicochem. Eng. Asp.* **2015**, *469*, 194–201. [CrossRef]
129. Gaydhane, M.K.; Kanuganti, J.S.; Sharma, C.S. Honey and curcumin loaded multilayered polyvinylalcohol/cellulose acetate electrospun nanofibrous mat for wound healing. *J. Mater. Res.* **2020**, *35*, 600–609. [CrossRef]
130. Liu, F.; Li, X.; Wang, L.; Yan, X.; Ma, D.; Liu, Z.; Liu, X. Sesamol incorporated cellulose acetate-zein composite nanofiber membrane: An efficient strategy to accelerate diabetic wound healing. *Int. J. Biol. Macromol.* **2020**, *149*, 627–638. [CrossRef] [PubMed]
131. Esmaeili, E.; Eslami-Arshaghi, T.; Hosseinzadeh, S.; Elahirad, E.; Jamalpoor, Z.; Hatamie, S.; Soleimani, M. The biomedical potential of cellulose acetate/polyurethane nanofibrous mats containing reduced graphene oxide/silver nanocomposites and curcumin: Antimicrobial performance and cutaneous wound healing. *Int. J. Biol. Macromol.* **2020**, *152*, 418–427. [CrossRef] [PubMed]
132. Beikzadeh, S.; Akbarinejad, A.; Swift, S.; Perera, J.; Kilmartin, P.A.; Travas-Sejdic, J. Cellulose acetate electrospun nanofibers encapsulating Lemon Myrtle essential oil as active agent with potent and sustainable antimicrobial activity. *React. Funct. Polym.* **2020**, *157*, 104769. [CrossRef]
133. Wang, X.; Huang, Z.; Miao, D.; Zhao, J.; Yu, J.; Ding, B. Biomimetic Fibrous Murray Membranes with Ultrafast Water Transport and Evaporation for Smart Moisture-Wicking Fabrics. *ACS Nano* **2019**, *13*, 1060–1070. [CrossRef]
134. Murray, C.D. The Physiological Principle of Minimum Work: I. The Vascular System and the Cost of Blood Volume. *Proc. Natl. Acad. Sci. USA* **1926**, *12*, 207–214. [CrossRef]

135. Guan, Q.-F.; Han, Z.-M.; Zhu, Y.; Xu, W.-L.; Yang, H.-B.; Ling, Z.-C.; Yan, B.-B.; Yang, K.-P.; Yin, C.-H.; Wu, H.; et al. Bio-Inspired Lotus-Fiber-like Spiral Hydrogel Bacterial Cellulose Fibers. *Nano Lett.* **2021**, *21*, 952–958. [CrossRef]
136. Liu, H.; Pang, B.; Tang, Q.; Müller, M.; Zhang, H.; Dervişoğlu, R.; Zhang, K. Self-Assembly of Surface-Acylated Cellulose Nanowhiskers and Graphene Oxide for Multiresponsive Janus-Like Films with Time-Dependent Dry-State Structures. *Small* **2020**, *16*, 2004922. [CrossRef]
137. Gevorkian, A.; Morozova, S.M.; Kheiri, S.; Khuu, N.; Chen, H.; Young, E.; Yan, N.; Kumacheva, E. Actuation of Three-Dimensional-Printed Nanocolloidal Hydrogel with Structural Anisotropy. *Adv. Funct. Mater.* **2021**, *31*, 2010743. [CrossRef]
138. Wang, M.; Tian, X.; Ras, R.H.A.; Ikkala, O. Sensitive Humidity-Driven Reversible and Bidirectional Bending of Nanocellulose Thin Films as Bio-Inspired Actuation. *Adv. Mater. Interfaces* **2015**, *2*, 1500080. [CrossRef]
139. Zhu, Q.; Jin, Y.; Wang, W.; Sun, G.; Wang, D. Bioinspired Smart Moisture Actuators Based on Nanoscale Cellulose Materials and Porous, Hydrophilic EVOH Nanofibrous Membranes. *ACS Appl. Mater. Interfaces* **2019**, *11*, 1440–1448. [CrossRef]
140. Duan, R.; Lu, M.; Tang, R.; Guo, Y.; Zhao, D. Structural Color Controllable Humidity Response Chiral Nematic Cellulose Nanocrystalline Film. *Biosensors* **2022**, *12*, 707. [CrossRef]
141. Wu, T.; Li, J.; Li, J.; Ye, S.; Wei, J.; Guo, J. A bio-inspired cellulose nanocrystal-based nanocomposite photonic film with hyper-reflection and humidity-responsive actuator properties. *J. Mater. Chem. C* **2016**, *4*, 9687–9696. [CrossRef]
142. Zhang, Z.-L.; Dong, X.; Fan, Y.-N.; Yang, L.-M.; He, L.; Song, F.; Wang, X.-L.; Wang, Y.-Z. Chameleon-Inspired Variable Coloration Enabled by a Highly Flexible Photonic Cellulose Film. *ACS Appl. Mater. Interfaces* **2020**, *12*, 46710–46718. [CrossRef]
143. Zhang, Z.-L.; Dong, X.; Zhao, Y.-Y.; Song, F.; Wang, X.-L.; Wang, Y.-Z. Bioinspired Optical Flexible Cellulose Nanocrystal Films with Strain-Adaptive Structural Coloration. *Biomacromolecules* **2022**, *23*, 4110–4117. [CrossRef]
144. Lin, F.; Wang, Z.; Shen, Y.; Tang, L.; Zhang, P.; Wang, Y.; Chen, Y.; Huang, B.; Lu, B. Natural skin-inspired versatile cellulose biomimetic hydrogels. *J. Mater. Chem. A* **2019**, *7*, 26442–26455. [CrossRef]
145. Hou, K.; Nie, Y.; Tendo Mugaanire, I.; Guo, Y.; Zhu, M. A novel leaf inspired hydrogel film based on fiber reinforcement as rapid steam sensor. *Chem. Eng. J.* **2020**, *382*, 122948. [CrossRef]
146. Spadaccini, C.M. Additive manufacturing and processing of architected materials. *Mater. Res. Soc. Bull.* **2019**, *44*, 782–788. [CrossRef]
147. Wang, D.; Chen, D.; Chen, Z. Recent Progress in 3D Printing of Bioinspired Structures. *Front. Mater.* **2020**, *7*, 286. [CrossRef]
148. Alexander, A.E.; Wake, N.; Chepelev, L.; Brantner, P.; Ryan, J.; Wang, K.C. A guideline for 3D printing terminology in biomedical research utilizing ISO/ASTM standards. *3D Print. Med.* **2021**, *7*, 8. [CrossRef]
149. *ISO/ASTM52900-21*; Additive Manufacturing—General Principles—Fundamentals and Vocabulary. West ASTM International: Conshohocken, PA, USA, 2022; p. 14. [CrossRef]
150. Kam, D.; Chasnitsky, M.; Nowogrodski, C.; Braslavsky, I.; Abitbol, T.; Magdassi, S.; Shoseyov, O. Direct Cryo Writing of Aerogels via 3D Printing of Aligned Cellulose Nanocrystals Inspired by the Plant Cell Wall. *Colloid Interfac* **2019**, *3*, 46. [CrossRef]
151. Mendes, B.B.; Gómez-Florit, M.; Hamilton, A.G.; Detamore, M.S.; Domingues, R.M.A.; Reis, R.L.; Gomes, M.E. Human platelet lysate-based nanocomposite bioink for bioprinting hierarchical fibrillar structures. *Biofabrication* **2020**, *12*, 015012. [CrossRef] [PubMed]
152. Jiang, J.; Oguzlu, H.; Jiang, F. 3D printing of lightweight, super-strong yet flexible all-cellulose structure. *Chem. Eng. J.* **2021**, *405*, 126668. [CrossRef]
153. Guo, J.; Li, Q.; Zhang, R.; Li, B.; Zhang, J.; Yao, L.; Lin, Z.; Zhang, L.; Cao, X.; Duan, B. Loose Pre-Cross-Linking Mediating Cellulose Self-Assembly for 3D Printing Strong and Tough Biomimetic Scaffolds. *Biomacromolecules* **2022**, *23*, 877–888. [CrossRef] [PubMed]
154. Liu, G.; Bhat, M.P.; Kim, C.S.; Kim, J.; Lee, K.-H. Improved 3D-Printability of Cellulose Acetate to Mimic Water Absorption in Plant Roots through Nanoporous Networks. *Macromolecules* **2022**, *55*, 1855–1865. [CrossRef]
155. Esmaeili, M.; Norouzi, S.; George, K.; Rezvan, G.; Taheri-Qazvini, N.; Sadati, M. 3D Printing-Assisted Self-Assembly to Bio-Inspired Bouligand Nanostructures. *Small* **2023**, *2023*, 2206847. [CrossRef]
156. Raviv, D.; Zhao, W.; McKnelly, C.; Papadopoulou, A.; Kadambi, A.; Shi, B.; Hirsch, S.; Dikovsky, D.; Zyracki, M.; Olguin, C.; et al. Active Printed Materials for Complex Self-Evolving Deformations. *Sci. Rep.* **2014**, *4*, 7422. [CrossRef]
157. Tibbits, S. The emergence of "4D printing". In Proceedings of TED Conferences. Available online: https://youtu.be/0gMCZFHv9v8 (accessed on 20 December 2022).
158. Tibbits, S.; McKnelly, C.; Olguin, C.; Dikovsky, D.; Hirsch, S. 4D printing and universal transformation. In Proceedings of the 34th Annual Conference of the Association for Computer Aided Design in Architecture, Los Angeles, CA, USA, 23–25 October 2014; pp. 539–548.
159. Momeni, F.; Hassani.N, S.M.; Liu, X.; Ni, J. A review of 4D printing. *Mater. Des.* **2017**, *122*, 42–79. [CrossRef]
160. 4D Printing Market, by Application (Aerospace and Defense, Healthcare, Automotive, Construction, Clothing, Utility, Others.), by Region-Forecast to 2030. 2023, p. 100. Available online: https://www.precedenceresearch.com/4d-printing-market (accessed on 4 January 2023).
161. Gladman, A.S.; Matsumoto, E.A.; Nuzzo, R.G.; Mahadevan, L.; Lewis, J.A. Biomimetic 4D printing. *Nat. Mater.* **2016**, *15*, 413–418. [CrossRef]

162. Siqueira, G.; Kokkinis, D.; Libanori, R.; Hausmann, M.K.; Gladman, A.S.; Neels, A.; Tingaut, P.; Zimmermann, T.; Lewis, J.A.; Studart, A.R. Cellulose Nanocrystal Inks for 3D Printing of Textured Cellular Architectures. *Adv. Funct. Mater.* **2017**, *27*, 1604619. [CrossRef]
163. Zhao, X.; Tekinalp, H.; Meng, X.; Ker, D.; Benson, B.; Pu, Y.; Ragauskas, A.J.; Wang, Y.; Li, K.; Webb, E.; et al. Poplar as Biofiber Reinforcement in Composites for Large-Scale 3D Printing. *ACS Appl. Bio Mater.* **2019**, *2*, 4557–4570. [CrossRef]
164. Correa, D.; Papadopoulou, A.; Guberan, C.; Jhaveri, N.; Reichert, S.; Menges, A.; Tibbits, S. 3D-Printed Wood: Programming Hygroscopic Material Transformations. *3D Print Addit. Manuf.* **2015**, *2*, 106–116. [CrossRef]
165. Le Duigou, A.; Correa, D.; Ueda, M.; Matsuzaki, R.; Castro, M. A review of 3D and 4D printing of natural fibre biocomposites. *Mater. Des.* **2020**, *194*, 108911. [CrossRef]
166. Correa, D.; Poppinga, S.; Mylo, M.D.; Westermeier, A.S.; Bruchmann, B.; Menges, A.; Speck, T. 4D pine scale: Biomimetic 4D printed autonomous scale and flap structures capable of multi-phase movement. *Philos. Trans. A Math. Phys. Eng. Sci.* **2020**, *378*, 20190445. [CrossRef]
167. Mulakkal, M.C.; Trask, R.S.; Ting, V.P.; Seddon, A.M. Responsive cellulose-hydrogel composite ink for 4D printing. *Mater. Des.* **2018**, *160*, 108–118. [CrossRef]
168. Wang, W.; Yao, L.N.; Zhang, T.; Cheng, C.Y.; Levine, D.; Ishii, H. Transformative Appetite: Shape-Changing Food Transforms from 2D to 3D by Water Interaction through Cooking. In Proceedings of the 2017 ACM Sigchi Conference on Human Factors in Computing Systems (Chi'17), Denver, CO, USA, 6–11 May 2017; pp. 6123–6132. [CrossRef]
169. Teng, X.; Zhang, M.; Mujumdar, A.S. 4D printing: Recent advances and proposals in the food sector. *Trends Food Sci. Technol.* **2021**, *110*, 349–363. [CrossRef]
170. Navaf, M.; Sunooj, K.V.; Aaliya, B.; Akhila, P.P.; Sudheesh, C.; Mir, S.A.; George, J. 4D printing: A new approach for food printing; effect of various stimuli on 4D printed food properties. A comprehensive review. *Appl. Food Res.* **2022**, *2*, 100150. [CrossRef]
171. Lai, J.; Ye, X.; Liu, J.; Wang, C.; Li, J.; Wang, X.; Ma, M.; Wang, M. 4D printing of highly printable and shape morphing hydrogels composed of alginate and methylcellulose. *Mater. Des.* **2021**, *205*, 109699. [CrossRef]

applied biosciences

Article

Are There Wheat Cultivars Allowing Enhanced Carbon Allocation to Soils?

Vincent Chaplot [1,2,*], Isack Mathew [1], Alistair Clulow [1] and Hussein Shimelis [1]

1 School of Agricultural, Earth and Environmental Sciences, University of KwaZulu-Natal, Pietermaritzburg 3209, South Africa
2 Laboratory of Oceanography and Climate, Experiments and Numerical Approaches (LOCEAN), UMR 7159, IRD/C NRS/UPMC/MNHN, IPSL, 75005 Paris, France
* Correspondence: vincent.chaplot@ird.fr

Abstract: The transfer of atmospheric carbon (C) in soils is a possible strategy for climate change mitigation and for restoring land productivity. While some studies have compared the ability of existing crops to allocate C into the soil, the genetic variations between crop genotypes have received less attention. The objective of this study was to compare the allocation to the soil of atmospheric C by genetically diverse wheat genotypes under different scenarios of soil water availability. The experiments were set up under open-field and greenhouse conditions with 100 wheat genotypes sourced from the International Maize and Wheat Improvement Centre and grown at 25% (drought stressed) and 75% (non-stressed) field capacity, using an alpha lattice design with 10 incomplete blocks and 10 genotypes per block. The genotypes were analyzed for grain yield (GY), plant shoot and root biomass (SB and RB, respectively) and C content, and stocks in plant parts. Additionally, ^{13}C pulse labeling was performed during the crop growth period of 10 selected genotypes for assessing soil C inputs. The average GY varied from 75 to 4696 g m^{-2} and total plant biomass (PB) from 1967 to 13,528 g m^{-2}. The plant C stocks ranged from 592 to 1109 g C m^{-2} (i.e., an 87% difference) under drought condition and between 1324 and 2881 g C m^{-2} (i.e., 117%) under well-watered conditions. Atmospheric C transfer to the soil only occurred under well-drained conditions and increased with the increase in the root to shoot ratio for C stocks (r = 0.71). Interestingly, the highest transfer to the soil was found for LM-26 and LM-47 ($^{13}C/^{12}C$ of 7.6 and 6.5 per mille, respectively) as compared to LM-70 and BW-162 (0.75; 0.85). More is to be done to estimate the differences in C fluxes to the soil over entire growing seasons and to assess the long-term stabilization of the newly allocated C. Future research studies also need to identify genomic regions associated with GY and soil C transfer to enable the breeding of "carbon-superior" cultivars.

Keywords: Acrisols; soil carbon; cultivars; cereals; water stress; global change

Citation: Chaplot, V.; Mathew, I.; Clulow, A.; Shimelis, H. Are There Wheat Cultivars Allowing Enhanced Carbon Allocation to Soils?. *Appl. Biosci.* **2023**, *2*, 115–135. https://doi.org/10.3390/applbiosci2010010

Academic Editor: Robert Henry

Received: 17 November 2022
Revised: 13 January 2023
Accepted: 17 February 2023
Published: 20 March 2023

1. Introduction

Climate change is mainly driven by the human-induced release of large quantities of carbon (C), as greenhouse gases (GHGs), to the atmosphere. The C was initially sequestered in soils and rocks by plants through photosynthesis and by other organisms. Evidence points to agriculture as a major source of GHGs starting several thousand years ago through the conversion of natural ecosystems (e.g., forests and grasslands) into croplands. Following this conversion, most of the C from the biomass and soil pools, the latter consisting of plant and animal residues at various stages of decomposition, enters into the atmosphere through oxidation by microbes. A total amount of 133 Pg of soil C is estimated to have been lost to the atmosphere [1].

The subsequent rise in atmospheric CO_2 concentrations has caused an increase in global temperatures, which has significantly affected precipitation patterns, leading to intermittent warmer and drier environments. This, in turn, has negatively affected agriculture. For instance, rising surface temperatures curtail crop production because crops are adapted to specific optimal temperature ranges, and deviations from the optimal range negatively affect biomass production, grain yield (GY) and grain quality [2]. Specifically, higher temperatures exacerbate drought stress by increasing transpiration and reducing crop water use efficiency [3]. In addition, higher atmospheric CO_2 concentrations reduce the nutritional quality of grains, thus impacting on the dietary requirements of billions of people across the world [2,4].

Several propositions have been made to mitigate climate change, such as agroforestry and changes in farming practices, e.g., from conventional to reduced tillage systems, but they are shrouded in controversy and not easy to adopt [5]. The low adoption rates could be partly because farmers, in their large majority, do not necessarily want to change their farming practices. Therefore, pragmatic solutions for transferring and storing the excess atmospheric C into soils that fit well into farmers' already-established production systems are still needed. One such solution could be the identification of genetic resources that are useful to farmers, while at the same time enhancing the amount of soil C. In particular, special attention to the ability of food and cash crops to allocate C in their rooting system and to the soil is needed. Studies have reported that 50–80% of soil C comes from root activity during growth and decomposition after senescence [6–8]. Moreover, a larger root system, is not only expected to increase root C inputs to the soil, but also to increase the water extraction capacity of plants [9], which is a key defense trait to counter potential climate change-driven water scarcity.

Mathew et al. [10], in their analysis of different crop types from 389 trials worldwide, reported that the highest C allocation to roots was in grasses (R:S = 1.19), followed by cereals (0.95), legumes (0.86), oil crops (0.85), and fiber crops (0.50). Further analysis pointed to a decrease in C allocation to roots from humid to dry climates, while, as C allocation to roots increases, C inputs to the soil rise ($r = 0.33$) [11].

Several studies have shown that R:S for annual cereals varies greatly. For example, Amanullah and Stewart [12] reported values of 0.41 and 0.29 for sorghum and maize, respectively. Yang et al. [13] reported a mean of 0.25 for maize and wheat, and Bolinder et al. [14] showed that the average R:S for winter wheat (0.2) was significantly lower than for oats (0.4) and barley (0.5). Variations of R:S within single species have also been reported. Fang et al. [15] reported a ratio of 1.13 for landrace wheat compared to 0.61 and 0.81 for two modern cultivars. Similarly, Siddique et al. [16] reported ratios ranging from 0.72 to 0.84 for modern wheat cultivars. However, Bolinger et al. (1997) [14] did not find differences between cultivars.

Despite several studies on plant C stocks, little is known about the potential ability of different cultivars of a single crop to transfer C to soils. Experimental verification of variations in soil C resulting from changes, such as the activity of different genotypes, may require long-term experiments (e.g., Paustian et al. [17]). Alternatively, the advent of isotopic CO_2 labeling allows for C transfers within the atmosphere–plant–soil system to be accessed over short periods of time, such as days or crop growing seasons. Isotopic labeling exploits the differences in the atomic signature of atmospheric CO_2 and commercially labeled CO_2 ($^{13}CO_2$ or $^{14}CO_2$) to estimate the transfer of C from one pool to another.

In numerous studies, the transfer of atmospheric C to soils has been investigated using C isotopes [8,18]. indicated that annual crops retained more C (45% of assimilated ^{13}C or ^{14}C) in shoots than grasses (34%), mainly perennials, and allocated 1.5 times less C below ground. Similarly, Kuzyakov and Domanski [19] reported that grasses allocate 2200 kg C $ha^{-1}yr^{-1}$ below ground, versus 1500 kg C ha^{-1} yr^{-1} for cereals.

Further, assimilation in aboveground organs [20] and the proportion of the belowground allocation being dependent on plant genetics and environmental conditions were shown by Rangel-Castro et al. [21], de Neergaard and Gorissen [22] and Brüggemann et al. [23].

The objective of the current study was to evaluate and compare the capacity to capture and transfer atmospheric C to roots among wheat cultivars with diverse rooting capabilities and drought tolerance for selecting "carbon superior" cultivars. Here we considered 100 wheat cultivars from the International Maize and Wheat Improvement Centre (CIM-

MYT). These were grown in the field and in a glasshouse and subjected to water-stressed and -non-stressed conditions, and a subset of 10 cultivars was selected for further C translocation studies. The results could be helpful to crop breeders breeding carbon-efficient cultivars, and to farmers who seek solutions for storing more C into soils without changing their current practices.

2. Materials and Methods

2.1. Selection of the Plant Material

One hundred wheat genotypes from the International Maize and Wheat Improvement Centre (CIMMYT) were chosen for their genetic variability in terms of rooting abilities and breeding history for drought tolerance. The performance of these genotypes in terms of GY, total biomass production, and biomass allocation to shoots and to roots was evaluated at the University of KwaZulu-Natal (UKZN), Pietermaritzburg, South Africa, under controlled conditions. One experiment was conducted in the field and another one in a greenhouse. Two water regimes were tested in order to enhance the generality of the results.

A subset of 10 genotypes was selected and thereafter planted in the greenhouse to evaluate through ^{13}C multiple pulse labeling variations in the ability to capture atmospheric C and transfer it to crop tissues and to the soil.

2.2. Field Experiment for Grain Yield and Biomass Production

The field experiment was carried out in 2017 during the winter season at the Ukulinga Research farm (Latitude: 29.667°; longitude: 30.406°; and altitude above sea level of 811 m). The 30-year mean annual temperature at Ukulinga is 18 °C and the mean annual precipication is 738 mm.

The soil was classified as Westleigh (Soil Classification Working Group) [24] or Plinthic Acrisols (WRB) [25] and was characterized by a surface *Orthic* A horizon and a subsurface plinthic B horizon. The A horizon (0–0.3 m) was clay–loamy with a clay content of 28% and an organic content of 2.6%. It was relatively loose (bulk density of 1.04 g cm^{-3}) and acidic (pH = 4.6) (Table 1).

Table 1. Selected soil properties of the surface A horizon (0–0.3 m) at the field trial (Ukulinga farm of the University of KwaZulu Natal, Pietermaritzburg, South Africa).

Bulk density	1.04
Phosphorus (g kg^{-1})	39
Potassium (g kg^{-1})	241
Calcium (g kg^{-1})	1.4
Magnesium (mg kg^{-1})	369
Electrical conductivity (cmol L^{-1})	11
pH KCl	4.6
Organic carbon (%)	2.6
Nitrogen (%)	0.23
Clay (%)	28

Soil tillage occurred before planting in May 2017 at the end of the dry season with the soil surface being immediately covered with a plastic sheet to control the moisture entering the soil. For planting following a 10 × 10 alpha lattice design with three replicates per genotype, small holes of about 5 cm in diameter were drilled every 10 cm along 30 cm spaced lines. The plot size was 0.24 m^2 per genotype per replication for each water regime. In total, the experiment was conducted in 96.0 m^2 of land. Three seeds were planted in each hole on May 15th. Each row was constituted by 10 genotypes and was further treated as an incomplete block. The application of 120 kg ha^{-1} of nitrogen, 30 kg ha^{-1} of phosphorus and 30 kg ha^{-1} of potassium was conducted prior to planting followed by 30 kg ha^{-1} of nitrogen at tillering to stem elongation stage, which is common practice in the area, and crop protection practices followed the recommendations by the Department of Agriculture, Forestry and Fisheries [26].

Drip irrigation was applied throughout the duration of the study for the two treatments. The two treatments were well-watered (75% FC) where the water applied was maintained at 75% field capacity, and water stress (25% FC) where water applied was maintained at 25% of FC. Irrigation was withheld 5 weeks after crop emergence in the drought stress treatment, and was further re-established after the appearance of wilting signs. Irrigation was subsequently withheld before anthesis to mimic drought stress at a key stage of the wheat growing cycle until harvest, which occurred on the 15th of September. Intermittent irrigation was provided to avoid permanent wilting under the drought stress treatment. The amount of water applied, hourly temperature, wind direction and wind speed, and soil water content from 0 to 1 m deep using TDRs every 10 cm depth at 3 locations in each watering treatment were recorded for the duration of the trial to determine water consumption.

2.3. Greenhouse Experiment for Grain Yield and Biomass Production

The first greenhouse experiment, which was performed on the 100 genotypes, was carried out from October 2016 to February 2017. The humidity in the greenhouse was between 55% and 65%, with an average of 11 h of natural day light and daily temperature between 20 and 30 °C. In the 10 × 10 alpha lattice design the 100 genotypes (with 3 replicates per genotype) were randomly assigned to minimize experimental errors associated with location. Five-liter plastic pots with holes at their bottom and were filled with soil material obtained from the 0–30 cm soil layer of the Ukulinga farm. In each pot, 10 seeds were planted and watered to field capacity. Two weeks after germination, the seedlings were adjusted to eight seedlings per pot by removing the two weakest seedlings. The plants were watered and fertilized (300 kg N ha^{-1} and 200 kg P_2O_5 ha^{-1}) using drip irrigation. Two water regimes were used: drought-stressed by maintaining soil moisture at 25% field, and a well-watered treatment that maintained a minimum soil moisture at 75% field capacity. Watering to 75% and 25% FC was determined by a soil moisture probe and weighing of a sample of 10 randomly selected pots daily. The drought stress treatments were initiated 6 weeks after growth crop in the greenhouse and maintained by intermittent irrigation to avoid permanent wilting.

2.4. Selection of the Subset of 10 Genotypes

Several agronomic traits were considered. These included (i) number of days from sowing to heading (DTH) when 50% of the plants had fully emerged spikes, (ii) number of days to maturity (DTM) when 50% of the plants were dry, (iii) number of productive tillers (NPT), (iv) grain yield (GY), (v) shoot and root biomass (SB, RB) at harvest after oven drying at 60 °C for 72 h and the resulting root to shoot (R:S) ratio for biomass, and (vi) thousand kernel weight (TKW).

Hierarchical clustering based on phenotypic data from the field and the glasshouse trials combined across water regimes was used to group the 100 genotypes into 10 sub-groups. A Dendrogram of similarities was derived using the Unweighted Pair Group Method with Arithmetic mean algorithm (UPGMA) and, based on this, the genotype with the highest GY and plant biomass (root + shoot) in each sub-group was selected for further carbon studies. In total, 10 genotypes were selected for carbon allocation analyses.

2.5. Assessment of C Transfer to the Soil

The subset of 10 genotypes was considered in a second greenhouse experiment dedicated to assess the variation among the genotypes to transfer atmospheric C into the soil. The genotypes were planted on 13 January 2018 with 3 pots per genotype and per water regime and laid out in a randomized complete block design. The 10 genotypes were treated similarly to the initial 100 genotypes. Similar water regimes (25 and 75% FC) were used and applied at similar stages as previously explained.

Four pulse labelings (i.e., enrichment of stable ^{13}C isotope in the atmosphere during short periods called pulses) were carried out starting from 4 weeks after emergence

(9 February, i.e., 27 days after sowing) until maturity (110 days). Each pulse labeling event was performed using a closed system, which consisted of an upper chamber made of plastic sheeting. The chamber was installed in the late afternoon of the day before isotope labeling. Extra fine earth was packed firmly around the base of the chamber to reduce gas leakage. The airtight chamber was 1.2 m long, 0.6 m wide and 1 m high, and was built in a greenhouse where the air temperature was 25 ± 5 °C. The inner surface of the chamber was smeared with anti-fog agent to reduce water vapor condensation during labeling, thus maintaining light intensity and reducing the dissolution of $^{13}CO_2$ in water drops forming on the chamber's inner plastic sheet.

The ^{13}C-labeling was achieved by replacing the ambient CO_2 in the chamber with $^{13}CO_2$, as common practice (e.g., Liu et al. [27]). The ambient CO_2 from the chamber was trapped in a 1 M NaOH solution in a beaker placed in the center of the chamber. This was performed until the CO_2 concentration, monitored by an infrared gas analyzer (EGM-1; Environmental Gas Monitor, PP Systems, Hitchin, UK), reached 10 ppmv. When the ambient CO_2 concentration reached 10 ppmv, approximately 5 g $Na_2{}^{13}CO_3$ (~97 atom%) powder was added into a second beaker containing 1 M H_2SO_4 solution to generate $^{13}CO_2$ to bring the carbon dioxide concentration back to 450 ppmv. More $Na_2{}^{13}CO_3$ powder was added to the H_2SO_4 solution when necessary to maintain the chamber carbon dioxide concentration at about 450 ppmv. Each labeling process lasted 6 h to ensure sufficiently high ^{13}C abundance in both shoot and root samples compared with unlabeled control samples. The air inside the chamber was circulated by a vertically mounted electric (12 V, 0.21 A) fan placed in the center of the chamber to ensure good distribution of the $^{13}CO_2$ within the chamber throughout the labeling session. All the chamber carbon dioxide was trapped into the NaOH solution at the end of labeling session before opening the chamber to prevent $^{13}CO_2$ assimilation by the non-labeled neighboring plants.

The content in ^{13}C in the soil represents the balance between 13C allocation to the soil by the genotypes during the pulse labeling events and C mineralization between the pulse labeling events and soil sampling. Since all wheat genotypes were treated equally, i.e., same ambient CO2 concentration, same $^{13}CO_2$ pulsing protocol, it is assumed that the differences in ^{13}C allocation to the soil constitute a fair predictor of overall C allocation to the soil, and we assume differences in the temporal dynamics of C allocation to the soil between genotypes, or in the ^{13}C signal of the C allocated, to be secondary, and if not secondary, to be out of scope of the present study.

2.6. Descriptive Variables

2.6.1. Primary Variables

Plant shoots of all experiments were harvested by clipping at the soil surface after crop maturity. Stems and leaves were pooled and used for C analysis. Soil and roots from the pots were separated manually, by stirring the soil–root mixture in a bucket to ensure complete disaggregation of the soil followed by sieving. The plants were cut off at the soil surface to separate above- from belowground biomass. The roots were then manually separated from the soil by gently washing off the soil under running tap-water and using a 2 mm sieve.

All plant and soil materials were dried in an oven at 60 °C for 72 h before weighing and grinding in ceramic mortar. The ground materials were passed through 2 mm sieves. The sieved samples were stored at room temperature prior to analysis for total C and ^{13}C contents.

Grain, shoot and root samples of the 10 selected genotypes were ground to <0.5 mm and analyzed for total C and N in triplicates using a LECO CNS-2000 Dumas dry matter combustion analyzer (LECO Corp., St. Joseph, MI, USA). The organic C content (OC_C) was used to estimate C stocks in the different plant parts as follows:

$$GC_S = GY \times GC_C \times b \tag{1}$$

where GC_S is the grain C stock in kg C m^{-2}, GY is the grain yield (g m^{-2}) and GC_C is the C concentration in the grain (g C kg^{-1}); b is a constant equal to 0.00001.

$$SC_S = SB \times SC_C \times b \tag{2}$$

where SC_S is the shoot C stock (kg C m^{-2}), SB is the shoot biomass (g m^{-2}) and SC_C is the C concentration in the shoots (g C kg^{-1}); b is a constant equal to 0.00001.

$$RC_S = RB \times RC_C \times b \tag{3}$$

where RC_S is the root C stock (kg C m^{-2}), RB is the root biomass (g m^{-2}) and RC_C is the C concentration in the roots (g C kg^{-1}); b is a constant equal to 0.00001.

The plant C stocks (PC_S) correspond to the sum of the stocks from the different plant parts (e.g., GC_S + SC_S + RC_S for C stocks).

The $^{13}C/^{12}C$ ratios in the bulk soil sampled immediately after harvest were analyzed using an isotope ratio mass spectrometer (Deltaplus, Finnigan MAT GmbH, Bremen, Germany) coupled with an elemental analyzer (NC 2500, ThermoQuest Italia S.p.A., Milan, Italy) by an interface (ConFlo II, Finnigan MAT GmbH, Bremen, Germany). The natural abundance of ^{13}C in the soil was expressed as ^{13}C (%) relative to Pee Dee Belemnite following Equations (4) and (5):

$$R_{sample} = \left(\frac{\delta^{13}C}{1000} + 1\right) \times 0.011237 \tag{4}$$

$$^{13}C(\%) = \left(\frac{R_{sample}}{R_{sample} + 1}\right) \times 100 \tag{5}$$

where R_{sample} is the isotope ratio ($^{13}C/^{12}C$) of a sample and 0.011237 is the ratio of $^{13}C/^{12}C$ in Pee Dee Belemnite; ^{13}C (%) represents the percent of ^{13}C atom in total C atoms in a given sample. Here we considered the difference in soil ^{13}C content before the multiple pulse labeling and after harvest.

2.6.2. Secondary Variables

The water-use efficiency of productivity, typically defined as the ratio of mass produced to the rate of evapotranspiration, was calculated for the selected variables: GY, SB, RB, GC_S, SC_S, RC_S, PC_S and ^{13}C (‰). Meteorological data, such as air temperature, air humidity, wind speed, and quantity of water applied through irrigation or natural rains, were recorded using a standard weather station located in the vicinity of the pots. These data were recorded each day to estimate crop evapotranspiration as follows:

$$ET = \Delta W + I + P + CR - D - R \tag{6}$$

where ET is crop evapotranspiration (mm day^{-1}), ΔW is the change in soil or pot water mass between two consecutive measurements of soil water content (g converted to mm day^{-1}), I is the amount irrigation water applied (mm day^{-1}), P is the amount of rainfall (mm day^{-1}), CR is the capillary rise from the water table to the crop root zone (mm day^{-1}), D is the downward drainage from the crop root zone (mm day^{-1}) and R is the surface runoff (mm day^{-1}). The soils at the open field experiment exhibited high infiltration by water and the water table was below 20 m, hence runoff and capillary contribution from groundwater were ignored. Crop water use efficiency for biomass, C stocks and ^{13}C (‰) was calculated as biomass or C stocks divided by crop total evapotranspiration.

2.6.3. Data Analysis

Descriptive statistics were estimated for the study variables. In addition, data were tested for normality using the Shapiro–Wilk test and homogeneity of variance, and subjected to an analysis of variance (ANOVA) using the general linear model procedure for unbalanced designs. A multivariate procedure for hierarchical clustering was performed based on phenotypic data from the field and greenhouse trials combined across water regimes to group the genotypes based on their similarities. Principal Component Analysis (PCA) was performed to depict the multiple relationships between plant biomass, C allocation and stocks, and C transfer to the soil. The PCA axes were generated using ^{13}C content in the different plant parts and the other variables of plant biomass and plant C, based on mean values across the field and greenhouse experiments.

3. Results

3.1. Variations in Grain Yield and Plant Biomass Amongst the 100 Genotypes

The overall average GY of the initial set of 100 genotypes, across the open-field and glasshouse trials for both water regimes, was 1387 g m^{-2} with a standard error of $\pm$ 84 g m^{-2} (Table 2). GY varied by a factor of 62, from a minimum of 75 g m^{-2} to a maximum of 4696 g m^{-2}, and showed a positively skewed distribution (Skew = 1.53). Total plant biomass (PB) among the 100 genotypes was much less variable than GY as values ranged from 1967 to 13,528 g m^{-2}, a difference of only 6.8 times, which was significant at $p < 0.001$. The 100 genotypes had an average R:S for biomass of 0.12, with values from 0.03 to 0.38 (Table 2).

Table 2. Summary statistics (min = minimum: max = maximum; Q1 = 1st quartile; Q3 = 3rd quartile; CV = coefficient of variation; SEM = standard mean error; skew = skewness; kurt = kurtosis) for grain yield (GY), selected morphological variables (SB: shoot biomass, RB: root biomass, PB: total plant biomass, R:S: root to shoot) for the 100 wheat genotypes grown in the field and in the glasshouse, and across water regimes.

	GY	SB	RB	PB	R:S
	---------------------g m^2 ---------------------				
Mean	1387	2498	305	4189	0.12
Median	1309	2332	263	3930	0.11
Min	75	1179	65	1976	0.03
Max	4696	8658	1219	13,529	0.38
Q1	959	1827	189	3026	0.09
Q3	1644	2908	365	4893	0.15
CV%	47	37	57	37	41
SEM	84	121	22	200	0.01
Skew	1.53	2.04	2.03	1.82	1.67
Kurt	4.35	7.15	5.76	5.66	4.39

In addition, the mean GY of the 10 selected genotypes in the glasshouse increased from 326 g m^{-2} at 25% FC to 414 g m^{-2} at 75% FC, which corresponded to a 27% increase (Table 3). In the open field, the overall increase was from 1700 gm^{-2} at 25% FC to 2062 g m^{-2} at 75% FC, i.e., a 21% increase. A similar trend of greater overall mean values of the 10 selected genotypes at 75% FC as compared to 25% FC was observed for all the biomass variables studied. For instance, the overall mean PB of the 10 genotypes decreased in the glasshouse from 6289 gm^{-2} at 75% FC to 2992 gm^{-2} at 25% FC, as did the overall mean R:S of the 10 genotypes, from 0.57 to 0.43 (Table 3). Overall, the impact of water regime was significant for all the studied variables (Table 4).

The dendrogram generated from UPGMA using the results from the two water regimes and conditions (field and greenhouse) revealed two major distinct clusters of the 100 wheat genotypes based on their similarity in agronomic performance (Figure 1). The first cluster comprised 97 genotypes, and was further divided into subgroups A and B. Sub-cluster A was further divided into four sub-sub-clusters comprising genotypes such as LM26, BW141,

BW140, LM70 and LM48. Sub-cluster B was further divided into two sub-sub-clusters with genotypes such as BW152, LM47, LM75, BW162 and LM71. Mean values of the selected phenotypic variables for each of the 10 clusters and for all experiments are displayed in Table 5. The highest grain yields were obtained for B1 (706 g m^{-2}), followed by B22 (695 g m^{-2}), while the poorest-performing cluster was C (215 g m^{-2}). The greatest plant biomass was also observed for B1 (2218 g m^{-2}) and the highest R:S ratio for C, at 0.82. In contrast, the lowest plant biomass was found for A12, and the lowest R:S (0.44) for A11, A12 and B21 (Table 5).

Table 3. Summary statistics (min = minimum: max = maximum; CV = coefficient of variation; SEM = standard mean error) for grain yield (GY), selected morphological variables (SB: shoot biomass, RB: root biomass, PB: total plant biomass, R:S: root to shoot), and plant carbon variables (SCC: shoot carbon content, RCC: root carbon content, SCS: shoot carbon stocks, RCS: root carbon stocks and PCS: plant carbon stocks) for the selected 10 wheat varieties grown in the field and glasshouse, and across water regimes.

	GY	SB	RB	PB	R:S	SCC	RCC	SCS	RCS	PCS
	----------------------g m^2 ----------------------					---------%--------		-------------g m^2 ---------------		
					Glasshouse					
25% Field capacity										
Mean	326	1873	792	2992	0.43	33	31	625	239	865
Median	266	1826	762	2918	0.44	34	31	621	225	854
Min	58	1349	368	2156	0.15	25	21	409	103	592
Max	869	2415	1263	3983	0.70	37	39	805	379	1109
CV%	68	24	36	24	34	19	23	25	34	23
SEM	28	57	37	94	0.02	1	1	20	10	26
75% Field capacity										
Mean	414	3726	2149	6289	0.57	33	31	1243	647	1890
Median	397	3440	1823	5508	0.57	34	31	1186	605	1716
Min	136	2779	1290	4531	0.37	25	21	851	333	1324
Max	811	5192	3575	9305	0.78	37	39	1822	1059	2881
CV%	50	26	38	30	26	19	23	27	35	28
SEM	27	127	104	247	0.02	1	1	44	29	69
					Field					
25% Field capacity										
Mean	1700	2507	302	4508	0.12	33	31	832	95	927
Median	1401	2534	277	4042	0.12	34	31	827	80	968
Min	621	1598	152	2475	0.08	25	21	534	45	582
Max	4488	3775	642	8100	0.19	37	39	1215	214	1409
CV%	53	27	39	34	24	8	16	26	47	27
SEM	116	87	15	199	0.004	0.36	0.62	28.10	5.79	32.30
75% Field capacity										
Mean	2062	3714	403	6179	0.12	34	31	1254	121	1375
Median	1720	3154	382	5462	0.10	35	31	1052	112	1170
Min	1118	1954	135	3555	0.06	30	21	675	53	728
Max	4383	8658	1006	13,529	0.26	37	39	2869	302	3022
CV%	44	42	44	39	43	6	15	42	43	39
SEM	117	202	23	313	0.01	0.26	0.61	67.57	6.69	69.93

3.2. *Variations Amongst the 10 Best Performing Genotypes*

The best-performing genotype in terms of plant biomass of each of the 10 clusters was selected for further investigation. The plant biomass variables are displayed in Figure 2 for the field experiments and in Figure 3 for the glasshouse experiments. In the field and under drought stress, BW152 was the highest grain-yielding amongst the 10 selected genotypes, with 2330 g m^{-2}, followed by BW162 (1748 g m^{-2}), LM75 (1716 g m^{-2}) and LM71 (1582 g m^{-2}), while LM26 was the worst one (1088 g m^{-2}) (Figure 2). Plant biomass ranged between 3471 g m^{-2} for LM26 and 6351 g m^{-2} for BW152, and the R:S varied between 0.087 for LM75 and 0.17 for BW152. Under 75% FC, BW141 ranked first for GY at 3061 g m^{-2}, and LM71 was second at 2554 g m^{-2}, immediately followed by LM75 (2530 g m^{-2}). Seven genotypes had a GY over 2000 g m^{-2}. BW141 had the highest PB, followed by BW152, LM26, LM70 and LM75 (Figure 2). R:S varied between 0.071 for BW141 and 0.17 for LM70.

Table 4. Mean squares after combined analysis of variance for phenotypic traits of 100 wheat genotypes evaluated across the two water regimes and the two environments (field and glasshouse). GY: grain yield; SB: shoot biomass weight; RB: root biomass weight; R:S: root to shoot ratio.

	DF	GY	SB	RB	R:S
Rep	1	1.54	3.2	13.9	0.07
Block	18	117.1 ***	236.4	18.5	0.10 ***
Genotype	99	101.7 ***	556.5	52.1 *	0.06 *
Water regime	1	43.7 *	43.1 *	2.8 *	1.12 *

Phenotypic coefficient of variation at 0.05 (*) and 0.001 (***), respectively.

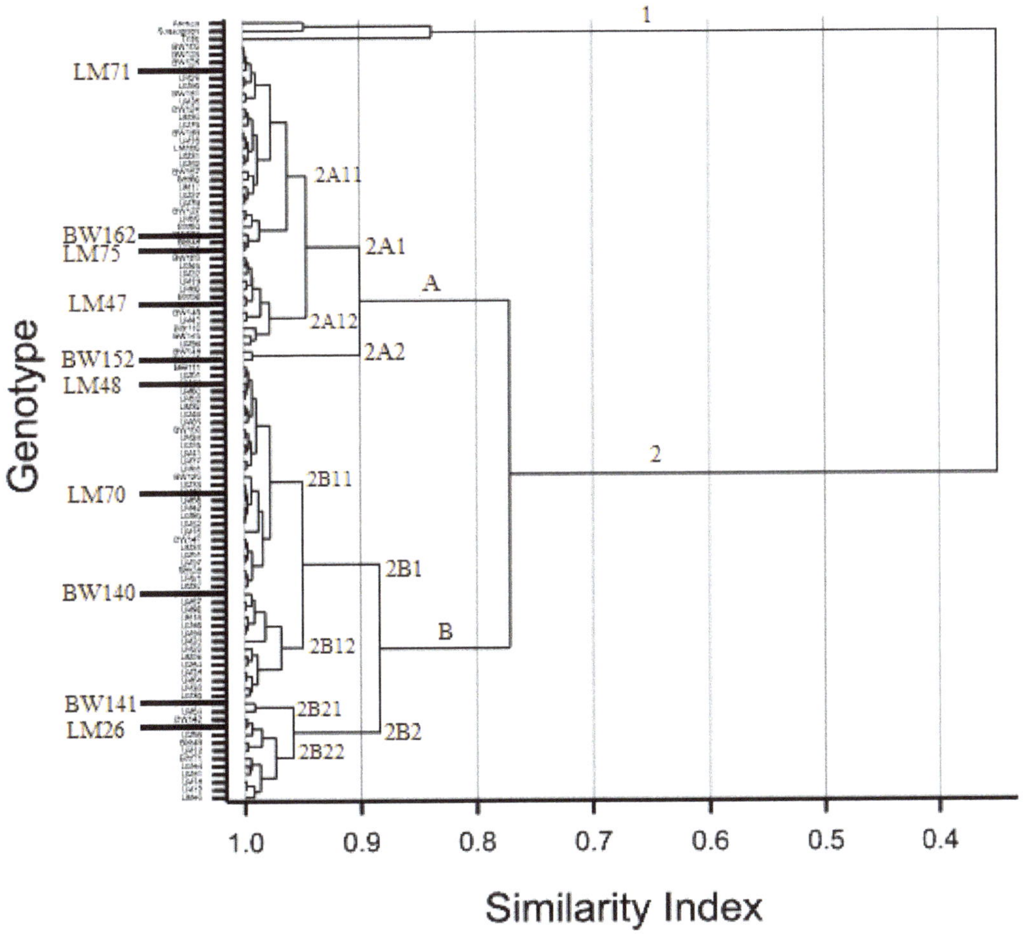

Figure 1. Dendrogram showing clusters according to phenotypic relatedness of 100 genotypes evaluated.

Table 5. Mean of selected phenotypic traits for the 10 clusters presented in Figure 1 and for the different water regimes and environments (field and glasshouse).

Cluster	Treatment	GY	SB	RB	PB	R:S
2A11	Overall	613.8	992.5	224.1	1830.4	0.44
	Drought	*480.9*	*777.6*	*187.1*	*1445.7*	*0.44*
	Well-watered	*746.7*	*1207.5*	*261.1*	*2215.2*	*0.45*
2A12	Overall	586.8	860.7	216.2	1663.6	0.44
	Drought	*385.5*	*678.1*	*166.0*	*1229.6*	*0.43*
	Well-watered	*771.3*	*1028.2*	*262.1*	*2061.5*	*0.44*
2A1	Overall	481.5	983.3	227.3	1686.3	0.45
	Drought	*317.3*	*795.1*	*186.3*	*1295.0*	*0.40*
	Well-watered	*649.6*	*1176.1*	*269.2*	*2087.1*	*0.49*
2A2	Overall	576.4	986.1	238.4	1795.3	0.46
	Drought	*432.5*	*824.9*	*197.6*	*1447.5*	*0.42*
	Well-watered	*717.8*	*1147.3*	*279.5*	*2143.0*	*0.50*
2B1	Overall	706.6	1159.1	352.9	2218.6	0.50
	Drought	*610.3*	*968.5*	*243.9*	*1822.7*	*0.43*
	Well-watered	*802.9*	*1349.8*	*461.9*	*2614.5*	*0.57*
2B21	Overall	568.4	1182.4	270.7	2005.7	0.44
	Drought	*413.4*	*929.9*	*223.8*	*1555.4*	*0.42*
	Well-watered	*719.1*	*1427.9*	*316.4*	*2443.7*	*0.46*
2B22	Overall	695.3	1125.9	264.8	2067.0	0.48
	Drought	*502.7*	*873.8*	*210.8*	*1567.2*	*0.45*
	Well-watered	*885.6*	*1376.5*	*318.7*	*2566.8*	*0.50*
Grand Mean		**603.0**	**1040.5**	**256.1**	**1893.4**	**0.46**

Greenhouse GY values were significantly lower than in the field, with values from 170 g m^{-2} for BW162 to 490 g m^{-2} for LM75 under 25% FC, and from 195 g m^{-2} for BW141 to 815 g m^{-2} for LM75 under 75% FC. BW162 and BW140 had the highest RB under well-watered conditions, while LM26 and BW162 ranked first for water-stressed conditions.

From Figure 3, we can see that LM75 was on average the most water-efficient genotype in terms of GY, shoot and plant biomass and carbon stocks, but ranked second in terms of water use efficiency for RB. In contrast, LM47, which ranked second for water use efficiency (WUE) in terms of GY, was the least efficient genotype for RB. BW162 and BW140 were the most water-efficient for RB.

3.3. Variations in Below- and Aboveground C Allocation among the Selected Genotypes

The average C content of the selected 10 genotypes was 34 $\pm$ 0.9% in the shoots (SC_C), which decreased to 30 $\pm$ 1.1% in the roots (RC_C) (Table 3), corresponding to a significant difference at $p < 0.05$. SC_C ranged from 30 to 37%, while RCc varied from 21 to 39%, with all the differences being significant at $p < 0.05$. Plant C stocks (PCs) exhibited large variations between the tested genotypes, with a mean value from all experiments of 1174$\pm$ 64 g m^{-2}. The values ranged from 582 g m^{-2} (in the field and under 25% FC) to 3022 g m^{-2} (field and 75% FC), i.e., a 5-fold difference. From Table 3, we also learn that the distribution of plant C stocks (PC_S) was positively skewed (skewness = 4.7), with most of the PC_S being found in shoots (91%) rather than in roots (9%).

In the field, plant C stocks (PCs) under 25% FC were the highest for BW 152 (1059 g C m^{-2}) and BW 141 (1004 g C m^{-2}), while four genotypes had values below 850 g C m^{-2} (BW140, LM26, LM75 and LM71) (Figure 4). PCs under 75% FC were the highest for BW141 (2260 g C m^{-2}), followed by LM70 (1712 g C m^{-2}), LM26 (1684 g C m^{-2}) and BW152 (1663 g C m^{-2}), the lowest values being found for LM71 (901). Under glasshouse conditions, PCs under 25% FC were the highest for BW162 (1043 g C m^{-2}) and LM70 (1012 g C m^{-2}), and the lowest for BW152 (679 g C m^{-2}), LM140 (732 g C m^{-2}) and LM71 (794 g C m^{-2})

(Figure 5). Under 75% FC, BW140 ranked first for PCs (2733 g C m^{-2}), followed by BW162 (2259 g C m^{-2}), LM70 (2277 g C m^{-2}) and LM75 (2072 g C m^{-2}), while all the other cultivars had significantly lower PCs.

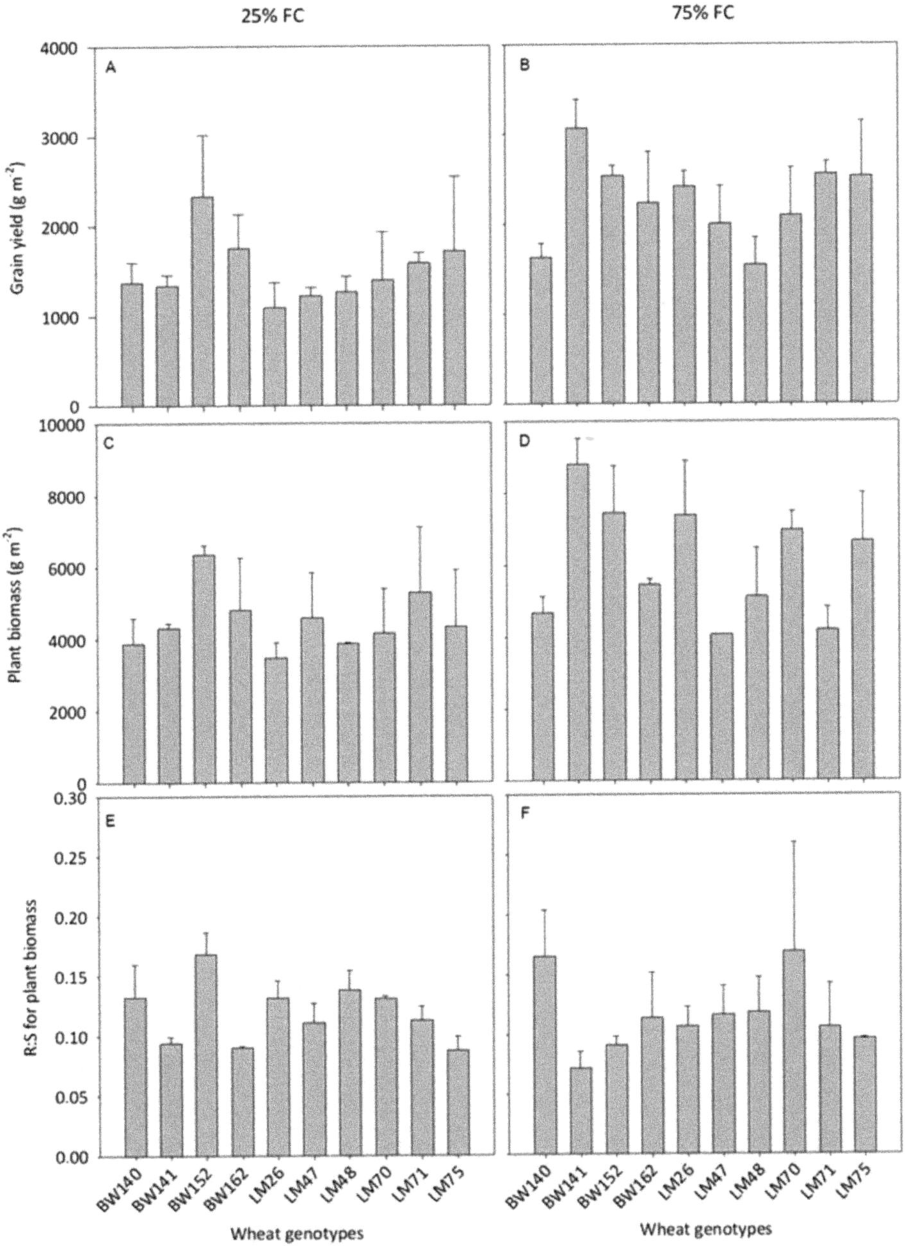

Figure 2. Variations in grain yield (**A**,**B**), plant biomass (**C**,**D**) root to shoot ratio for plant biomass (**E**,**F**) due to genotype and soil water availability (25 and 75% field capacity) under field conditions. Data are mean and standard errors (n = 9).

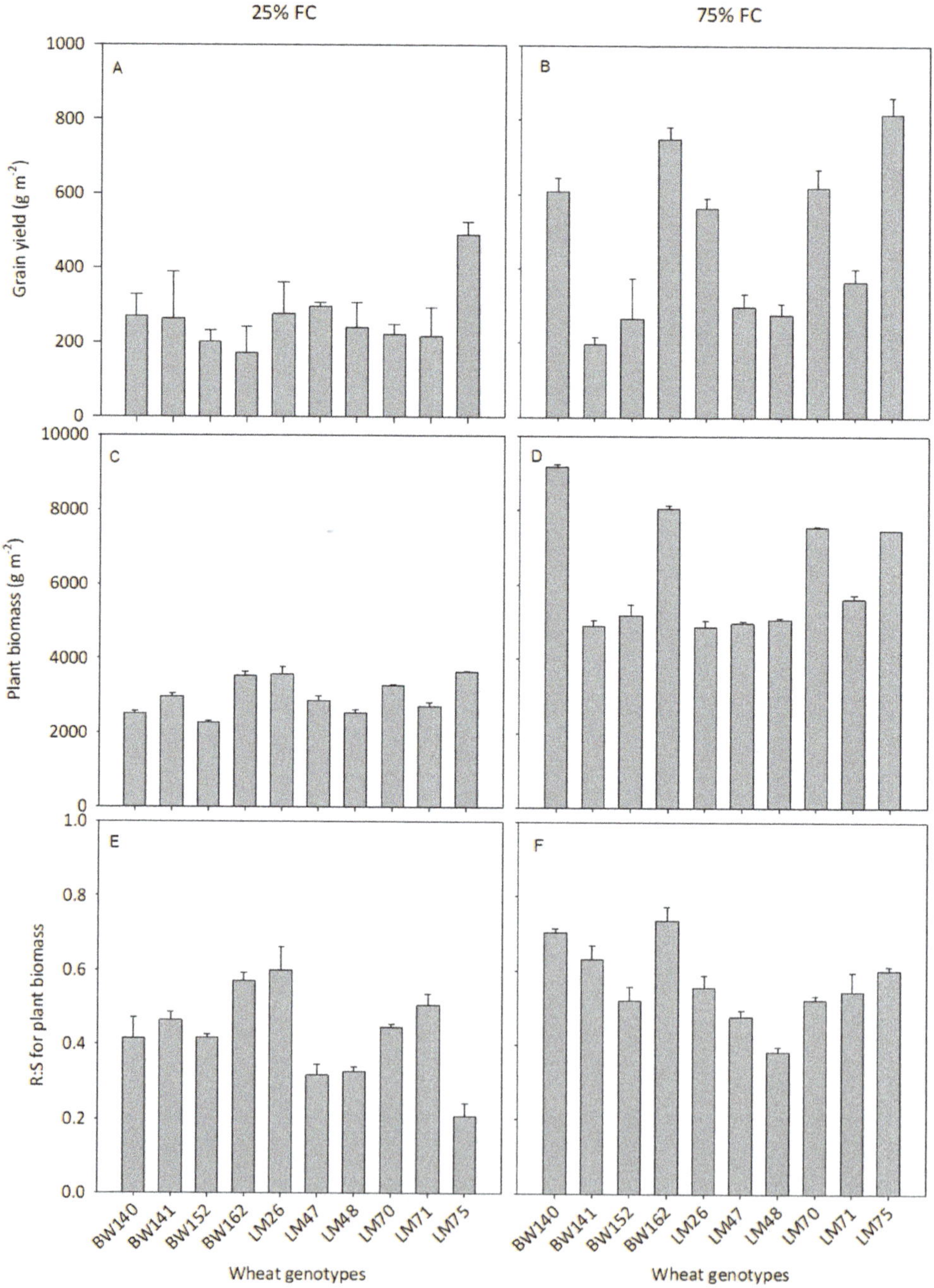

Figure 3. Variation in grain yield (**A**,**B**), plant biomass (**C**,**D**) root to shoot ratio for plant biomass (**E**,**F**) due to genotype and soil water availability (25 and 75% field capacity) under greenhouse conditions. Data are mean and standard errors (n = 9).

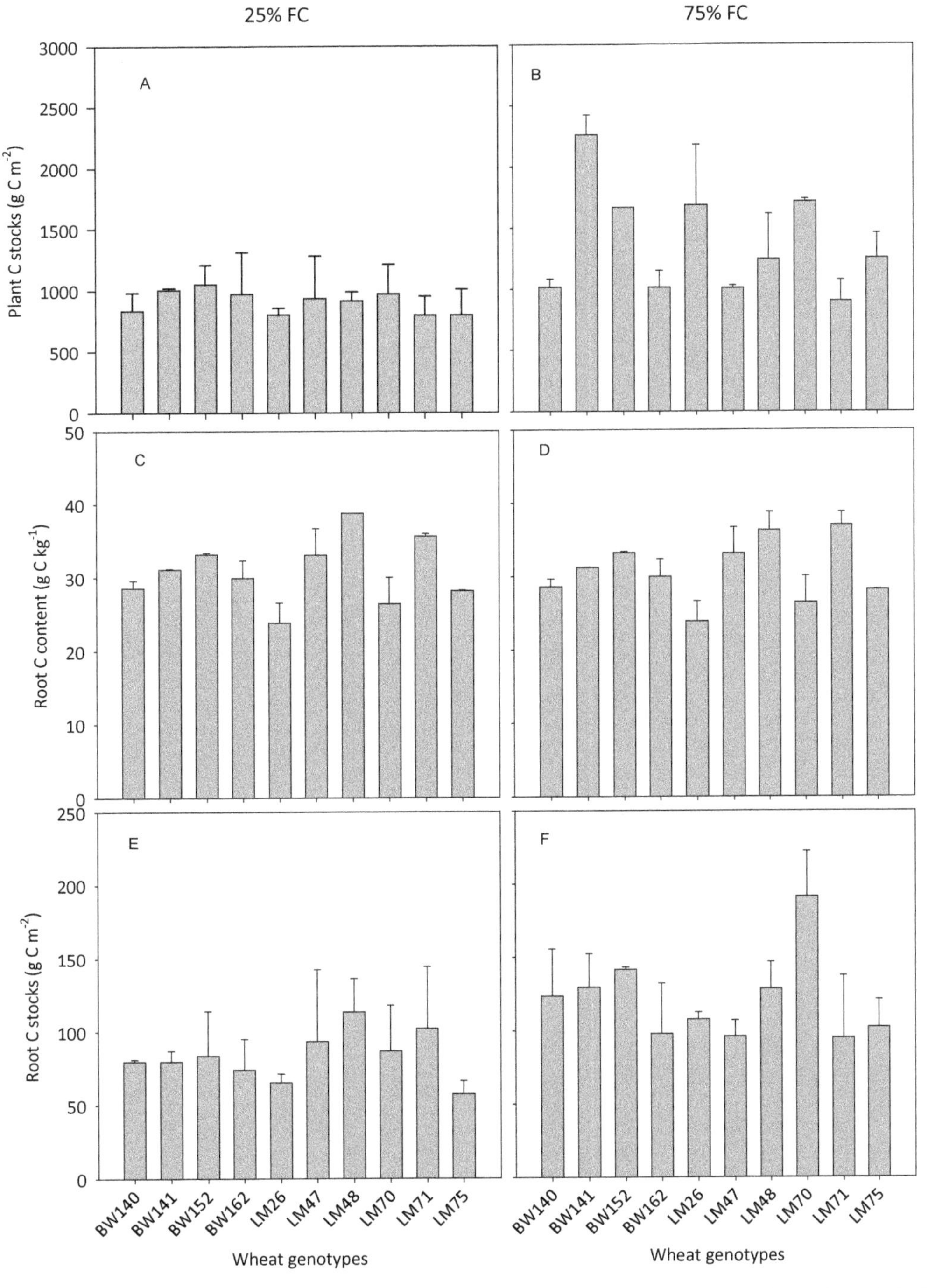

Figure 4. Variation in plant C stocks (**A**,**B**), root C content (**C**,**D**) and root C stocks (**E**,**F**) due to genotype and soil water availability (25 and 75% field capacity) under field conditions. Data are mean and standard errors (n = 9).

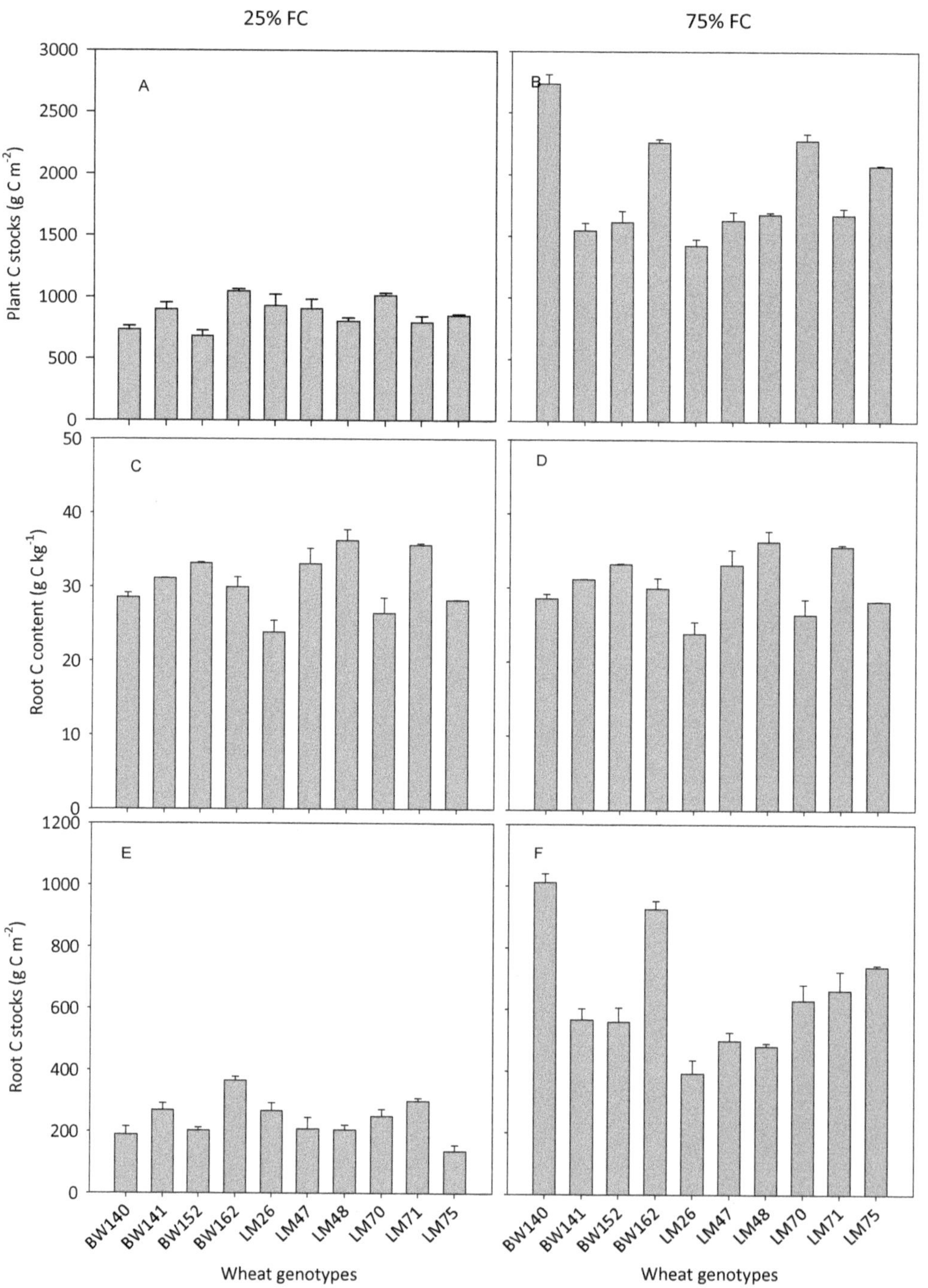

Figure 5. Variation in plant C stocks (**A**,**B**), root C content (**C**,**D**) and root C stocks (**E**,**F**) due to genotype and soil water availability (25 and 75% field capacity) under greenhouse conditions. Data are mean and standard errors (n = 9).

Excess ^{13}C in the soil after harvesting as compared to the bulk soil at sowing was significantly higher under well-watered than under dry conditions (3.0 ± 0.16‰ vs. 0.011 ± 0.0018‰) (Figure 6). Under water-stressed conditions, excess ^{13}C ranged between 0‰ for LM-75 and 0.059‰ for LM-48, but even though these differences were significant, they remained very close to zero (Figure 4). In contrast, the soil ^{13}C content of the well-watered treatment ranged between 0.75 ± 0.21‰ for LM-70 and 7.58 ± 0.058‰ for LM-26, which corresponded to a 1006% difference, significant at the $p < 0.05$ level (Figure 4). Contents over 3‰ also appeared for LM-71 (3.1‰), BW-140 (3.5‰), and LM-47 (6.5‰), while contents below 3‰ occurred for LM-75 (2.9‰), LM-48, BW-152 (1.6‰), BW-141 (1.3‰) and BW-162 (0.84‰) (Figure 6).

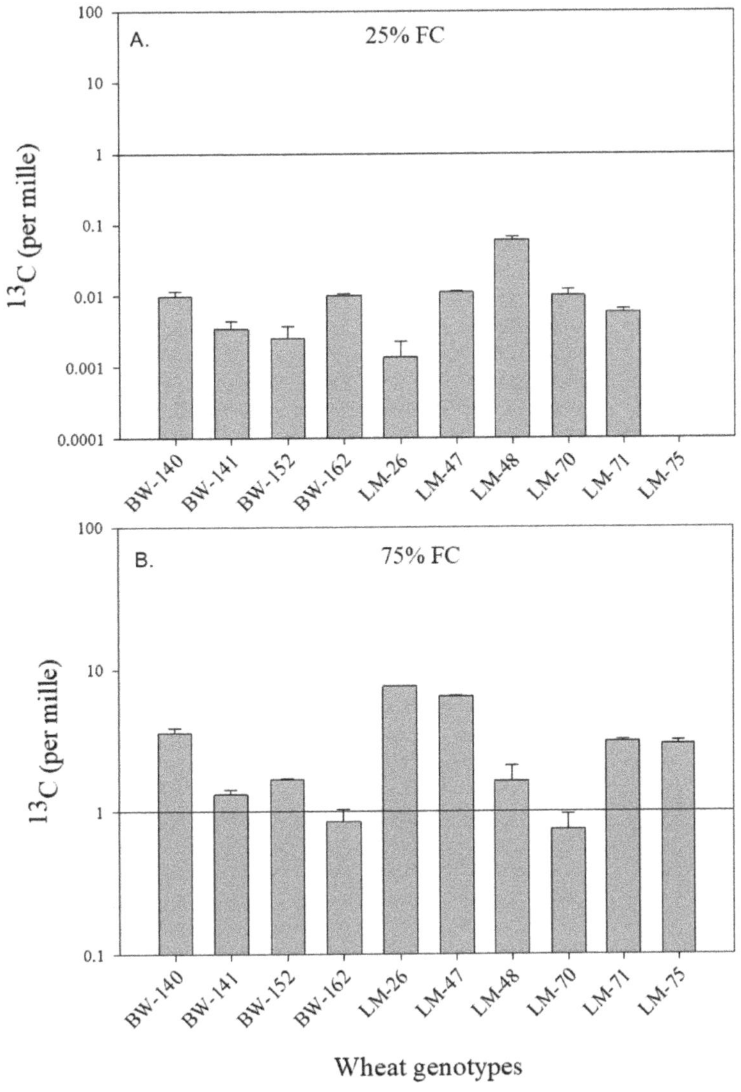

Figure 6. Excess ^{13}C content in the soil after harvest as compared to before pulse labeling for drought (**A**) and well-watered (**B**) treatments and for the studied wheat genotypes. Data are mean and standard errors (n = 3).

4. Discussion

4.1. Large Variations in Below- and Aboveground C Allocation among the Selected Genotypes

The results point to a maximum 102% difference in wheat C stocks between the studied cultivars, the poorest-performing one storing 592 g C m^{-2} under drought conditions and 1324 g C m^{-2} under well-watered conditions, while the best-performing cultivar allocated 1109 and 2881 g C m^{-2}, respectively.

Such a difference in the ability of cultivars to allocate C in their body was unexpected. Our investigations show that the "carbon superior" cultivars had higher total biomass (grain, shoot and root) and greater concentrations of C in the biomass.

Large variations in total biomass production among cultivars agree with reports by Nevo and Chen [28], Waines et al. [29], and more recently by Akman et al. [30] and Fang et al. [15], and can be explained by differences in genotypic characteristics leading to variations in the production and the storage of assimilates by photosynthesis.

The higher shoot biomass C of some cultivars can most likely be explained by a greater ability for C uptake, and the ability of the phloem to transport the C compounds belowground, probably as a result of a greater pressure applied to the flow system and a higher size and continuity of the "tubes" that allow C to be transported over long distances (Liu et al. [27]). Wide differences were also observed in the ability of the genotypes to allocate C into their roots, as shown by a 3.7-times difference between the lowest and the highest root C stocks. This translated into wide variations in R:S for C stocks (CV of 34% with values between 0.15 and 0.70), which was also most unexpected, as wheat genotypes have been shown so far to only slightly vary in their biomass allocation to roots with, for instance, Fang et al. [15] pointing to ratios between 1 and 1.36.

Such considerable variations in root traits among genotypes have been previously suggested [31,32], and the higher allocation of biomass and plant C in the roots of the selected CIMMYT genotypes might be explained by the fact that these have been developed for heat and drought tolerance, for which the root system is key. The impacts of drought stress on sensitive genotypes include stomatal closure, high root death, compromised cell membranes and reduced mesophyll conductance [33,34]. These changes may cause leakages of cell sap, reduced enzyme activity, reduced photosynthetic activity and ultimately the low transportation of photosynthates to plant organs and rhizodeposition [35]. However, these effects are less pronounced in drought-tolerant genotypes, which could explain the significantly higher biomass productivity and C content in the CIMMYT heat- and drought-tolerant genotypes compared to the other genotypes in the test population. Lavinsky et al. [36] found that drought-tolerant maize reduced starch and total soluble sugar production, while increasing the production of phenolics, which helped in avoiding excessive losses of water and the collapse of the xylem vessels. This ensured that photosynthesis was maintained at a higher rate compared to the drought-sensitive genotypes, whose photosynthetic capacity decline, as evidenced by reductions in lignin content, grain yield and biomass allocation to roots and shoots. These mechanisms could also have been used by the CIMMYT heat- and drought-tolerant varieties, since maize and wheat share a common ancestry and their physiology is largely governed by syntenic genes. In contrast, Aljazairi [37] indicated that modern genotypes tend to invest more C in their spikes than the traditional genotypes, which invested more C in non-reproductive shoot tissues such as roots, thus probably explaining the low variability in RB observed in previous studies. Interestingly, van der Graaff et al. [38] reported an R:S of 0.6 for the wild genotype, as compared to 0.3 for a modern one. Not only did the present study point to a higher range of R:S values, but it also indicates the higher C translocation to roots and potentially to soils than previously though [27].

The average root C content was 33% in shoots and 31% in roots, which are low, as Poaceas generally exhibit shoot C contents in the 40–45% range. A meta-analysis study on world data showed that the mean C content was 45% in shoots and 41% in roots [10]. However, the values in the present study differed highly, for instance in the glasshouse from 25% to 37% in shoots, and from 21% to 39% in roots. The low C content values were lower than our expectations. However, deviations from expected values in mineral content have been reported as a function of soil or environmental conditions affecting nutrient uptake and plant development. Gavito et al. [39] found that soil temperature, atmospheric CO_2 concentration and soil nitrogen content had significant interactive effects on N and P acquisition and content in plant biomass. Rengel [40] also indicated that C and other mineral contents in crops change with differences in genotype and soil conditions.

The existence of large C content differences between crop cultivars and between crop parts (stems vs. roots) is similar to reports by Corneo et al. [31], and might be explained by different abilities to produce C compounds, such as lignin of high C:N. In support, Luquet et al. [41] showed for instance that the lignin content in stems is positively correlated with plant height, because lignin has a high tensile strength, which is required to support tall plants. The differences in plant height among the test genotypes would indicate that they have different lignin contents, and thus the C content would vary widely.

4.2. Variations in the Ability of Wheat Genotypes to Transfer Atmospheric C into the Soil

Several reasons might be given to explain the differences in C translocation to soils by the studied cultivars, where a 1006% maximum difference was observed. These are likely to involve differences in the supply of assimilates to roots as mentioned above, but also differences in: (1) the ability of roots to exudate C to the rhizosphere while limiting respiration; (2) the capacity of exudates to be stabilized, either adsorbed by clay particles or assimilated by mycorrhizal fungi and other rhizosphere microorganisms from the rhizosphere [42]. A further process (3), which mostly occurs at later stages, after harvest, is the ability of root tissues to become stable organic matter following decomposition [43,44].

Several other reasons might explain the differences in C transfer into the soil of the different genotypes. Hütsch et al. [45] showed that cereal root exudates, which are 80% water-soluble (64% carbohydrates, 22% amino acids and 14% organic acids), are rapidly (1–2 days) stabilized into water-insoluble forms and bound preferentially to clay particles. Warembourg and Estelrich [46] also indicated that only 13% to 21% of total root exudates are found in the soil matrix as water-insoluble or stable organic matter. The present study, which involved soil sampling after harvest, weeks after the last pulse labeling occurred, thus most likely identifies the presence in the soil of water-insoluble forms of organic matter. Genotypes such as LM47 and LM48 showed a superior ability to produce water-insoluble organic matter in the soil, which we estimated to be about 30% higher than the least efficient genotypes. The underlying reasons are still unknown. Do these cultivars produce higher amounts of root exudates? Are these exudates of higher stability (high C:N)? Do exudates support a rhizobiome that is enhancing the biological, chemical and physical C stabilization in soils? What genetic factors are associated with this? These are important research questions to be investigated. Moreover, increased C transfer to the soil, as we estimated for certain genotypes, might induce a detrimental effect for soil organic matter. Indeed, the production of root exudates by the best C genotypes could, indeed, stimulate microbial activity and accelerate the turnover of soil organic matter [47], a process known as the "priming effect", where soil organic matter (SOM) decomposition is stimulated by the addition of labile substrates to the soil. However, this process might be limited because exudate stabilization is quick, and a large fraction of root exudates become stable organic matter [46]. Finally, the large amounts of dead root and shoot material produced by LM48 and LM47 might also become important sources of soil organic matter, hence their contribution to total soil organic matter should also be further investigated.

A high level of C translocation to the soil was observed for LM47, which was also one of the highest grain-yielding cultivars, ranking 2nd out of the 10 studied cultivars. This points out that in spite of water supply, this cultivar not only showed an efficient production of sugars but also an optimized transfer belowground and to the soil.

4.3. *Impact of Soil Water Availability on Plant C Storage and C Transfer to the Soil*

Overall, water stress decreased wheat's ability to allocate C into the soil. There was an average 2.2-fold decrease in plant C stocks following water stress, and the decrease was more acute for root C stocks (2.7-fold) than shoot C stocks (1.9-fold). Such a decrease in plant C stocks is not explained by changes in C content in body parts, but by a decrease in biomass. As water becomes less available, plants tend to respond by reducing the stem height and diameter, as well as the dry weight [48], with cell elongation being more impacted than cell division [49]. As the content of water in stem cells decreases, there is an increase in the content of soluble sugars to maintain tissue turgor, and consequently, less of it is transported belowground as reserves in roots [41]. Such an adaptation mechanism to drought is likely to explain the sharper decrease in root C stocks than in shoot C stocks following water stress.

In addition, not only did water stress decrease root C stocks, but it also tended to significantly decrease C transfers to the soil, and to lessen the differences between cultivars. In water-stressed conditions, the excess soil ^{13}C content only differed by less than half a percent among the genotypes, while the difference increased to 21% under well-watered conditions. This is likely to be due to a shortage of exudate supply to the soil for all cultivars as water becomes scarce.

From the PCA, there was a trend for C transfer to the soil to increase with R:S. This may be explained by the fact that the more a plant builds roots, the more carbon is exudated to the soil, thus calling for breeding strategies enhancing the R:S ratio. LM47 could thus be considered as a "carbon superior" wheat cultivar since it induced the highest C transfer to the soil. It interestingly exhibited the highest root C content amongst the cultivars, while it ranked second in terms of GY. The latter is a very promising result, pointing out that the objectives of grain production and soil C sequestration are not necessarily antinomic.

The present study showed that drought stress significantly affected C translocation to roots and to the soil. Indeed, not only is the production of photosynthesis products reduced, but its transport from leaf to roots is too, as previously demonstrated [50].

5. Conclusions

Three main conclusions can be drawn from this study that investigated the ability of 100 wheat cultivars from the CIMMYT gene bank to transfer atmospheric carbon (C) into soils. The first conclusion is that wheat genotypes highly differed in their ability to allocate C into the soil, with differences of up to 1006%. The second conclusion is that C allocation to soil was highly correlated with soil water availability, with greater allocation under well-watered soil conditions. The third conclusion is that soil C transfer tended to increase with the increase in plant biomass and biomass allocation to roots, and this was not systematically to the detriment of grain yield. These findings point out the enormous potential of the available genetic resources in wheat to meet the multiple objectives of grain and biomass production, the sustainable use of soils, and climate mitigation and adaptation. Such an improved understanding of the links between plant characteristics and soil C storage may open up opportunities for wheat breeding with multiple objectives in mind. Research on the mechanisms of the decomposition of roots, as well as of aboveground post-harvest biomass, with management that addresses farmers' needs, is also required. As the results were obtained under clay–loam conditions in South Africa, confirmation of the trends should be made under different soil conditions and climates. In addition, further research on the factors that control the transfer and storage of atmospheric C into soils is still important. Amongst the emerging research questions that may need to be addressed in the near future are: How does a cultivar's physiology affect the transfer of carbohydrates to the soil and their incorporation into stable organic matter? What are the exact differences in C input to the soil amongst the cultivars at the level of the entire wheat-growing cycle? What role do exudates play in C translocation to soils, as well as their quantity and quality?

Author Contributions: Conceptualization and methodology, V.C. and H.S.; investigation, data curation, I.M. and A.C.; writing, review and editing by all authors. All authors have read and agreed to the published version of the manuscript.

Funding: This research was funded by the South African Water Research Commission (WRC) Project No. K5-2721-4.

Institutional Review Board Statement: The study did not require ethical approval.

Informed Consent Statement: Not applicable.

Data Availability Statement: All available data are present in the resent paper.

Acknowledgments: The authors appreciate the funding from the South African Water Research Commission (WRC) Project No. K5-2721-4 and the National Research Foundation (NRF) of South Africa. This manuscript is based upon a report to the funders entitled "WATER USE EFFICIENCY AND CARBON SEQUESTRATION POTENTIAL OF INDIGENOUS CROPS". We would also like to acknowledge technical support from The African Centre for Crop Improvement (ACCI).

Conflicts of Interest: The authors declare no conflict of interest.

References

1. Sanderman, J.; Hengl, T.; Fiske, G.J. Soil carbon debt of 12,000 years of human land use. *Proc. Natl. Acad. Sci. USA* **2017**, *114*, 9575–9580. [CrossRef]
2. Pandey, Y.; Vyas, R.P.; Kumar, J.; Singh, L.; Singh, H.C.; Yadav, P.C. Heritability, Correlation and Path Coefficient Analysis for Determining Interrelationships among Grain Yield and Related characters in Maize (*Zea mays* L.). *Int. J. Pure App. Biosci.* **2017**, *5*, 595–603. [CrossRef]
3. Ray, J.D.; Gesch, R.W.; Sinclair, T.R.; Allen, L.H. The effect of vapor pressure deficit on maize transpiration response to drying soil. *Plant Soil* **2002**, *239*, 113–121. [CrossRef]
4. Smith, M.R.; Myers, S.S. Impact of anthropogenic CO_2 emissions on global human nutrition. *Nat. Clim. Chang.* **2018**, *8*, 834. [CrossRef]
5. Minasny, B.; Malone, B.P.; McBratney, A.B.; Angers, D.A.; Arrouays, D.; Chambers, A.; Chaplot, V.; Chen, Z.-S.; Cheng, K.; Bhabani, S.; et al. Soil carbon 4 per mille. *Geoderma* **2017**, *292*, 59–86. [CrossRef]
6. Rasse, D.P.; Rumpel, C.; Dignac, M.-F. Is soil carbon mostly root carbon? Mechanisms for a specific stabilization. *Plant Soil* **2005**, *269*, 341–356. [CrossRef]
7. Yang, Y.; Fang, J.; Ji, C.; Ma, W.; Mohammat, A.; Wang, S.; Smith, P. Widespread decreases in topsoil inorganic carbon stocks across China's grasslands during 1980s–2000s. *Glob. Chang. Biol.* **2012**, *18*, 3672–3680. [CrossRef]
8. Hirte, J.; Leifeld, J.; Abiven, S.; Oberholzer, H.-R.; Mayer, J. Below ground carbon inputs to soil via root biomass and rhizodeposition of field-grown maize and wheat at harvest are independent of net primary productivity. *Agric. Ecosyst. Environ.* **2018**, *265*, 556–566. [CrossRef]
9. Lynch, J.P.; Wojciechowski, T. Opportunities and challenges in the subsoil: Pathways to deeper rooted crops. *J. Exp. Bot.* **2015**, *66*, 2199–2210. [CrossRef]
10. Mathew, I.; Shimelis, H.; Mutema, M.; Chaplot, V. What crop type for atmospheric carbon sequestration: Results from a global data analysis. *Agric. Ecosyst. Environ.* **2017**, *243*, 34–46. [CrossRef]
11. Mathew, I.; Shimelis, H.; Mutema, M.; Minasny, B.; Chaplot, V. Crops for increasing soil organic carbon stocks—A global meta analysis. *Geoderma* **2020**, *367*, 114230. [CrossRef]
12. Amanullah, Jr.; Stewart, B.A. Dry matter partitioning, growth analysis and water use efficiency response of oats (*Avena sativa* L.) to excessive nitrogen and phosphorus application. *J. Agr. Sci. Tech.* **2013**, *15*, 479–489.
13. Yang, C.H.; Chai, Q.; Huang, G.B. Root distribution and yield responses of wheat/maize intercropping to alternate irrigation in the arid areas of northwest China. *Plant Soil Environ.* **2010**, *56*, 253–262. [CrossRef]
14. Bolinder, M.A.; Angers, D.A.; Dubuc, J.P. Estimating shoot to root ratios and annual carbon inputs in soils for cereal crops. *Agric. Ecosyst. Environ.* **1997**, *63*, 61–66. [CrossRef]
15. Fang, Y.; Du, Y.; Wang, J.; Wu, A.; Qiao, S.; Xu, B.; Chen, Y. Moderate drought stress affected root growth and grain yield in old, modern and newly released cultivars of winter wheat. *Front. Plant Sci.* **2017**, *8*, 672. [CrossRef] [PubMed]
16. Siddique, K.H.M.; Belford, R.K.; Tennant, D. Root: Shoot ratios of old and modern, tall and semi-dwarf wheats in a Mediterranean environment. *Plant Soil* **1990**, *121*, 89–98. [CrossRef]
17. Paustian, K.; Lehmann, J.; Ogle, S.; Reay, D.; Robertson, G.P.; Smith, P. Climate-smart soils. *Nature* **2016**, *532*, 49–57. [CrossRef]
18. Pausch, J.; Kuzyakov, Y. Carbon input by roots into the soil: Quantification of rhizodeposition from root to ecosystem scale. *Glob. Chang. Biol.* **2018**, *24*, 1–12. [CrossRef]
19. Kuzyakov, Y.; Domanski, G. Carbon input by plants into the soil. Review. *J. Plant Nutr. Soil Sci.* **2000**, *163*, 421–431. [CrossRef]
20. Kaštovská, E.; Šantrůčková, H. Fate and dynamics of recently fixed C in pasture plant-soil system under field conditions. *Plant Soil* **2007**, *300*, 61–69. [CrossRef]

21. Rangel-Castro, J.I.; Killham, K.; Ostle, N.; Nicol, G.W.; Anderson, I.C.; Scrimgeour, C.M. Stable isotope probing analysis of the influence of liming on root exudates utilization by soil microorganisms. *Environ. Microbiol.* **2005**, *7*, 828–838. [CrossRef] [PubMed]
22. De Neergaard, A.; Gorissen, A. Carbon allocation to roots, rhizodeposits and soil after pulse labeling: A comparison of white clover (*Trifolium repens* L.) and perennial ryegrass (*Lolium perenne* L.). *Biol. Fertil. Soils* **2004**, *39*, 228–234. [CrossRef]
23. Brüggemann, E.N.; Gessler, A.; Kayler, Z.; Keel, S.G.; Badeck, F.; Barthel, M.; Boeckx, P.; Buchmann, N.; Brugnoli, E.; Esperschütz, J.; et al. Carbon allocation and carbon isotope fluxes in the plant-soil-atmosphere continuum: A review. *Biogeosciences* **2011**, *8*, 3457–3489. [CrossRef]
24. Soil Classification Working Group. *Soil Classification—A Taxonomic System for South Africa*; Department of Agricultural Development: Pretoria, South Africa, 1991.
25. IUSS Working Group WRB. *World Reference Base for Soil Resources. A Framework for International Classification, Correlation and Communication*; World Soil Resources Report No. 103; FAO: Rome, Italy, 2006.
26. DAFF. *Wheat Production Guideline*; Department of Agriculture, Forestry and Fisheries: Pretoria, South Africa, 2010.
27. Liu, X.; Bonhomme, J.; Merbach, I.; Kümmel, S.; Richnow, H.H. Uptake of a-HCH by wheat from the gas phase and translocation to soil analyzed by a stable carbon isotope labeling experiment. *Chemosphere* **2021**, *264*, 128489. [CrossRef] [PubMed]
28. Nevo, E.; Chen, G. Drought and salt tolerances in wild relatives for wheat and barley improvement. *Plant Cell Environ.* **2010**, *33*, 670–685. [CrossRef] [PubMed]
29. Waines, J.G. *Determination of Optimum Root and Shoot Size in Bread Wheat for Increased Water and Nutrient Use Efficiency and Grain Yield*; Report to California Wheat Commission: GH 2011–2012 Experiments; California Wheat Commission: Woodland, CA, USA, 2012.
30. Akman, H.; Akgün, N.; Tamkoç, A. Comparison of root and shoot traits of different wheat species and wild wheat relatives: Does feature of shoot biomass have positive and significant relationships with grain yield and root traits? *Rev. Fac. Agron.* **2017**, *34*, 428–447.
31. Corneo, P.E.; Suenaga, H.; Kertesz, M.A.; Dijkstra, F.A. Effect of twenty four wheat genotypes on soil biochemical 734 and microbial properties. *Plant Soil.* **2016**, *404*, 141–155. [CrossRef]
32. Comas, L.H.; Becker, S.R.; Cruz, V.M.V.; Byrne, P.F.; Dierig, D.A. Root traits contributing to plant productivity under drought. *Front. Plant Sci.* **2013**, *4*, 442. [CrossRef]
33. Grassi, G.; Magnani, F. Stomatal, mesophyll conductance and biochemical limitations to photosynthesis as affected by drought and leaf ontogeny in ash and oak trees. *Plant Cell Environ.* **2005**, *28*, 834–849. [CrossRef]
34. Henry, A.; Doucette, W.; Norton, J.; Bugbee, B. Changes in crested wheatgrass root exudation caused by flood, drought, and nutrient stress. *J. Environ. Qual.* **2007**, *36*, 904–912. [CrossRef]
35. Canarini, A.; Merchant, A.; Dijkstra, F.A. Drought effects on Helianthus annuus and Glycine max metabolites: From phloem to root exudates. *Rhizosphere* **2016**, *2*, 85–97. [CrossRef]
36. Lavinsky, A.O.; Magalhães, P.C.; Ávila, R.G.; Diniz, M.M.; De Souza, T.C. Partitioning between primary and secondary metabolism of carbon allocated to roots in four maize genotypes under water deficit and its effects on productivity. *Crop J.* **2015**, *3*, 379–386. [CrossRef]
37. Aljazairi, S.; Arias, C.; Nogues, S. Carbon and nitrogen allocation and partitioning in traditional and modern wheat genotypes under pre-industrial and future CO_2 conditions. *Plant Biol.* **2015**, *17*, 647–659. [CrossRef] [PubMed]
38. Van der Graaff, E.; Laux, T.; Rensing, S.A. The WUS homeobox-containing (WOX) protein family. *Genome Biol.* **2009**, *10*, 248. [CrossRef] [PubMed]
39. Gavito, M.E.; Curtis, P.S.; Mikkelsen, T.N.; Jakobsen, I. Interactive effects of soil temperature, atmospheric carbon dioxide and soil N on root development, biomass and nutrient uptake of winter wheat during vegetative growth. *J. Exp. Bot.* **2001**, *52*, 1913–1923. [CrossRef]
40. Rengel, Z. The role of crop residues in improving soil fertility. In *Nutrient Cycling in Terrestrial Ecosystems*; Springer: Berlin/Heidelberg, Germany, 2007; pp. 183–214.
41. Luquet, D.; Perrier, L.; Clément-Vidal, A.; Jaffuel, S.; Verdeil, J.L.; Roques, S.; Soutiras, A.; Baptiste, C.; Fabre, D.; Bastianelli, D. Genotypic covariation of traits underlying sorghum stem biomass production and quality and its regulation by water availability: Insight from studies at organ and tissue levels. *GCB Bioenergy* **2018**, *11*, 444–462. [CrossRef]
42. Kuzyakov, Y.; Friedel, J.K.; Stahr, K. Review of mechanisms and quantification of priming effects. *Soil Biol. Biochem.* **2000**, *32*, 1485–1498. [CrossRef]
43. Whipps, J.M.; Budge, S.P. Screening for sclerotial mycoparasites of *Sclerotinia sclerotiorum*. *Mycol. Res.* **1990**, *94*, 607–612. [CrossRef]
44. Kuzyakov, Y.; Cheng, W. Photosynthesis controls of rhizosphere respiration and organic matter decomposition. *Soil Biol. Biochem.* **2001**, *33*, 1915–1925. [CrossRef]
45. Hutsch, B.W.; Augustin, J.; Merbach, W. Plant rhizodeposition—An important source for carbon turnover in soils. *J. Plant Nutr. Soil Sci.* **2002**, *165*, 397–407. [CrossRef]
46. Warembourg, F.R.; Estelrich, D.H. Towards a better understanding of carbon flow in the rhizosphere: A time-dependent approach using carbon-14. *Biol. Fertil. Soils* **2000**, *30*, 528–534. [CrossRef]
47. Phillips, R.P.; Finzi, A.C.; Bernhardt, E.S. Enhanced root exudation induces microbial feedbacks to N cycling in a pine forest under long-term CO_2 fumigation. *Ecol. Lett.* **2011**, *14*, 187–194. [CrossRef] [PubMed]

48. Perrier, L.; Rouan, L.; Jaffuel, S.; Clement-Vidal, A.; Roques, S.; Soutiras, A.; Luquet, D. Plasticity of sorghum stem biomass accumulation in response to water deficit: A multiscale analysis from internode tissue to plant level. *Front. Plant Sci.* **2017**, *8*, 1516. [CrossRef] [PubMed]
49. Lacube, S.; Fournier, C.; Palaffre, C.; Millet, E.J.; Tardieu, F.; Parent, B. Distinct controls of leaf widening and elongation by light and evaporative demand in maize. *Plant Cell Environ.* **2017**, *40*, 2017–2028. [CrossRef]
50. Barthel, M.; Hammerle, A.; Sturm, P.; Baur, T.; Gentsch, L.; Knohl, A. The diel imprint of leaf metabolism on the $\delta^{13}C$ signal of soil respiration under control and drought conditions. *New Phytol.* **2011**, *192*, 925–938. [CrossRef]

applied biosciences

MDPI

Article

Tissue Microarray Lipidomic Imaging Mass Spectrometry Method: Application to the Study of Alcohol-Related White Matter Neurodegeneration

Isabel Gameiro-Ros [1,†], Lelia Noble [2,†], Ming Tong [3,†], Emine B. Yalcin [2] and Suzanne M. de la Monte [2,3,4,*]

1 Department of Pharmacology and Therapeutics, Faculty of Medicine, Autonomous University of Madrid, 28029 Madrid, Spain
2 Department of Pathology and Laboratory Medicine, Rhode Island Hospital, Alpert Medical School of Brown University, Providence, RI 02903, USA
3 Department of Medicine, Rhode Island Hospital, Alpert Medical School of Brown University, Providence, RI 02903, USA
4 Departments of Neurology & Neurosurgery, Rhode Island Hospital, Alpert Medical School of Brown University, Providence, RI 02903, USA
* Correspondence: suzanne_delamonte_md@brown.edu; Tel.: +1-401-444-7364
† These authors contributed equally to this work.

Abstract: Central nervous system (CNS) white matter pathologies accompany many diseases across the lifespan, yet their biochemical bases, mechanisms, and consequences have remained poorly understood due to the complexity of myelin lipid-based research. However, recent advances in matrix-assisted laser desorption/ionization-imaging mass spectrometry (MALDI-IMS) have minimized or eliminated many technical challenges that previously limited progress in CNS disease-based lipidomic research. MALDI-IMS can be used for lipid identification, semi-quantification, and the refined interpretation of histopathology. The present work illustrates the use of tissue micro-arrays (TMAs) for MALDI-IMS analysis of frontal lobe white matter biochemical lipidomic pathology in an experimental rat model of chronic ethanol feeding. The use of TMAs combines workload efficiency with the robustness and uniformity of data acquisition. The methods described for generating TMAs enable simultaneous comparisons of lipid profiles across multiple samples under identical conditions. With the methods described, we demonstrate significant reductions in phosphatidylinositol and increases in phosphatidylcholine in the frontal white matter of chronic ethanol-fed rats. Together with the use of a novel rapid peak alignment protocol, this approach facilitates reliable inter- and intra-group comparisons of MALDI-IMS data from experimental models and could be extended to human disease states, including using archival specimens.

Keywords: MALDI; white matter; lipidomic; alcohol; tissue micro-array; mass spectrometry; central nervous system

Citation: Gameiro-Ros, I.; Noble, L.; Tong, M.; Yalcin, E.B.; de la Monte, S.M. Tissue Microarray Lipidomic Imaging Mass Spectrometry Method: Application to the Study of Alcohol-Related White Matter Neurodegeneration. *Appl. Biosci.* **2023**, 2, 173–193. https://doi.org/10.3390/applbiosci2020013

Academic Editor: Robert Henry

Received: 9 January 2023
Revised: 28 March 2023
Accepted: 29 March 2023
Published: 4 April 2023

1. Introduction

1.1. White Matter Pathology in Neurodegeneration

White matter (WM) atrophy is an important and consistent feature of many chronic central nervous system (CNS) diseases, including alcohol-related brain degeneration (ARBD) [1–3], Alzheimer's disease (AD) [4–6], vascular dementias [7,8], and frontotemporal lobar degeneration [9], yet the mechanisms have been vastly under-investigated due to technical challenges posed by myelin lipidomic research. WM is mainly composed of axons that project neuronal connections to different brain regions and myelin, which provides the insulation and support needed to optimize axonal functions. Damage to or the loss of axons disrupts structural and pathway communications. The degeneration or loss of myelin compromises the efficiency of neurotransmission needed for cognitive and motor functions [10–17] and renders axons vulnerable to toxic, metabolic, and inflammatory injury that can lead to sustained or permanent functional deficits [18].

1.2. Alterations in Myelin Lipid Composition with Disease

A unique property of CNS myelin is its very high dry mass of lipid (70–85%) compared with that of protein (15–30%). Major WM lipids include cholesterol, glycosphingolipids, sulfatides, gangliosides, phospholipids, and sphingomyelin [19]. Chronic disease states leading to WM degeneration are marked by altered expression and metabolism of phospholipids and sulfatides [20–22]. Alterations in membrane phospholipid composition perturb lipid raft and receptor functions [23,24]. The aberrant or reduced expression of glycerosphingolipids, sphingomyelins, and sulfatides broadly impairs neuronal and glial functions, including plasticity, neuronal conductivity, memory, myelin maintenance, protein trafficking, adhesion, glial-axonal signaling, insulin secretion, and oligodendrocyte survival [20,25,26]. The degradation of sulfatide increases ceramide abundance [27,28], which promotes neuroinflammation, reactive oxygen species formation, and apoptosis, and impairs cellular survival and metabolic signaling [25].

1.3. Strategy for Increasing Knowledge of Disease-Specific WM Pathology

Although imbalances in phospholipids or sphingolipids represent biochemical correlates of WM degeneration, the breadth of knowledge in this field is limited. Thus far, relatively few lipids have been characterized in relation to disease, and the available information about the specificity, time course, and reversibility of CNS myelin lipid pathology is modest. Progress in this field has been impeded by the lack of accessible tools for the simultaneous analysis of multiple samples, including different sources, such as diseased versus control brains, and correlations of biochemical parameters with histopathology. Herein, we describe the use of tissue microarrays (TMAs) coupled with matrix-assisted laser desorption/ionization-imaging mass spectrometry (MALDI-IMS) to characterize the effects of ARBD on frontal lobe WM lipid profiles in an established experimental rat model of chronic ethanol consumption.

1.4. Benefits of Tissue Microarrays in Research

TMAs enable the simultaneous in situ analysis of multiple tissue samples for correlations with histopathology and intra- and inter-group comparisons. The TMA approach has been applied to studies of brain microvascular pathology in neurodegeneration [29], gene expression changes in malignancies [30,31], and proteomic mass spectrometry [32]. The unique feature of this report is the utilization of TMA technology for untargeted disease-oriented lipidomic research. Our methodological approach was designed to overcome limitations of data acquisition and interpretation encountered with standard protocols that are not focused on lipidomics research. Furthermore, although we have successfully utilized standard MALDI-IMS procedures [33–35], the need to ensure the uniform handling and simultaneous analysis of samples to soundly investigate the pathophysiology of diseases, either experimental or human, drew us to re-strategize by developing a TMA-MALDI-Lipidomics protocol. The only currently feasible alternative would be to extract lipids for the simultaneous analysis of samples deposited onto a MALDI target plate [36], but that approach abrogates correlative in situ studies of tissue pathology.

2. Materials and Methods

2.1. Overview

The methods herein detail our TMA-MALDI-Lipidomics protocol, including the critical reagents and procedural steps needed to generate TMAs that are suitable for lipidomic research. The example illustrated is an established experimental rat model of chronic alcohol exposure that causes WM degeneration together with deficits in spatial learning and memory tasks as demonstrated using the Morris Water Maze [37]. This section includes a brief description of the model and a summary of the technical details pertaining to the acquisition of MALDI-IMS data. The methods corresponding to both the rat model and MALDI-IMS have been detailed in earlier publications [21,38,39].

2.2. Experimental Model

Chronic ethanol feeding of adult Long Evans rats causes significant deficits in spatial learning and memory, along with atrophy and degeneration of WM [38,40,41]. To generate the model, 4-week-old Long Evans rats were pair-fed for 6 weeks with commercial isocaloric liquid diets (Research Diets, Inc., New Brunswick, NJ, USA) that contained 0% or 36% (caloric content) ethanol [21,42]. The diets were prepared daily according to the manufacturer's protocol. Immediately after sacrifice via isoflurane inhalation, the brains were harvested to obtain a standardized slice of both frontal lobes by making an initial cut just anterior to the temporal tips and a second cut 3 mm further anterior. The right and left 3 mm-thick frontal lobe slices were laid flat in separate pre-labeled Tissue-Tek acetyl polymer square mesh slotted embedding and processing cassettes (Electron Microscopy Sciences, Hatfield, PA, USA). One slice was quickly frozen on dry ice and stored at −80 °C, and the other was immersion-fixed in 10% neutral buffered formalin and stored at 4 °C. The use of experimental animals for this research was approved by the Lifespan Institutional Animal Care and Use Committee, Providence RI, USA.

2.3. TMA Generation

2.3.1. Overview

TMAs were generated to enable the batch processing and analysis of multiple samples on a single slide. The TMAs should optimally include 2 or 3 replicate control and experimental samples, randomly positioned to assess the reproducibility of results and perform robust intra- and inter-group comparisons [43,44]. The main considerations for generating TMAs are as follows: (a) nature of the embedding compound; (b) tissue state (fresh-frozen or formalin-fixed); and (c) technical steps used to efficiently collect and transfer samples for constructing the arrays.

2.3.2. Selection of the Embedding Compound

Embedding compounds must enable optimum tissue sectioning and minimize background signals. In addition, they should be easy to cut and withstand repeated cryo-sectioning. Embedding compounds that generate high backgrounds or diminish the MALDI-IMS signal-to-noise ratios are unsuitable. Since the suitability and effects of embedding compounds can vary with the tissue type, source, and preparation, preliminary studies must be conducted to optimize experimental conditions. The sources and preparations of four embedding compounds considered in developing the MALDI-TMA method are provided below.

a. Tissue-Plus Optimal Cutting Temperature embedding compound (Tissue-Tek O.C.T.) from Sakura Finetek USA Inc., Torrance, CA, USA, was purchased and used according to the manufacturer's instructions.
b. Two percent carboxymethylcellulose (CMC) gel was prepared by dissolving high-viscosity carboxymethlycellulose sodium salt (Sigma-Aldrich, St. Louis, MO, USA) in sterile, deionized water and storing the product at 4 °C for up to 2 weeks.
c. Modified O.C.T. (mOCT) was prepared as described [45]. In brief, polyvinyl alcohol (PVA) 6–98 (10 g) was heat-solubilized in Hank's Balanced Salt Solution (HBSS; 100 mL), and after cooling to room temperature, polypropylene glycol (PPG) 2000 (8 mL) and sodium azide (100 mg) were added and vortex-mixed to form a thin milky white gel. The mOCT was stored at room temperature for up to 3 months and vortex mixed just prior to each use.

d. Gelatin (10–15%) is the fourth potential embedding compound that ultimately was not tested for this study [46–48] due to very low signals obtained with WM tissue, as previously reported [46]. Unfortunately, the gelatin problem could not be resolved because it would have been impracticable to precisely peel the embedding compound away from the TMA cores.

2.3.3. Tissue Sample Preparation for MALDI-IMS Lipidomic Studies

1. Only fresh-frozen or formalin-fixed tissues are suitable.
2. Paraffin-embedded samples are unsuitable because alcohol-containing solvents used for tissue processing destroy and solubilize lipids.
3. For sample testing, 3 mm-diameter fresh-frozen or formalin-fixed frontal lobe WM tissue cores from adult Long Evans rats (2 control and 2 alcohol-fed) were embedded in O.C.T., CMC, or mOCT.

2.3.4. TMA Construction

The workflow diagram in Figure 1 depicts the required steps for generating sample recipient blocks, coring the samples, and assembling the TMAs.

(1) Generating TMA recipient blocks:
    (a) Plan the array.
        1. Select a validated and reproducible system for making TMA recipient blocks.
            a. We used reusable silicone molds from Arraymold (Salt Lake City, UT, USA).
        2. Decide on the number of samples to be included and core diameter requirements for the TMA.
            a. Arraymold recipient block template configurations: 1 mm cores = 170 samples; 1.5 mm cores = 150 samples; 2 mm cores = 70 samples; 3 mm cores = 40 samples; 4 mm cores = 15 samples; 5 mm cores = 15 samples.
        3. Plan to include either fresh-frozen or formalin-fixed tissue in a single TMA.
            a. Different tissue preparations can impact MALDI-IMS signal intensities.
            b. Large differences in MALDI-IMS signal intensities significantly distort results.
        4. Construct an asymmetric grid map to designate sample core insertion sites. Asymmetric mapping:
            a. Establishes orientation of the TMAs.
            b. Designates sample addresses to prevent misidentification at later analytical stages.
    (b) Evenly distribute the embedding compound across the mold, then allow the compound to solidify in a cryostat chamber set to −20 °C.
        1. For this protocol, we used a Leica CM 3050 cryostat microtome (Leica Biosystems, Wetzlar, Germany).
    (c) Carefully release the frozen recipient block by peeling off the silicone mold.
    (d) Store the recipient block at −80 °C in an air-tight container for up to 3 months. Commercial air-tight laboratory plastic containers were used to store TMA slides organized and held in molded plastic cork-lined slide storage boxes that were further sealed in plastic bags.

(2) Sample preparation and coring:
    (a) Keep samples in their original cassettes at least until the tissue cores have been harvested and transferred to the recipient block.
    (b) Select a 1 mm to 5 mm diameter re-usable Arraymold coring tool or a disposable surgical biopsy punch that corresponds to the recipient block.

1. Match the Arraymold coring tool size to the commercial Arraymold to enable snug fitting of the specimens and minimize gaps.

(c) For fresh-frozen tissue, retrieve samples from the −80 °C freezer and equilibrate for 15–30 min in a −20 °C cryostat chamber.

(d) For formalin-fixed tissue, prepare samples 2 or 3 days in advance by thoroughly rinsing them (in their original cassettes) in phosphate buffered saline (PBS) at 4 °C for 48 h with gentle platform agitation.

(3) Assembling the TMA:

(a) Transfer the recipient block from the −80 °C freezer to the cryostat chamber (−20 °C) and equilibrate for 15–30 min before use.

1. Keep the recipient block in the cryostat chamber until all cored samples have been transferred.

(b) According to the grid map, fill designated blank wells with embedding compound only.

(c) For fresh-frozen samples, just prior to generating the cores, semi-thaw the tissue slices, on a bed of wet ice, one by one. Use aluminum foil to separate tissue from wet ice.

1. Obtain the core (gentle pressure may be required) and immediately transfer it to the recipient block taking care to avoid thawing.
2. Use a cryostat-chilled metal spatula to tap the core flush with the recipient block's surface.

(d) For formalin-fixed samples, blot the tissue dry with lint-free laboratory grade paper wipes.

1. Obtain the core and immediately transfer it to the recipient block.
2. Alternatively, core the tissue and momentarily leave it in place but slightly elevated above the surrounding tissue to apply a unique Microdot (1 μL) orientation pattern to the sample edges using surgical biopsy ink (MarginMarker; Vector Surgical, Waukesha, WI, USA) [33], and then transfer it to the recipient block.
3. Use gentle tapping with a pre-chilled metal spatula to fully insert the core.

(e) To control for reproducibility of the results, include duplicate or triplicate samples spatially dispersed across the TMA.

(f) Carefully fill gaps between tissue cores and well walls with supplemental embedding compound, particularly if the frozen cores fragment.

(g) Apply a smooth flat skim coat of embedding compound across the entire surface of the TMA.

(h) Place the completed TMA in a labeled Tissue-Tek cassette and store in an air-tight container at −80 °C for up to 3 months.

This step-by-step protocol enables anyone with experience handling tissue to generate TMAs suitable for MALDI-IMS. The storage times indicated above and in subsequent sections enable TMAs to be generated in stages and batch-analyzed. With practice and possibly the assistance of a second person, the workflow should be rapid and efficient. For example, undergraduate students and technicians readily mastered the protocol and currently enable its routine use in the laboratory. However, it is imperative to minimize the time allotted for coring and transferring the fresh-frozen cores to recipient blocks since the sample temperatures should not be permitted to rise above −20 °C for prolonged periods due to potential degradation. The coring and transfer process should take 30 s or less per sample. Complete sample thawing should be avoided. Formalin-fixed tissue cores are less problematic as the sample integrity is preserved by fixation.

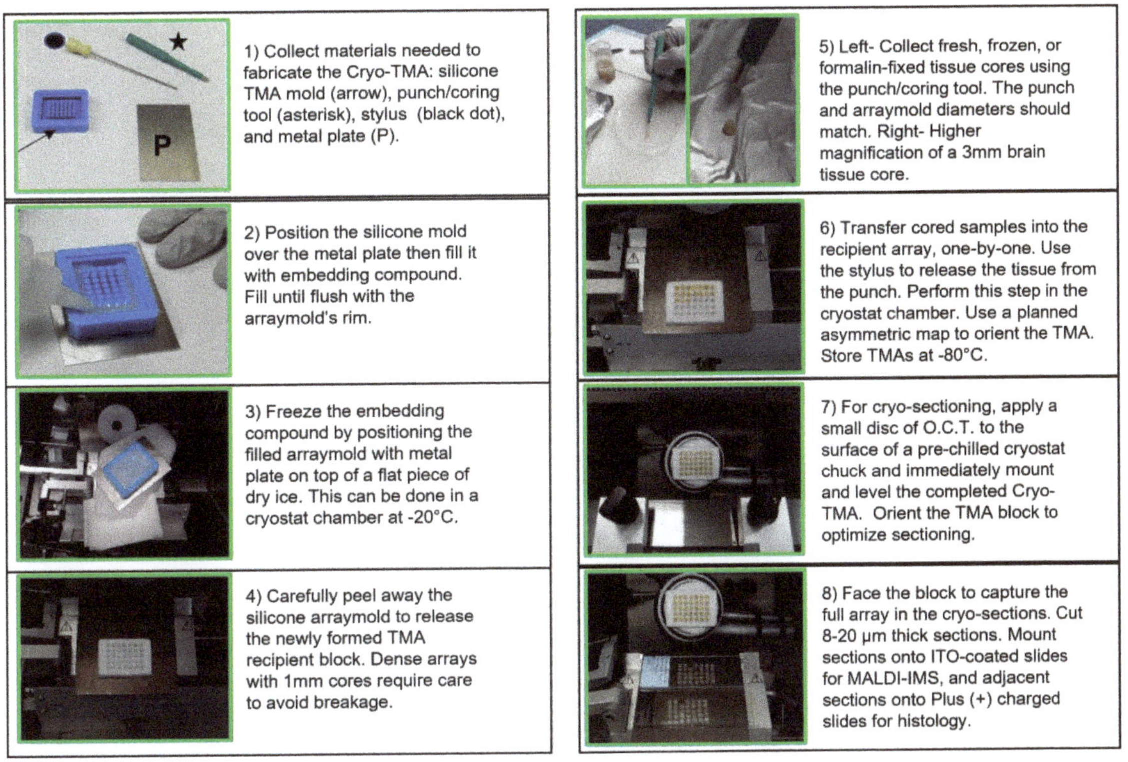

Figure 1. TMA/MALDI-IMS workflow. The major steps used to generate Cryo-TMAs for MALDI-IMS are depicted in captioned Panels 1–8. (1) Tools needed to generate the Cryo-TMAs include a silicone arraymold (arrow), a metal plate (P), a coring/punch tool (asterisk), and a stylus for pushing cored tissue into the arraymolds (black dot). (2) Fill the arraymold with embedding compound. (3) Freeze the embedding compound on a bed of dry ice inside a cryostat chamber (−20 °C or lower). (4) Release the newly formed frozen recipient block from the arraymold and store it in an air-tight container at −80 °C. (5) Core tissue samples with a punch. Equilibrate frozen tissue to −20 °C in a cryostat chamber prior to sampling. (6) Transfer tissue cores to the recipient array block according to a pre-planned asymmetric map to maintain orientation. Apply a skim coat of embedding compound to secure the tissue cores in place. Store completed TMAs at −80 °C. (7) Mount the TMA block onto a chuck and equilibrate it to the ambient cryostat chamber's temperature (−20 °C) prior to sectioning. (8) Trim to fully face the block (TMA) and section (8–20 μm thickness). Mount sections onto ITO-coated slides for MALDI-IMS and at least one additional section onto a Plus (+) charged slide for histology. Store TMA slides at −80 °C until ready for MALDI-IMS.

2.4. TMA Sectioning and Slide Preparation for MALDI-IMS

1. Equilibrate the frozen TMA (stored at −80 °C) to −18 °C in a cryostat microtome chamber for 20–30 min prior to sectioning.
2. Mount the block onto a cryostat chuck to optimize sectioning of the full TMA, including all tissue cores.
3. Section the TMA at a thickness between 8 μm and 20 μm using a clean, fresh disposable blade for each TMA.
    a. Generate two to four sets of 4 TMA sections.
    b. Use the first 3 adjacent sections/set for MALDI-IMS and the fourth for histologic staining.

   c. Label and number each slide in the order of sectioning. Label with a pencil or permanent marking pen.
4. For MALDI-IMS, thaw-mount the cryosections onto indium tin oxide (ITO)-coated slides (Delta Technologies, Loveland, CO, USA).
5. Desiccate the slides designated for MALDI-IMS at room temperature in a sealed chamber, and then, either store them in an air-tight container at −80 °C for up to 3 months or immediately proceed with sample sublimation.
6. For histology co-registration with images acquired through MALDI-IMS, thaw-mount cryosections onto Plus-charged glass slides (Thermo Fisher Scientific, Plainville, MA, USA) and air-dry. Either store in an air-tight container (up to 3 months) or immediately proceed with staining protocol below.
   a. Fix tissue sections in 10% neutral buffered formalin.
   b. Rinse several times (10–15 dips) in distilled water.
   c. Stain with Gil's Hematoxylin (Thermo Fisher Scientific, Plainville, MA, USA) according to the manufacturer's instructions.
   d. Dehydrate tissue sections in graded ethanol solutions (50%, 70%, 95%, 95%, 100%, 100%) for 30 s each, clear in two changes of xylenes (30 s each), and then, preserve under coverglass with Per Mount mounting medium (MilliporeSigma, Burlington, MA, USA) or a comparable product. Store the stained slides at room temperature in a dust-free slide box/holder.
   e. Scan the slides to generate 3600 DPI resolution images (Epson's Perfection V850 Scanner, Los Alamitos CA, USA) just prior to MALDI-IMS data acquisition.

2.5. Matrix Application

The matrix application methods for MALDI-IMS lipidomics with cryosections have been described in detail elsewhere [21,38,39]. Since the goal of this manuscript is to provide a method for generating TMAs for MALDI-IMS lipidomics, the further steps are summary-listed rather than detailed.

1. Equilibrate stored frozen TMA slides to room temperature.
2. Rinse in aqueous buffer, such as 50 mM ammonium formate (pH 6.4), to increase lipid ion signal intensities [49].
3. Vacuum-dry slides for 30 min to promote tissue adhesion to slides.
4. Sublime the slides with a suitable matrix using a commercial apparatus, such as that from Chemglass Life Sciences (Vineland, NJ, USA).
5. Choose a matrix appropriate for negative (NIM) and/or positive ion mode (PIM) imaging.
   a. NIM imaging is optimum for detecting most phospholipids and sulfatides.
   b. PIM imaging is most suited for detecting ceramides, sphingomyelin, phosphatidylcholine, and cholesterol.
   c. Use 2,5-dihydroxybenzoic acid (DHB; Sigma-Aldrich Co, St. Louis, MO, USA), 208 $\mu g/cm^2$, as a matrix for NIM or PIM [49–52].
      i. Alternative matrices for NIM or PIM imaging should be considered to optimize the results. See references [53–56].
6. After sublimation, add external mass-calibration standards (Peptide Calibration Standard II, Bruker Daltonics, Bremen, Germany) by depositing 1 µL of a standard peptide mixture with 15 mg/mL α-cyano-4-hydroxycinnamic acid (HCCA) as a matrix, as recommended by the manufacturer.
   a. The mass range is from 377 Da to 2463 Da.
   b. Enables mass accuracy determinations for phospholipids and sphingolipids.

2.6. MALDI-IMS

The methods for performing MALDI-IMS lipidomics have been described and are widely available [21,38,39]. The major steps used to analyze data for this article are summarized below.

1. TMA slides sublimed with DHB as the matrix were imaged in the negative and positive ion modes using a reflectron geometry MALDI-time-of-flight (TOF)/TOF mass spectrometer (Ultraflextreme, Bruker Daltonics, Bremen, Germany).
2. Data acquisition (restricted to a mass range of 600–1200 Da) was performed by focusing a Smartbeam II Nd:YAG laser with a spatial resolution of ~100 μm^2 [50,51,57].
3. Regions of interest were selected based on co-registration with adjacent hematoxylin-stained slides.
4. Data sequence preparation, normalization to total ion counts, and visualization were carried out using FlexImaging software (v 4.0, Bruker Daltonics, Bremen, Germany).
5. Data processing, which included normalization, baseline correction, peak defining, and recalibration, was performed with ClinProTools v3.0 (Bruker Daltonics, Bremen, Germany).
6. Signals corresponding to specific m/z values were visualized using pseudo-colored intensities.
7. Statistical analysis, including Principal Component Analysis (PCA), was carried out using ClinProTools v3.0 (Bruker Daltonics, Bremen, Germany).
8. NIM lipid identification was accomplished through a comparison of precursor and product ion m/z values with corresponding data in the LIPID MAPS prediction tools database (https://www.lipidmaps.org/tools/structuredrawing/GP_p_form.php, accessed on 8 December 2022).
    a. Using tandem mass spectrometry (MS/MS), analytes were fragmented, and their product ions were collected in the MS/MS spectra.
    b. The parent ion and all fragments were used to search the LIPID MAPS database and assign structure/identity.
    c. However, for low-intensity lipid ions, the MS/MS spectra were not informative, necessitating tentative assignments made by matching our m/z values with published data [54,55,58].
9. PIM lipid identification was more challenging due to the presence of multiple adducts (H^+, Na^+, K^+, etc.), resulting in complex spectra. Structural identification was often difficult with TOF since the same ion could appear as multiple adducts. For these studies, we did not perform MS/MS to definitively identify ambiguous lipids detected in the PIM. Instead, tentative assignments were made using the literature where the same m/z was detected under similar conditions [52–55,59–63].
10. The Rapid Peak Alignment Method (RPAM) [64] was used to simultaneously process the MALDI data across the TMA.
    a. In brief, RPAM replaces manual peak alignments and reduces the data processing time for 24 samples from 10 or more hours to approximately 90 min [64].
    b. RPAM data can be transferred to statistical packages for analysis.
    c. The RPAM algorithm greatly facilitates intra- and inter-group comparisons of lipid ion expression and abundance [64].

2.7. Statistical Analyses and Graphics

1. Data were exported to Excel for re-organization.
2. Excel was used to generate data bar plots for illustrating inter-group percentage differences in lipid expression.
3. Graphpad Prism 9 (San Diego, CA, USA) and Number Cruncher Statistical Systems (NCSS) (Kaysville, UT, USA) software were used to generate graphs and analyze data with Student T and Wilcoxon Signed Rank tests. Results were corrected for a 5% false discovery rate (FDR).

3. Results

Experimental data were used to illustrate the utility of TMAs for MALDI-IMS lipidomics. Critical comments include the effects of different embedding compounds, fresh-frozen versus formalin-fixed tissue, and feasibility for the analysis of disease effects.

3.1. Embedding Compound Qualitative Differences

1. mOCT produced very low background signals and exhibited ample firmness for reproducible cryostat sectioning.
2. The commercial OCT embedding compound was unsuitable for MALDI-IMS lipidomics due to high background signals. Although this limitation was known [46], commercial OCT was tested to provide evidence of its unsuitability for lipidomics and definitively discourage its use despite ready availability.
3. Two percent CMC produced very low background noise but proved too soft to generate consistent replicate and flat cryo-sections of brain TMAs.
    a. We also tested 4% CMC, which produced very low background signals but was unsuitable for TMAs due to extreme brittleness, particularly for brain tissue sectioning.

Conclusions

Select an embedding compound that optimizes the signal-to-noise ratio for specific tissues.

Select embedding compound preparations that reproducibly generate high-quality TMA cryosections.

3.2. Effects of Embedding Compound on Disease Characterization with Fresh Frozen Tissue TMAs (Figure 2)

1. Paired control and ethanol rat frontal lobe WM cores (3 mm) were embedded in mOCT and 2% CMC (hybrid recipient block—see Figure 2 legend).
2. Data analysis focused on sulfatide ST(42:2), m/z 888.772, which was identified as previously described [21,38,39,50]. A mixture of calibration standards with m/z values spanning the range of analytes of interest and applied to the MALDI target was used to visualize the tissue distribution and relative intensity of each ion at every pixel using a pseudo-color scale. In the NIM, pseudo-colored images demonstrated the following:
    a. Brighter (more intense) MALDI-IMS signals for cores of frontal lobe WM from the same animals embedded in 2% CMC compared to those with mOCT.
    b. The ethanol exposure-associated higher signal intensities were more conspicuous for samples embedded in 2% CMC compared to those with mOCT.
    c. However, the 2% CMC embedding compound posed challenges for generating multiple replicate TMA sections.

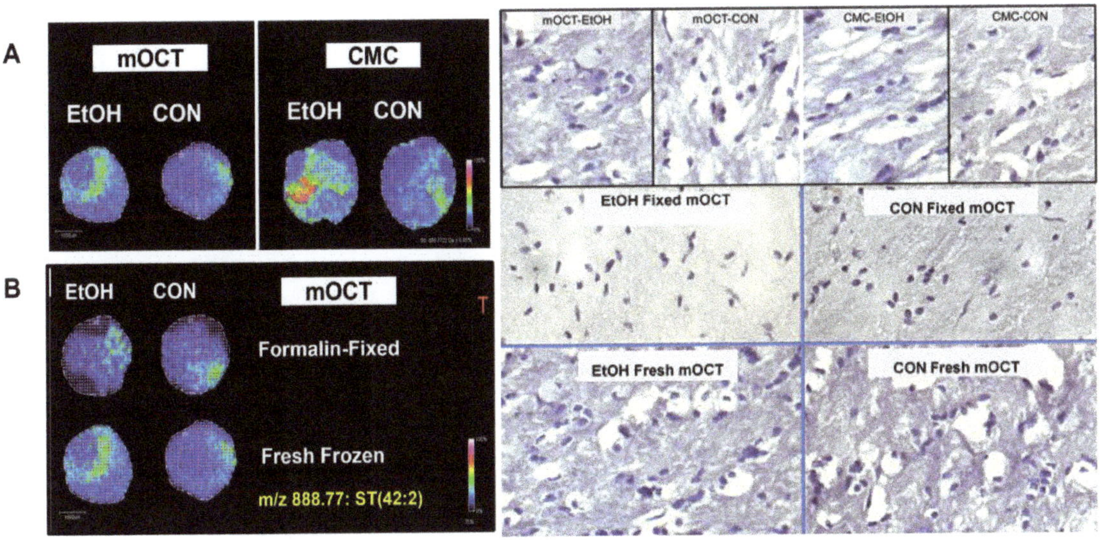

Figure 2. Example images illustrating the impact of the embedding compound and tissue preparation on the characterization of altered ST(42:2) expression following chronic ethanol exposure. ST(42:2) is a sulfatide with an m/z of 888.8. Frontal lobe WM cores (3 mm diameter) from adult Long Evans rats maintained for 6 weeks on isocaloric liquid diets that contained 36% (EtOH) or 0% (CON) caloric ethanol were used to generate TMAs in hybrid recipient blocks constructed with mOCT in one half and 2% CMC in the other half. To achieve this, two glass slides were stood together in the center of the arraymold, aligned with its shorter axis as the embedding compounds solidified on either side of the dam. Cryosections (8 µm-thick) of the TMAs were sublimated with DHB and imaged via MALDI-IMS in the negative ion mode (NIM). Signal intensities were pseudo-colored based on an internal standard reference scale. (**A**) Comparisons between mOCT and 2% CMC in fresh-frozen tissue. (**B**) Comparisons between formalin-fixed and fresh-frozen EtOH and CON samples embedded in mOCT. The panel to the right shows hematoxylin-stained histological sections of white matter in the cores imaged on the left. The small round-to-oval blue dot-like structures are nuclei of glial cells. The large white spaces associated with fresh tissue sections correspond to freeze artifacts common to white matter, regardless of the embedding compound. Note the absence of similar artifacts in fixed tissue (original magnification, ×400 for all histology images).

Conclusions

Interpret results obtained with samples embedded in the same compound and analyzed simultaneously under identical conditions.

Use an embedding compound that enables the optimum detection of lipids with a broad range of m/z values (corresponding to the calibration standards) in both control and experimental/diseased samples.

Consider technical feasibility challenges in planning experiments.

3.3. Fresh Frozen vs. Formalin-Fixed Tissue TMAs

1. NIM MALDI-IMS was performed on paired fresh-frozen and formalin-fixed frontal lobe WM cores embedded in an mOCT arraymold recipient block.
    a. By slicing the brains in the coronal plane at the temporal tips, 3 mm cores of central white matter were obtained. The small amounts of peripheral contamination from the adjacent cortex were easy to exclude from the analysis by marking the co-registered images.

2. Signal intensities were similar for the same samples and tended to be higher (brighter) for fresh-frozen compared than for formalin-fixed tissue cores embedded in mOCT (Figure 2B).
3. Example graphed results from individual paired samples analyzed via NIM MALDI-IMS depict peak profiles corresponding to the 21 lipids detected in the fresh-frozen and fixed samples (Figure 3).

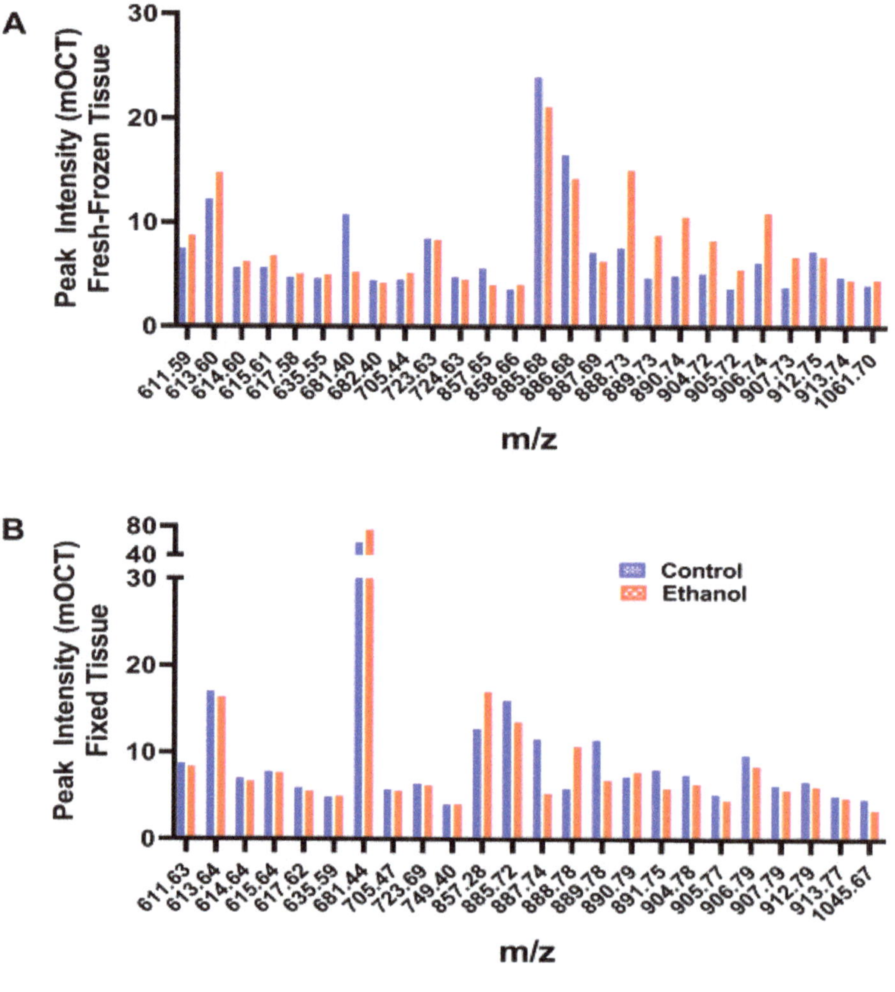

Figure 3. Peak intensity spectra corresponding to lipids detected in mOCT-embedded fresh-frozen and formalin-fixed frontal WM from control or ethanol-exposed rats. The TMAs were generated with paired tissue cores (3 mm diameter) from the same brains (n = 4 rats/group) (see Methods). Data were acquired via NIM MALDI-IMS. Unique peaks were excluded from these comparisons. Note the extensive overlap of signal intensities within a group, i.e., (**A**) frozen or (**B**) fixed, and conspicuous peak profile distinctions based on tissue preservation (Panel (**A**) vs. Panel (**B**)).

4. Results were nearly identical for control and ethanol brains that were either fresh-frozen (Figure 3A) or formalin-fixed (Figure 3B).
5. However, the signal intensities (peak heights) corresponding to individual lipids differed for formalin fixed versus fresh-frozen tissue samples (Figure 3A,B).

Conclusions

TMAs generated with formalin-fixed tissue are suitable for MALDI-IMS lipidomics.

The finding that formalin-fixed brain tissue could be used for MALDI-IMS is consistent with earlier reports [33,34,65].

Due to non-identical peak profiles with the same embedding compound, as previously reported [65], formalin-fixed and fresh-frozen sample results should not be compared.

The suitability of formalin-fixed tissue for TMA MALDI-IMS lipidomics expands opportunities for large-scale retrospective studies, including the analysis of human diseases.

Although formalin fixation with paraffin embedding is suitable for metabolomics [66], it markedly distorts lipidomics [60]. However, formalin fixation alone with proper and simultaneously analyzed controls enables reliable inter-group comparisons. Correspondingly, we detected nearly identical lipid expression profiles but differences in peak intensities (lipid abundances) in ethanol-exposed versus control samples (see Figure 3). We strongly discourage the use of paraffin-embedded samples due to the unpredictable and often substantial removal of lipids in the dehydration and clearing steps, rendering the samples unsuitable for MALDI-IMS lipidomics [60].

Formalin fixation generally does not interfere with lipidomic profiling through MALDI [67]. However, prolonged formalin fixation can adversely affect the structural integrity of lipids that contain primary amines via cross-linking [60]. Therefore, formalin fixation should be optimized by the following: (1) keeping the specimen thickness to 3 mm or less; (2) using neutral-buffered formalin; (3) limiting fixation times to 48 or 72 h; (4) storing postfixed tissue in PBS containing 0.01% sodium azide at 4 °C up to a maximum of 3 weeks.

DHB is generally used for positive- and negative-mode imaging [54,55], but some accounts have reported lower lipid detection sensitivity with DHB compared to that with other matrices for negative-mode imaging [68]. However, after experimenting with different matrices, we found that DHB provides broad lipid coverage with many sphingolipids and phospholipids detected in positive and negative ion modes within the mass range of interest, i.e., m/z = 600–1200 [21,33,50].

3.4. *Embedding Compound Effects on Lipid Peak Profiles (Figure 4)*

1. NIM MALDI-IMS detected 140 shared lipid peaks (m/z 600–1200) in TMAs generated with control and ethanol-exposed fresh-frozen or formalin-fixed rat frontal WM cores embedded in mOCT or 2% CMC (n = 4 rats/group).
2. Principal Component Analysis (PCA) plots generated with ClinProTools, v3 demonstrated clear separation of control and ethanol fresh frozen versus formalin-fixed samples embedded in mOCT (Figure 4A) or CMC (Figure 4B).
3. PCA plots demonstrated the effects of ethanol, which were better distinguished with the 2% CMC embedding compound than with mOCT.
4. These studies demonstrate the effects of the treatment/disease model but the more prominent effects of tissue processing and the embedding compound.
5. Data generated with cores sampled from the same brains but spatially positioned in different regions of the TMA had a less than 5% mean coefficient of variation.

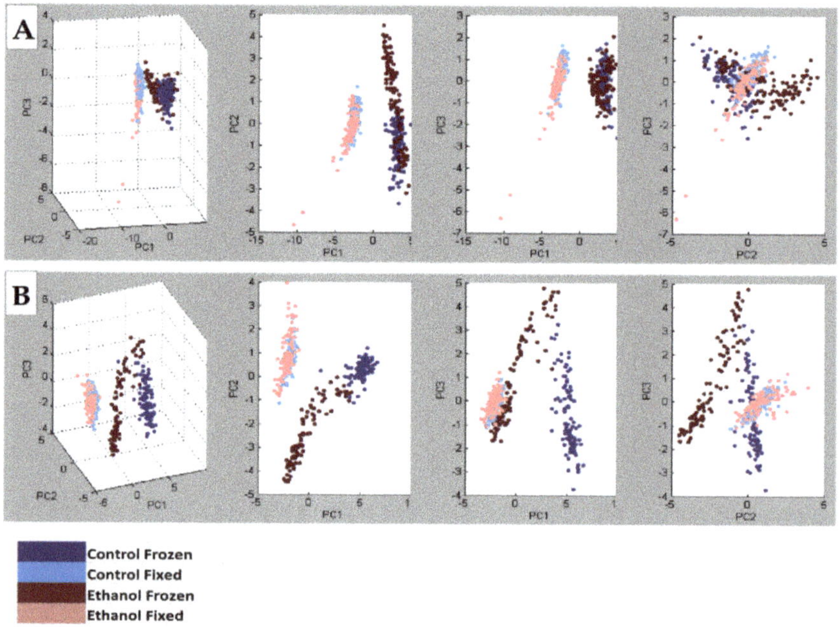

Figure 4. Principal Component Analysis (PCA) plots corresponding to the full spectra of lipids detected in fresh-frozen or formalin-fixed frontal lobe samples from control and ethanol-exposed rats (See Supplementary Figure S1). TMAs generated with (**A**) mOCT or (**B**) 2% CMC as the embedding medium were analyzed via MALDI-IMS with NIM signal acquisitions. PCA plots were generated with ClinProTools, v3. Note the clear separation of control and ethanol fresh-frozen versus formalin-fixed samples embedded in mOCT or CMC. However, with data combined, the overlap of signals was more tightly related to tissue processing (freezing of fixation) than experimental treatment (ethanol or control diet) with either mOCT or CMC as the embedding compound.

Conclusions

TMAs are suitable for characterizing normal and disease-related lipid profiles in tissue.

Since formalin-fixed tissue is suitable for detailed MALDI-IMS lipidomics, the approach could be used to analyze pathological shifts in lipid expression and responses to treatment in large-scale experimental and human disease states.

Tissue processing and selection of embedding compounds must be uniform for reliable data analysis via TMA-MALDI-IMS.

3.5. Illustrated Application of MALDI-IMS TMA Methodology

1. Chronic + Binge Ethanol Exposure model [69]: Adult Long Evans rats were maintained for 8 weeks on isocaloric Lieber-DeCarli liquid diets containing 24% or 0% caloric ethanol liquid diets (BioServ, Frenchtown, NJ, USA), and during the last 3 weeks, they were gavage-binged with 2 g/kg ethanol or saline in the liquid diet (2.5 mL total volume) on Tuesdays, Thursdays, and Saturdays.
2. Fresh-frozen WM cores (2 mm) from six rats per group were used to generate mOCT-embedded TMAs.
3. The NIM and PIM frontal white matter MALDI-IMS lipidomics spectra were distinct (Figure 5). The control and ethanol samples mainly differed with respect to selected peak intensities, rather than specific lipids detected within the NIM or PIM.

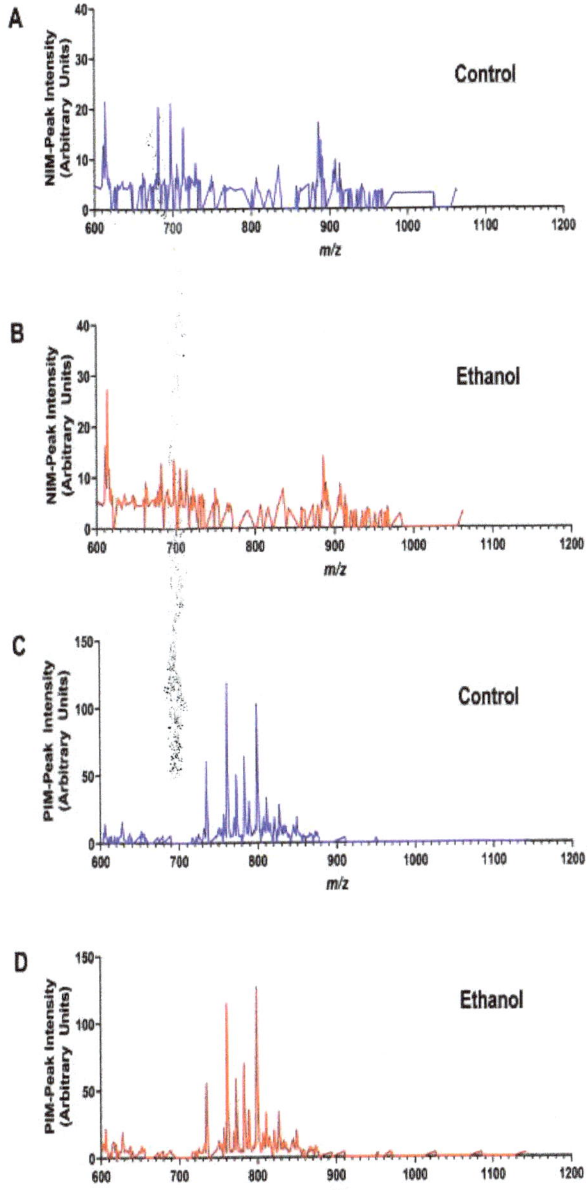

Figure 5. MALDI-IMS NIM and PIM spectra from a TMA constructed with fresh-frozen, 2 mm cores of frontal lobe white matter embedded in mOCT. Spectra correspond to relative intensities (abundances) of lipid ions between m/z 600 and 1200 detected by MALDI (**A**,**B**) NIM and (**C**,**D**) PIM. Data were acquired by rasterizing across the TMA WM cores of (**A**,**C**) control and (**B**,**D**) chronic ethanol-fed rats. The peak intensities (arbitrary units) reflect averaged results from four rats per group.

4. Further MALDI-IMS analysis was focused on phosphatidylinositol (PI) data (Figure 6A) acquired in NIM and phosphatidylcholine (PC) data (Figure 6B) acquired in the PIM

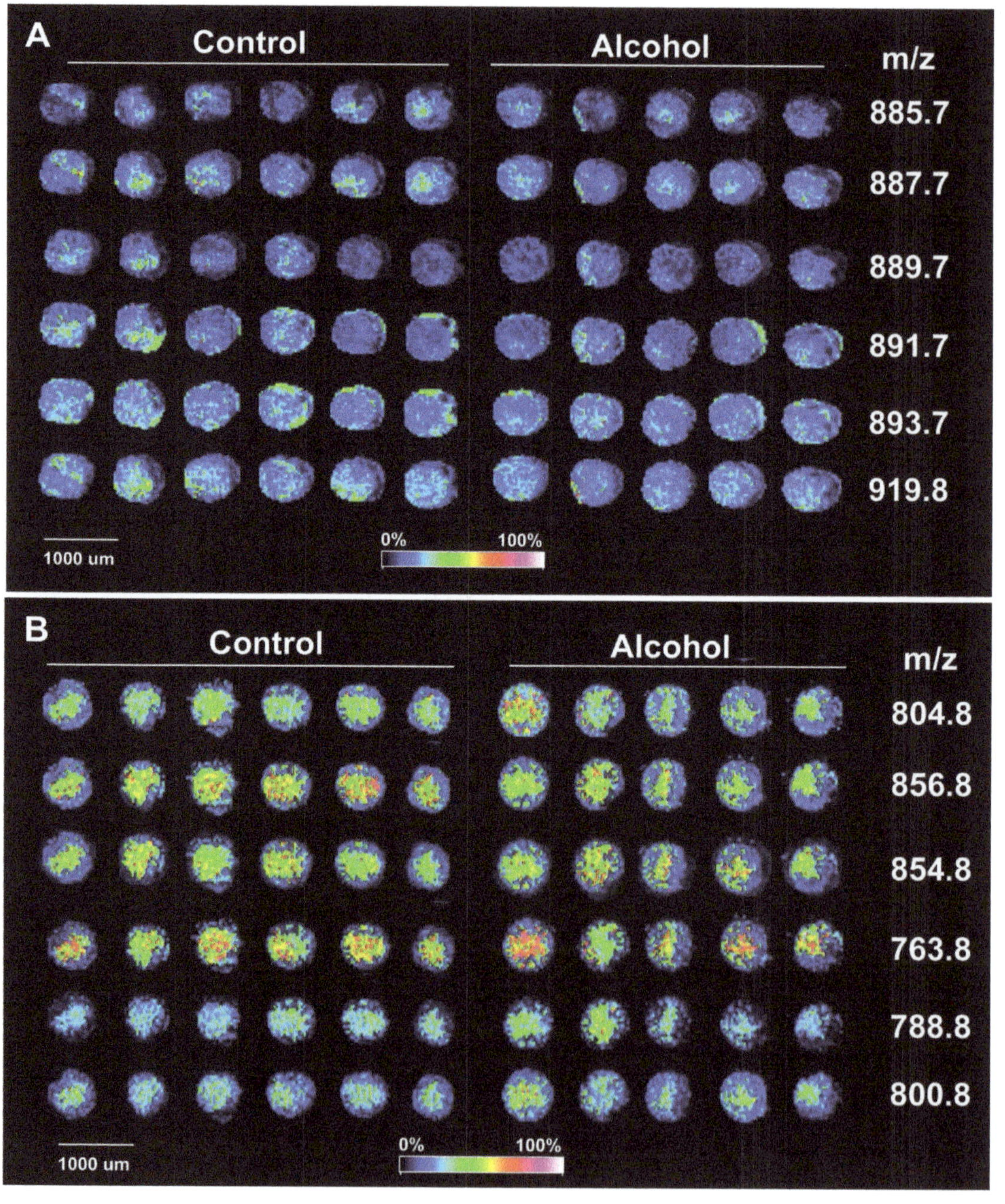

Figure 6. MALDI-IMS TMAs of frontal lobe white matter. MALDI-IMS TMAs generated with fresh frozen frontal lobe white matter cores from control and ethanol-fed rats were embedded in mOCT recipient blocks and imaged in the (**A**) negative or (**B**) positive ion mode. Lipid ion signals were pseudo-colored to reflect signal intensities. (**A**) Example NIM results correspond to phosphatidylinositols, and the (**B**) PIM results correspond to phosphatidylcholines. See Supplementary Figure S2 for representative histologic images.

1. Table 1 lists the 10 PIs and 11 PCs identified via tandem mass spectrometry (MS/MS) coupled with LIPID MAPS database searches and published reports [39].

Table 1. Phosphatidylinositols and phosphatidylcholines detected in frontal lobe white matter via MALDI-IMS.

Code	Phosphatidylinositols-NIM	Ionization Form	*m/z*
a	PI(34:1)	$[M^-H]^-$	**835.7**
b	PI(36:4)	$[M^-H]^-$	**857.6**
c	PI(O-38:3)/PI(P-38:2)	$[M^-H]^-$	**873.2**
d	PI(38:5)	$[M^-H]^-$	**883.7**
e	PI(38:4)	$[M^-H]^-$	**885.7**
f	PI(38:3)	$[M^-H]^-$	**887.7**
g	PI(38:2)	$[M^-H]^-$	**889.7**
h	PI(38:1)	$[M^-H]^-$	**891.7**
i	PI(38:0)	$[M^-H]^-$	**893.7**
j	PI(40:1)/PI(P-41:0)/LPIM2(18:2)	$[M^-H]^-$	**919.8**
	Phosphatidylcholines-PIM	**Ionization Form**	***m/z***
a	(1) PC(36:1); (2) pPC(36:4); (3) PC(34:6)	(1) $[M^+H]^+$; (2) $[M^+Na]^+$; (3) $[M^+K]^+$	**788.8**
b	(1) PC(36:7); (2) PC(O-36:0); (3) PC(34:1)	(1) $[M^+Na]^+$; (2) $[M^+Na]^+$; (3) $[M^+K]^+$	**798.8**
c	(1) PC(38:7); (2) PC(36:4)	(1) $[M+H]^+$; (2) $[M+Na]^+$	**804.8**
d	(1) PC(40:9); (2) PC(O-40:2); (3) PC(38:6); 4) PC(36:0)	(1) $[M^+H]^+$; (2) $[M^+H]^+$; (3) $[M^+Na]^+$; (4) $[M^+K]^+$	**828.8**
e	(1) PC(42:10); (2) PC(40:7); (3) PC(38:1)	(1) $[M^+H]^+$; (2) $[M^+Na]^+$; (3) $[M^+K]^+$	**854.8**
f	(1) PC(42:9); (2) PC(40:6)	(1) $[M^+H]^+$; (2) $[M^+Na]^+$	**856.8**
g	(1) PC(O-38:2); (2) PC(36:6); (3) PC(34:0)	(1) $[M^+H]^+$; (2) $[M^+Na]^+$; (3) $[M+K]^+$	**800.8**
h	PC(32:0)	$[M^+H]^+$	**734.8**
i	PC(34:0)	$[M^+H]^+$	**762.8**
j	PC(34:1)	$[M^+H]^+$	**760.8**
k	PC(36:3)	$[M^+K^-N(CH_3)_3]^+$	**763.8**

Phosphatidylinositols and phosphatidylcholines detected in frontal lobe white matter via MALDI-IMS. Cores (2 mm) of fresh-frozen frontal white matter from control and ethanol-fed Long Evans rats were embedded in mOCT ($n = 6$ samples/group). The TMAs, sublimated with DHB, were imaged in the negative ion mode (NIM) and positive ion mode (PIM). The 10 PI and 11 PC lipid ions detected in all samples are listed with their ionization forms and m/z values. Codes (first column) correspond to graphed results displayed in Figure 7.

2. Inter-group differences in the mean peak intensity for each lipid were assessed statistically with Student t-tests. The calculated percentage differences in the ethanol relative to control samples are depicted in databar plots alongside the p-values for the individual t-tests (Figure 7A,C).
3. However, the overall effects of ethanol on PI (Figure 7B) and PC (Figure 7D) expression were further assessed using the Wilcoxon Signed Rank Test to determine if the median percentage differences differed significantly from 0.00. The test demonstrated overall significantly reduced PI ($p = 0.002$) and increased PC ($p = 0.003$) expression in ethanol-exposed frontal WM.

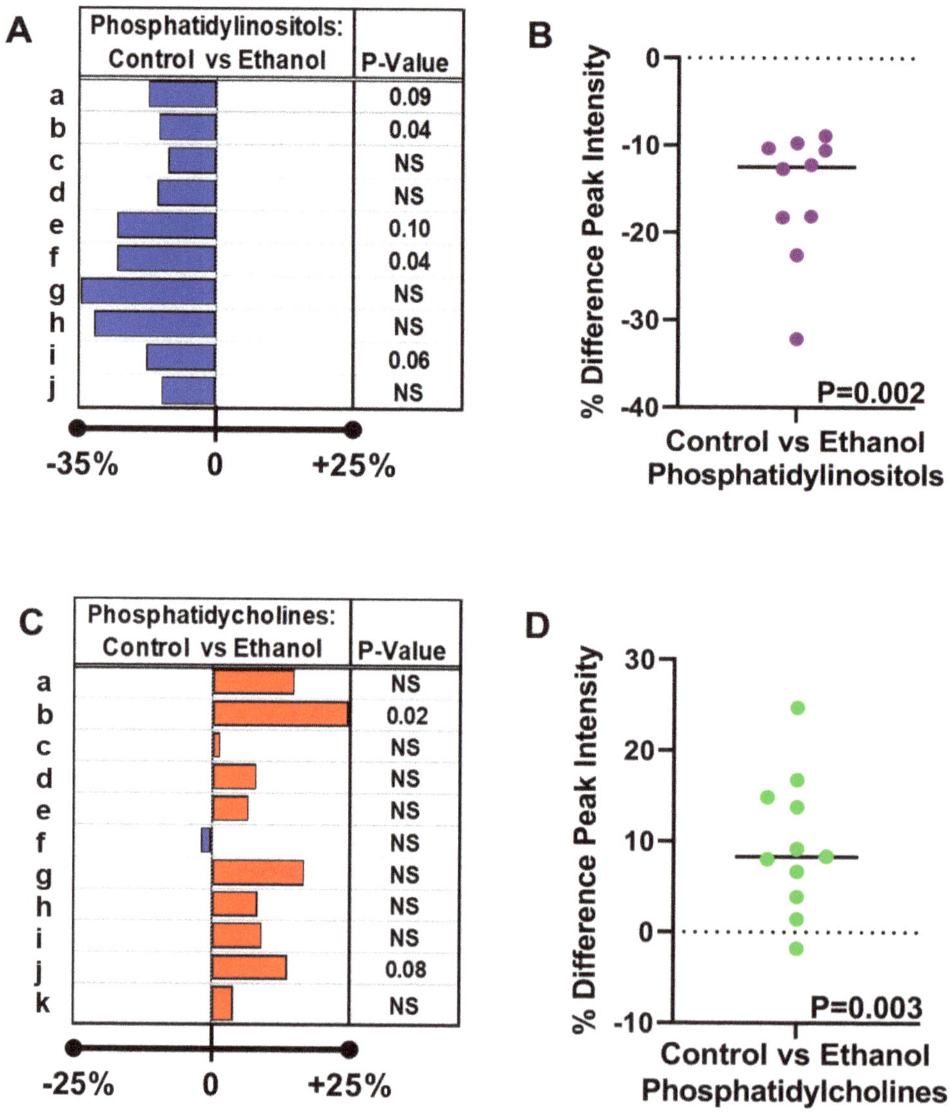

Figure 7. Databar plots were generated to reflect percentage differences in frontal lobe white matter levels of (**A**) phosphatidylinositol (PI) and (**C**) phosphatidylcholine (PC) expression resulting from chronic dietary ethanol exposure. This study utilized a chronic+binge ethanol feeding model with six rats per group. A TMA generated with fresh-frozen tissue embedded in mOCT was imaged. Ten PIs were detected in the NIM and 11 PCs in the PIM (see Table 1). The calculated paired percentage differences in lipid expression were used for the Databar plots, such that ethanol-associated reductions are represented by blue bars to the left and increases by red bars to the right. Inter-group differences were assessed by performing Student *t*-tests with a 5% false discovery rate. Significant differences ($p < 0.05$) and trend effects ($<0.05 < p < 0.10$) were determined using Prism Graphpad v9.4. NS = not statistically significant. (**B**,**D**) Scatter plots displaying the distribution of differences in (**B**) PI and (**D**) PC expression between ethanol and control frontal lobe white matter samples. The horizontal bar corresponds to the median. The hypothesis tested was that the inter-group differences equaled 0. The Wilcoxon Signed Rank test was statistically significant for the PI and PC grouped results.

Conclusions

TMAs are suitable for characterizing disease-related alterations in brain WM lipid expression.

Additional studies showed that data generated with cores sampled from the same brains but spatially positioned in different regions of the TMA had a less than 5% mean coefficient of variation.

4. Discussion

- This work provides a practical, high throughput approach for generating TMAs that are suitable for MALDI-IMS lipidomic studies.
- Although the efforts were focused on brain white matter biochemical histology, the methods can be successfully applied to a broad range of tissues.
- The demonstrated feasibility of using formalin-fixed tissue expands opportunities to conduct large-scale MALDI-IMS lipidomic studies of archival specimens to study the effects of disease in humans and experimental models.
- Although the TMA approach necessitates the inclusion of smaller samples for analysis compared to that ordinarily used for MALDI-IMS and therefore bears the risk of missing data, the streamlined higher throughput strategy enabling uniform sample handling and processing and simultaneous imaging of replicate, positive, and negative control samples under the same conditions outweigh potential limitations related to smaller sample sizes. Moreover, the TMA approach bolsters overall scientific rigor.
- Potential limitations posed by the need to analyze multi-sample lipid profile TMA datasets can be resolved using the open-access rapid peak alignment method (RPAM) [64].
- The use of TMAs for MALDI-IMS could potentially facilitate the better characterization of white matter myelin-associated pathologies that correlate with disease progression or responses to treatment, particularly with respect to neurodegeneration.

Supplementary Materials: The following supporting information can be downloaded at: https://www.mdpi.com/article/10.3390/applbiosci2020013/s1, Figure S1: MALDI-IMS spectra corresponding to (A,B,E,F) fresh frozen or (C,D,G,H) formalin-fixed frontal lobe white matter samples from (A,C,E,G) control and (B,D,F,H) ethanol-fed rats. 3-mm tissue cores were embedded in a hybrid (A–D) mOCT/(E–H) 2% CMC TMA and following DHB sublimation coating, the samples were imaged in the negative ion mode (NIM) along with calibration standards used to assess the relative abundance (peak intensity-arbitrary units) of each lipid ion. Each spectrum represents averaged results from 2 rats per group. See Figure S2; Figure S2: Example optical images of Hematoxylin-stained TMA frontal white matter cores from control and chronic ethanol-fed rats corresponding to the 4th core from the left in each group shown in Figure 6. Hematoxylin labels nuclei. The cores were photographed at different magnifications to reveal the glial-predominant and microvascular composition of white matter. Example oligodendrocytes with dot-like nuclei and astrocytes with oval nuclei are depicted to the left of the O's or A's. Microvessels with narrow lumens are marked with arrows. The red squares show the regions of higher magnification depicted in the immediately below panels.

Author Contributions: All authors contributed to the research efforts required to generate this manuscript and reviewed this manuscript. I.G.-R., L.N. and M.T. contributed equally to method development. E.B.Y. supervised the MALDI and contributed primary data. S.M.d.l.M. conceived the idea, contributed to the research design, performed data analysis, and wrote and revised the manuscript. All authors have read and agreed to the published version of the manuscript.

Funding: This research was funded by grants from the National Institutes of Health-National Institute of Alcohol Abuse and Alcoholism: AA-011431 and AA-024018.

Institutional Review Board Statement: The animal study protocol was approved by the Lifespan Institutional Animal Care and Use Committee (IACUC) of Rhode Island Hospital (CMTT000615, initially approved 11 February 2015) for studies involving animals.

Informed Consent Statement: Not applicable.

Data Availability Statement: Not applicable.

Acknowledgments: The authors acknowledge William Pelit, Department of Chemistry, Brown University, for assisting with the retrieval of MALDI spectra from data files.

Conflicts of Interest: The authors declare no conflict of interest.

References

1. Sutherland, G.T.; Sheedy, D.; Kril, J.J. Neuropathology of alcoholism. *Handb. Clin. Neurol.* **2014**, *125*, 603–615. [CrossRef]
2. de la Monte, S.M. Disproportionate atrophy of cerebral white matter in chronic alcoholics. *Arch. Neurol.* **1988**, *45*, 990–992. [CrossRef] [PubMed]
3. Harper, C.G.; Smith, N.A.; Kril, J.J. The effects of alcohol on the female brain: A neuropathological study. *Alcohol. Alcohol.* **1990**, *25*, 445–448. [PubMed]
4. de la Monte, S.M. Quantitation of cerebral atrophy in preclinical and end-stage Alzheimer's disease. *Ann. Neurol.* **1989**, *25*, 450–459. [CrossRef]
5. Englund, E.; Brun, A. White matter changes in dementia of Alzheimer's type: The difference in vulnerability between cell compartments. *Histopathology* **1990**, *16*, 433–439. [CrossRef]
6. Englund, E.; Brun, A.; Alling, C. White matter changes in dementia of Alzheimer's type. Biochemical and neuropathological correlates. *Brain J. Neurol.* **1988**, *111*, 1425–1439. [CrossRef] [PubMed]
7. Ruchoux, M.M.; Maurage, C.A. CADASIL: Cerebral autosomal dominant arteriopathy with subcortical infarcts and leukoencephalopathy. *J. Neuropathol. Exp. Neurol.* **1997**, *56*, 947–964. [CrossRef]
8. Wallin, A.; Blennow, K. Pathogenetic basis of vascular dementia. *Alzheimer Dis. Assoc. Disord.* **1991**, *5*, 91–102. [CrossRef] [PubMed]
9. Larsson, E.; Passant, U.; Sundgren, P.C.; Englund, E.; Brun, A.; Lindgren, A.; Gustafson, L. Magnetic resonance imaging and histopathology in dementia, clinically of frontotemporal type. *Dement. Geriatr. Cogn. Disord.* **2000**, *11*, 123–134. [CrossRef]
10. de la Monte, S.M.; Kril, J.J. Human alcohol-related neuropathology. *Acta Neuropathol.* **2014**, *127*, 71–90. [CrossRef]
11. Schmidt, K.S.; Gallo, J.L.; Ferri, C.; Giovannetti, T.; Sestito, N.; Libon, D.J.; Schmidt, P.S. The neuropsychological profile of alcohol-related dementia suggests cortical and subcortical pathology. *Dement. Geriatr. Cogn. Disord.* **2005**, *20*, 286–291. [CrossRef]
12. Elofson, J.; Gongvatana, W.; Carey, K.B. Alcohol use and cerebral white matter compromise in adolescence. *Addict. Behav.* **2013**, *38*, 2295–2305. [CrossRef]
13. Jacobus, J.; Squeglia, L.M.; Bava, S.; Tapert, S.F. White matter characterization of adolescent binge drinking with and without co-occurring marijuana use: A 3-year investigation. *Psychiatry Res.* **2013**, *214*, 374–381. [CrossRef]
14. Pfefferbaum, A.; Lim, K.O.; Desmond, J.E.; Sullivan, E.V. Thinning of the corpus callosum in older alcoholic men: A magnetic resonance imaging study. *Alcohol. Clin. Exp. Res.* **1996**, *20*, 752–757. [CrossRef]
15. Chanraud, S.; Martelli, C.; Delain, F.; Kostogianni, N.; Douaud, G.; Aubin, H.J.; Reynaud, M.; Martinot, J.L. Brain morphometry and cognitive performance in detoxified alcohol-dependents with preserved psychosocial functioning. *Neuropsychopharmacol. Off. Publ. Am. Coll. Neuropsychopharmacol.* **2007**, *32*, 429–438. [CrossRef] [PubMed]
16. Pfefferbaum, A.; Rosenbloom, M.J.; Adalsteinsson, E.; Sullivan, E.V. Diffusion tensor imaging with quantitative fibre tracking in HIV infection and alcoholism comorbidity: Synergistic white matter damage. *Brain A J. Neurol.* **2007**, *130*, 48–64. [CrossRef] [PubMed]
17. Estruch, R.; Nicolas, J.M.; Salamero, M.; Aragon, C.; Sacanella, E.; Fernandez-Sola, J.; Urbano-Marquez, A. Atrophy of the corpus callosum in chronic alcoholism. *J. Neurol. Sci.* **1997**, *146*, 145–151. [CrossRef] [PubMed]
18. Papp-Peka, A.; Tong, M.; Kril, J.J.; De La Monte, S.M.; Sutherland, G.T. The Differential Effects of Alcohol and Nicotine-Specific Nitrosamine Ketone on White Matter Ultrastructure. *Alcohol Alcohol.* **2017**, *52*, 165–171. [CrossRef]
19. Quarles, R.H.; Macklin, W.B.; Morell, P. *Myelin Formation, Structure and Biochemistry*, 6th ed.; Siegel, G.J., Ed.; Elsevier: Philadelphia, PA, USA, 2006.
20. Takahashi, T.; Suzuki, T. Role of sulfatide in normal and pathological cells and tissues. *J. Lipid Res.* **2012**, *53*, 1437–1450. [CrossRef]
21. Yalcin, E.B.; Nunez, K.; Tong, M.; de la Monte, S.M. Differential Sphingolipid and Phospholipid Profiles in Alcohol and Nicotine-Derived Nitrosamine Ketone-Associated White Matter Degeneration. *Alcohol. Clin. Exp. Res.* **2015**, *39*, 2324–2333. [CrossRef]
22. Roux, A.; Muller, L.; Jackson, S.N.; Baldwin, K.; Womack, V.; Pagiazitis, J.G.; O'Rourke, J.R.; Thanos, P.K.; Balaban, C.; Schultz, J.A.; et al. Chronic ethanol consumption profoundly alters regional brain ceramide and sphingomyelin content in rodents. *ACS Chem. Neurosci.* **2015**, *6*, 247–259. [CrossRef] [PubMed]
23. Mollinedo, F.; Gajate, C. Lipid rafts as major platforms for signaling regulation in cancer. *Adv. Biol. Regul.* **2015**, *57*, 130–146. [CrossRef] [PubMed]
24. Farooqui, A.A.; Ong, W.Y.; Horrocks, L.A.; Chen, P.; Farooqui, T. Comparison of biochemical effects of statins and fish oil in brain: The battle of the titans. *Brain Res. Rev.* **2007**, *56*, 443–471. [CrossRef]
25. Kolesnick, R.N.; Kronke, M. Regulation of ceramide production and apoptosis. *Annu. Rev. Physiol.* **1998**, *60*, 643–665. [CrossRef]
26. Ohanian, J.; Ohanian, V. Sphingolipids in mammalian cell signalling. *Cell. Mol. Life Sci. CMLS* **2001**, *58*, 2053–2068. [CrossRef]
27. Eckhardt, M. The role and metabolism of sulfatide in the nervous system. *Mol. Neurobiol.* **2008**, *37*, 93–103. [CrossRef]

28. Vos, J.P.; Lopes-Cardozo, M.; Gadella, B.M. Metabolic and functional aspects of sulfogalactolipids. *Biochim. Biophys. Acta* **1994**, *1211*, 125–149. [CrossRef] [PubMed]
29. Sjobeck, M.; Haglund, M.; Persson, A.; Sturesson, K.; Englund, E. Brain tissue microarrays in dementia research: White matter microvascular pathology in Alzheimer's disease. *Neuropathology* **2003**, *23*, 290–295. [CrossRef]
30. Luu, M.; Sabo, E.; de la Monte, S.M.; Greaves, W.; Wang, J.; Tavares, R.; Simao, L.; Wands, J.R.; Resnick, M.B.; Wang, L. Prognostic value of aspartyl (asparaginyl)-beta-hydroxylase/humbug expression in non-small cell lung carcinoma. *Hum. Pathol.* **2009**, *40*, 639–644. [CrossRef] [PubMed]
31. Casadonte, R.; Longuespee, R.; Kriegsmann, J.; Kriegsmann, M. MALDI IMS and Cancer Tissue Microarrays. *Adv. Cancer Res.* **2017**, *134*, 173–200. [CrossRef]
32. Cole, L.M.; Clench, M.R. Mass spectrometry imaging tools in oncology. *Biomark. Med.* **2015**, *9*, 863–868. [CrossRef]
33. de la Monte, S.M.; Kay, J.; Yalcin, E.B.; Kril, J.J.; Sheedy, D.; Sutherland, G.T. Imaging mass spectrometry of frontal white matter lipid changes in human alcoholics. *Alcohol* **2018**, *67*, 51–63. [CrossRef] [PubMed]
34. O'Rourke, M.B.; Smith, C.C.; De La Monte, S.M.; Sutherland, G.T.; Padula, M.P. Higher Mass Accuracy MALDI-TOF/TOF Lipid Imaging of Human Brain Tissue in Alzheimer's Disease. *Curr. Protoc. Mol. Biol.* **2019**, *126*, e86. [CrossRef]
35. Yalcin, E.; Kay, J.; De la Monte, S.M. Mass Spectrometry Imaging of White Matter Lipid Profile Changes with Time Course of Ethanol Exposure and Withdrawal in Adult Long Evans Rats. In *Proceedings of the Csaba Horvath Memorial Symposium, New Haven, CT, USA, 27 October 2016*; Yale University: New Haven, CT, USA, 2016.
36. Gallucci, G.M.; Tong, M.; Chen, X.; Stonestreet, B.S.; Lin, A.; de la Monte, S.M. Rapid Alterations in Cerebral White Matter Lipid Profiles After Ischemic-Reperfusion Brain Injury in Fetal Sheep as Demonstrated by MALDI-Mass Spectrometry. *Pediatr. Dev. Pathol.* **2019**, *22*, 344–355. [CrossRef]
37. Homans, C.; Yalcin, E.B.; Tong, M.; Gallucci, G.; Bautista, D.; Moriel, N.; Monte, S.d.l. Therapeutic Effects of Myriocin in Experimental Alcohol-Related Neurobehavioral Dysfunction and Frontal Lobe White Matter Biochemical Pathology. *J. Behav. Brain Sci.* **2022**, *12*, 23–42. [CrossRef] [PubMed]
38. Yalcin, E.B.; McLean, T.; Tong, M.; de la Monte, S.M. Progressive white matter atrophy with altered lipid profiles is partially reversed by short-term abstinence in an experimental model of alcohol-related neurodegeneration. *Alcohol* **2017**, *65*, 51–62. [CrossRef]
39. Yalcin, E.B.; de la Monte, S.M. Review of matrix-assisted laser desorption ionization-imaging mass spectrometry for lipid biochemical histopathology. *J. Histochem. Cytochem.* **2015**, *63*, 762–771. [CrossRef]
40. Tong, M.; Gonzalez-Navarrete, H.; Kirchberg, T.; Gotama, B.; Yalcin, E.B.; Kay, J.; de la Monte, S.M. Ethanol-Induced White Matter Atrophy Is Associated with Impaired Expression of Aspartyl-Asparaginyl-beta-Hydroxylase (ASPH) and Notch Signaling in an Experimental Rat Model. *J. Drug Alcohol. Res.* **2017**, *6*, 236033. [CrossRef] [PubMed]
41. Tong, M.; Yu, R.; Silbermann, E.; Zabala, V.; Deochand, C.; de la Monte, S.M. Differential Contributions of Alcohol and Nicotine-Derived Nitrosamine Ketone (NNK) to White Matter Pathology in the Adolescent Rat Brain. *Alcohol Alcohol.* **2015**, *50*, 680–689. [CrossRef] [PubMed]
42. Tong, M.; Longato, L.; Ramirez, T.; Zabala, V.; Wands, J.R.; de la Monte, S.M. Therapeutic reversal of chronic alcohol-related steatohepatitis with the ceramide inhibitor myriocin. *Int. J. Exp. Pathol.* **2014**, *95*, 49–63. [CrossRef] [PubMed]
43. Lowe, J.M.; Menendez, D.; Bushel, P.R.; Shatz, M.; Kirk, E.L.; Troester, M.A.; Garantziotis, S.; Fessler, M.B.; Resnick, M.A. p53 and NF-kappaB coregulate proinflammatory gene responses in human macrophages. *Cancer Res.* **2014**, *74*, 2182–2192. [CrossRef]
44. Meitner, P.A.; Resnick, M.B. LCM assisted biomarker discovery from archival neoplastic gastrointestinal tissues. *Methods Mol. Biol.* **2011**, *755*, 165–180. [CrossRef] [PubMed]
45. Berry, K.A.; Li, B.; Reynolds, S.D.; Barkley, R.M.; Gijon, M.A.; Hankin, J.A.; Henson, P.M.; Murphy, R.C. MALDI imaging MS of phospholipids in the mouse lung. *J. Lipid Res.* **2011**, *52*, 1551–1560. [CrossRef]
46. Gill, E.L.; Yost, R.A.; Vedam-Mai, V.; Garrett, T.J. Precast Gelatin-Based Molds for Tissue Embedding Compatible with Mass Spectrometry Imaging. *Anal. Chem.* **2017**, *89*, 576–580. [CrossRef]
47. Wang, Y.; Hummon, A.B. MS imaging of multicellular tumor spheroids and organoids as an emerging tool for personalized medicine and drug discovery. *J. Biol. Chem.* **2021**, *297*, 101139. [CrossRef] [PubMed]
48. Liu, X.; Flinders, C.; Mumenthaler, S.M.; Hummon, A.B. MALDI Mass Spectrometry Imaging for Evaluation of Therapeutics in Colorectal Tumor Organoids. *J. Am. Soc. Mass Spectrom.* **2018**, *29*, 516–526. [CrossRef] [PubMed]
49. Angel, P.M.; Spraggins, J.M.; Baldwin, H.S.; Caprioli, R. Enhanced sensitivity for high spatial resolution lipid analysis by negative ion mode matrix assisted laser desorption ionization imaging mass spectrometry. *Anal. Chem.* **2012**, *84*, 1557–1564. [CrossRef]
50. Yalcin, E.B.; Nunez, K.; Cornett, D.S.; de la Monte, S.M. Differential Lipid Profiles in Experimental Steatohepatitis: Role for Imaging Mass Spectrometry as a Diagnostic Aid. *J. Drug Alcohol Res.* **2015**, *4*, 1–11. [CrossRef]
51. Yalcin, E.B.; Nunez, K.; Tong, M.; Cornett, S.D.; de la Monte, S.M. MALDI-IMS Detects Differential White Matter Degeneration-Associated Lipid Profiles in Rat Models of Chronic Alcohol, Tobacco Nitrosamine, or Both Exposures. *J. Am. Soc. Mass Spectrom.* **2015**, *26*, 95.
52. Soltwisch, J.; Kettling, H.; Vens-Cappell, S.; Wiegelmann, M.; Muthing, J.; Dreisewerd, K. Mass spectrometry imaging with laser-induced postionization. *Science* **2015**, *348*, 211–215. [CrossRef]
53. Lemaire, R.; Wisztorski, M.; Desmons, A.; Tabet, J.C.; Day, R.; Salzet, M.; Fournier, I. MALDI-MS direct tissue analysis of proteins: Improving signal sensitivity using organic treatments. *Anal. Chem.* **2006**, *78*, 7145–7153. [CrossRef]

54. Shanta, S.R.; Zhou, L.H.; Park, Y.S.; Kim, Y.H.; Kim, Y.; Kim, K.P. Binary matrix for MALDI imaging mass spectrometry of phospholipids in both ion modes. *Anal. Chem.* **2011**, *83*, 1252–1259. [CrossRef] [PubMed]
55. Gode, D.; Volmer, D.A. Lipid imaging by mass spectrometry—A review. *Analyst* **2013**, *138*, 1289–1315. [CrossRef] [PubMed]
56. Thomas, A.; Charbonneau, J.L.; Fournaise, E.; Chaurand, P. Sublimation of new matrix candidates for high spatial resolution imaging mass spectrometry of lipids: Enhanced information in both positive and negative polarities after 1,5-diaminonapthalene deposition. *Anal. Chem.* **2012**, *84*, 2048–2054. [CrossRef]
57. Jackson, S.N.; Wang, H.Y.; Woods, A.S. In situ structural characterization of glycerophospholipids and sulfatides in brain tissue using MALDI-MS/MS. *J. Am. Soc. Mass Spectrom.* **2007**, *18*, 17–26. [CrossRef]
58. Dreisewerd, K.; Lemaire, R.; Pohlentz, G.; Salzet, M.; Wisztorski, M.; Berkenkamp, S.; Fournier, I. Molecular profiling of native and matrix-coated tissue slices from rat brain by infrared and ultraviolet laser desorption/ionization orthogonal time-of-flight mass spectrometry. *Anal. Chem.* **2007**, *79*, 2463–2471. [CrossRef] [PubMed]
59. Carter, C.L.; Jones, J.W.; Farese, A.M.; MacVittie, T.J.; Kane, M.A. Inflation-Fixation Method for Lipidomic Mapping of Lung Biopsies by Matrix Assisted Laser Desorption/Ionization-Mass Spectrometry Imaging. *Anal. Chem.* **2016**, *88*, 4788–4794. [CrossRef]
60. Dannhorn, A.; Swales, J.G.; Hamm, G.; Strittmatter, N.; Kudo, H.; Maglennon, G.; Goodwin, R.J.A.; Takats, Z. Evaluation of Formalin-Fixed and FFPE Tissues for Spatially Resolved Metabolomics and Drug Distribution Studies. *Pharmaceuticals* **2022**, *15*, 1307. [CrossRef]
61. Fournier, I.; Salzet, M.; Meriaux, C.; Franck, J. Matrices for Mass Spectrometry Imaging. Patent World Intellectual Property. Patent Application No. PCT/1B2009/056037.
62. Afshinnia, F.; Rajendiran, T.M.; Soni, T.; Byun, J.; Wernisch, S.; Sas, K.M.; Hawkins, J.; Bellovich, K.; Gipson, D.; Michailidis, G.; et al. Impaired beta-Oxidation and Altered Complex Lipid Fatty Acid Partitioning with Advancing CKD. *J. Am. Soc. Nephrol. JASN* **2018**, *29*, 295–306. [CrossRef]
63. Meriaux, C.; Franck, J.; Wisztorski, M.; Salzet, M.; Fournier, I. Liquid ionic matrixes for MALDI mass spectrometry imaging of lipids. *J. Proteom.* **2010**, *73*, 1204–1218. [CrossRef]
64. Pelit, W.; de la Monte, S.M.; Yalcin, E.B. Rapid Peak Alignment for MALDI-TOF Lipid Analysis. *J. Bioanal. Methods Tech.* **2021**, *1*, 104. [PubMed]
65. Carter, C.L.; McLeod, C.W.; Bunch, J. Imaging of phospholipids in formalin fixed rat brain sections by matrix assisted laser desorption/ionization mass spectrometry. *J. Am. Soc. Mass Spectrom.* **2011**, *22*, 1991–1998. [CrossRef] [PubMed]
66. Buck, A.; Ly, A.; Balluff, B.; Sun, N.; Gorzolka, K.; Feuchtinger, A.; Janssen, K.P.; Kuppen, P.J.; van de Velde, C.J.; Weirich, G.; et al. High-resolution MALDI-FT-ICR MS imaging for the analysis of metabolites from formalin-fixed, paraffin-embedded clinical tissue samples. *J. Pathol.* **2015**, *237*, 123–132. [CrossRef] [PubMed]
67. Neef, S.K.; Winter, S.; Hofmann, U.; Murdter, T.E.; Schaeffeler, E.; Horn, H.; Buck, A.; Walch, A.; Hennenlotter, J.; Ott, G.; et al. Optimized protocol for metabolomic and lipidomic profiling in formalin-fixed paraffin-embedded kidney tissue by LC-MS. *Anal. Chim. Acta* **2020**, *1134*, 125–135. [CrossRef]
68. Leopold, J.; Popkova, Y.; Engel, K.M.; Schiller, J. Recent Developments of Useful MALDI Matrices for the Mass Spectrometric Characterization of Lipids. *Biomolecules* **2018**, *8*, 173. [CrossRef] [PubMed]
69. Yalcin, E.B.; Tong, M.; Homans, C.; de la Monte, S.M. Myriocin Treatment Reverses Alcohol-Induced Alterations in Polyunsaturated Fatty Acid-Containing Phospholipid Expression in the Liver. *Nutr. Metab. Insights* **2022**, *15*, 11786388221082012. [CrossRef]

MDPI AG
Grosspeteranlage 5
4052 Basel
Switzerland
Tel.: +41 61 683 77 34

Applied Biosciences Editorial Office
E-mail: applbiosci@mdpi.com
www.mdpi.com/journal/applbiosci

www.ingramcontent.com/pod-product-compliance
Lightning Source LLC
Chambersburg PA
CBHW061444210926
51850CB00008B/1609